LEAD MANUFACTURING IN BRITAIN

Lead Manufacturing in Britain

a History

D.J. ROWE

CROOM HELM
London & Canberra

© 1983 Associated Lead
Croom Helm Ltd, Provident House, Burrell Row,
Beckenham, Kent BR3 1AT
Croom Helm Australia, PO Box 391,
Manuka, ACT 2603, Australia

British Library Cataloguing in Publication Data

Rowe, D.J.
    The British lead manufacturing industry 1778-1982.
    1. Lead industry and trade – Great Britain – History
    I. Title
    338.4'76694'0941   HD9539.L4

    ISBN 0-7099-2250-7

Printed and bound in Great Britain
by Billing & Sons Limited, Worcester.

CONTENTS

TABLES

In Memory of M.L. Walker

x

Relationship of Walker and Parker partners in Walkers, Parker & Co.

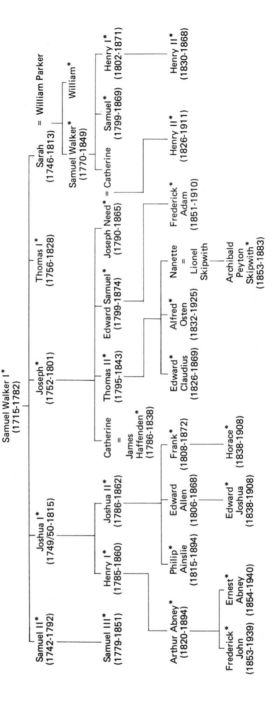

*Those marked with an asterisk were partners in the firm.. The table does not show a number of members of the Walker family who did not become partners in the lead concern.

xi

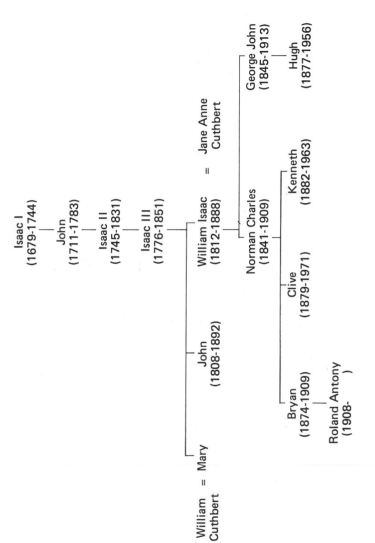

Cookson Family Tree

William = Mary
Cuthbert

Isaac I
(1679-1744)

John
(1711-1783)

Isaac II
(1745-1831)

Isaac III
(1776-1851)

William Isaac   =   Jane Anne
(1812-1888)   Cuthbert

John
(1808-1892)

Norman Charles
(1841-1909)

George John
(1845-1913)

Bryan
(1874-1909)

Clive
(1879-1971)

Kenneth
(1882-1963)

Hugh
(1877-1956)

Roland Antony
(1908-   )

PREFACE

In 1978, on the occasion of the two hundredth anniversary of the founding of the first of its many constituent companies, Associated Lead Manufacturers Ltd commissioned this book as an account of the development of the company and the lead manufacturing industry.

During the past four years I have been indebted to many people within the company for assistance in the research on which this narrative is based. It is not possible to mention all those current and ex-employees who have given of their time, memories and expertise in talking to me about the company and lead manufacturing. To select some of the many for mention would be invidious and I therefore name none but am, nevertheless, most grateful for the assistance and kindness which I have universally met – without their help the book would be poorer since interviews help to flesh out a story of which documents often form only the bare bones.

Outside Associated Lead circles completely I should like to acknowledge help from many institutions and individuals. Many public libraries and record offices have given me considerable assistance but I am especially grateful to Miss M.E. Williams, Bristol City Archivist and Mr F. Manders, Local History Librarian, Newcastle City Libraries, both of whom have put up with many unreasonable requests for information; Mr R. Potts of Tyne and Wear Archives drew up a hand-list of the initial company holding of records (and this may be the appropriate point at which to praise Associated Lead Manufacturers Ltd for having preserved a considerable range of records dating back to the eighteenth century and for being willing to consult an archivist about their preservation – it is to be hoped that other companies will follow this example); Mr A.A. Cole, ICI Librarian, made available the typescript history of British Associated Lead Manufacturers Ltd; Mr D.S. Derbyshire, of Goodlass Wall & Co. Ltd, made available a number of lead company records which had ended up with that firm. My colleagues Mr W.A. Campbell and Dr S.M. Linsley have read and commented constructively on much of the typescript, as has Dr L. Willies, while Mr M.L. Walker read it in its entirety and made many useful criticisms. A number of other specific debts is acknowledged at the appropriate point in the notes but I am, finally, particularly grateful to the several secretaries at Associated Lead's Elswick and Anzon's Howdon offices who have efficiently and speedily produced the various typescripts which have preceded this book.

D.J. Rowe

                    but thou, thou meagre lead,
Which rather threat'nest than dost promise aught,
Thy plainness moves me more than eloquence,
And here choose I.

            Shakespeare, *The Merchant of Venice*, Act III, Sc. II

1 THE LEAD MANUFACTURING INDUSTRY FROM ANCIENT TIMES TO THE EIGHTEENTH CENTURY

'Lead is the vilest and most abject of metals.'
Le Febvre[1]

Uses of lead from ancient times to the eighteenth century; methods of production of white lead, red lead, sheet lead, lead pipe and lead shot; geographical location of lead manufacturing in Britain in the eighteenth century.

Lead is one of the seven metals known from antiquity and it was regarded as the least of them according to the alchemists' classification which ran from gold, the chief of metals, down to lead. The alchemists named the metals after the planets – the sun was gold, the moon silver, Jupiter tin, Mars iron, Venus copper, Mercury quicksilver and Saturn lead.[2] Although lead was the basest of metals it was believed to be the oldest of them and played an important part in the alchemical experiments to find the philosophers' stone, which would transmute base metals into gold. The alchemists were, however, working along a blind alley and it was those who concentrated their activities on manufacturing lead who were to realise gold from it. John Gay would have made eighteenth-century attenders at performances of the *Beggar's Opera* laugh with the lines which he put into the mouth of his highwayman:

See the Ball I hold!
Let the Chymists toil like Asses,
Our Fire their Fire surpasses,
And turns all our Lead to Gold.[3]

Lead bullets, of course, were only one of a myriad uses of lead which had developed from classical times, for although lead was base it had an important range of uses.[4] It was the first metal to be extracted from an ore, probably around 9000 BC, since it could be smelted at low temperatures (not much over 300°C against more than 1000°C required to smelt copper).[5] Among the earliest references to lead is a number in the *Old Testament*, while the hanging gardens of Babylon, one of the seven wonders of the ancient world, are said to have been floored with sheets of lead to retain moisture for the plants growing there.[6] From Greek and especially Roman times[7] lead came to be used for a variety of products which included coins, seals, water pipes, spouts and cisterns, presses, anchors and weights of all kinds, roofing sheets, sling bullets and later cannonballs, the lining of coffins and sometimes the whole coffin, ornaments, statues and pewter for plates and mugs (in Roman times an alloy of roughly 50 per cent lead and 50 per cent tin), medical remedies, pigments and cosmetics.

One of the attractions to the Romans of holding the British Isles was the variety of minerals available and under Roman rule lead mining was developed to a much greater extent than before in various parts of the country. Excavations of Roman sites have produced lead objects, cisterns, pipes, statues, etc. and examples of Roman pig lead have been found.[8] As in so many other areas of technology, it is likely that the manufacture of lead declined when the Romans left this country and only began to recover in the early Middle Ages. Certainly from around 1200 it is possible to begin to trace the development of the mining industry with some certainty, although the manufacturing industry does not seem to have been of great significance and it may well be that much pig lead was exported to other European countries which were more advanced in industrial affairs.

Lead is an unusual metal in that its chemical uses have been as widespread as its metallurgical ones and it is important here to outline the chief distinctions. In metallic form, the production of sheet and pipe for instance, reference is made to blue lead products, while the chemical products are also referred to by their distinguishing colours: red lead (lead oxide, sometimes called minium, Pb_3O_4), orange lead (another oxide with the same chemical composition as red lead but with differing physical characteristics including colour), yellow lead (usually used to describe litharge, lead monoxide, PbO) and white lead (basic lead carbonate, $2PbCO_3.Pb(OH)_2$, that is to say two molecules of lead carbonate with one of lead hydroxide, sometimes referred to as hydrocerussite).[9] After the Roman period, with a decline in the concern for cleanliness, the major demand from cisterns and piping for lead had been considerably reduced and, while its use for building purposes remained the largest single demand for lead, it was increasingly required for its chemical properties as a pigment. Of these chemicals, white and red lead were the major products and their pigmentary properties had been known since classical times, although without, of course, any understanding of their chemical formulation. There has been considerable debate among historians interested in the subject as to whether the white lead or ceruse, as it was called, would have been the equivalent of a modern lead carbonate. From about the third century BC Greek and then Roman authors give accounts of the manufacture of ceruse which are all very similar in the process they describe.[10] Pliny, for instance, gives the following account:

> Psimithium, or ceruse, is another product of leadworks, the best quality being made in Rhodes. It is prepared by placing very thin lead shavings over a vessel of the sourest vinegar, which distils on to, and drops down from the lead. The portion of the latter which falls into the vinegar is dried, powdered and sifted. It is then moistened again with vinegar, made into tablets and dried in the summer sun.[11]

The problem about descriptions such as this is that they do not give enough information to make it clear, from a chemical view point, that lead carbonate was being produced. The chemical process was for the vapour of acetic acid (vinegar) to attack the oxidised outer surface of the metal and produce basic lead acetate, which in its turn would be converted to basic lead carbonate by the action of moist carbonic acid gas thus freeing the acetic acid to convert further metallic lead. It is not clear in the classical accounts as to where the carbon dioxide came from, although it is possible that the vinegar was sour wine, still containing debris from the wine-making process, grape skins, lees, etc. which would continue to ferment and produce carbon dioxide.

One further problem still seems to remain. While historians have accepted that the accounts of the classical writers are deficient, they have argued that they show an understanding of the basic requirements of a process to obtain basic lead carbonate.[12] This is not, however, true of Pliny's account which is one of the later of classical descriptions. If the product of the process fell into the vinegar (even if it had been lead carbonate) it would be converted into lead acetate which is soluble and of no use as a pigment. It is, therefore, not entirely certain that classical manufacture was producing a white lead which would be of the same chemical formula as a modern lead carbonate, although a white powder of lead was clearly produced which was used as a pigment both for the painting of wooden objects, especially ships, as a preservative, and also as a cosmetic. St Jerome, writing in AD 384, commented scathingly on women who

> should offend a Christian's eyes, who paint their cheeks with rouge and their eyes with belladonna; whose faces are covered with powder

> and so disfigured by excessive whiteness that they look like idols.

and subsequently he made it clear to what the whiteness was due, describing Gentile widows who 'are wont to paint their faces with rouge and white lead'.[13]

After these early beginnings it is clear that little in the way of further development took place for some hundreds of years. By about 1000 AD, however, the use of dung (stable manure), as a means of producing heat to cause the vaporisation of the acetic acid, in the white lead process was known. If lead carbonate had been produced in the classical period it is likely that it was only done during the summer when hot Mediterranean weather would have provided a sufficiently high temperature for the process to work. The introduction of dung was obviously an important one which allowed manufacture to take place in colder periods and in less fortunately endowed countries. Of possibly greater importance, although contemporaries would have been unaware of the chemistry involved, the introduction of dung provided a satisfactory source of carbon dioxide, which was released during its fermentation. The basis of the process for the manufacture of white lead, which was to last until the middle of the twentieth century, was complete. Also by about 1000 AD the use of linseed oil as a medium for spreading white lead as a paint was known.

For a long time the process appears to have remained confined to the northern Mediterranean and until the eighteenth century, when it was overtaken by the Dutch, the Venetian industry was the major centre for production and certainly the area in which the best quality white lead was produced. In Venice the process was developed on a large scale and from there it was borrowed by the Dutch, finally arriving in this country to be known, rather misleadingly, as the 'old Dutch' or 'stack'[14] process. As developed in Venice the process is described very thoroughly by Sir Philiberto Vernatti in 1678. Thin plates of lead were cast, rolled up without the surfaces touching, and then placed in pots containing vinegar, although ensuring that the lead did not touch the vinegar. A bed of horse-dung was made in the stack and covered with 400 pots which were then covered with boards and then three further similar layers were built on top. The stack was then closed for three weeks after which it was dismantled and the corroded lead removed. The account shows that the process was well established and that the Venetians were aware of the necessity for heat, for keeping the lead separate from the vinegar and for space for air movement around

the lead to be corroded, something which was not scientifically established until the early twentieth century. Nevertheless the account goes
on to show that the process was by no means foolproof:

> Accidents to the work are; that two pots alike ordered and set one
> by the other, without any possible distinction of advantage, shall
> yield, the one thick and good flakes, the other few and small or
> none, which happeneth in greater quantities, even over whole beds
> sometimes.[15]

Although Vernatti laid out the problems of the process in the seventeenth century they were to remain unsolved for more than 200 years
until developing science could give an explanation and enable the process to be more carefully monitored. Those historians who argue that
the Industrial Revolution was partly a result of scientific contribution
to industry can take no comfort from the manufacture of white lead,
which remained an entirely empirical process.

It was, however, a process of increasing importance as growth of
population and industrial development placed demands for pigments
which made white lead probably the most important single use for lead.
From the Mediterranean the process spread and became established in
the rapidly growing trading nations – firstly Holland and then England.[16]
We can be sure of its manufacture in England by 1622 when Christopher
Eland registered a patent for a new method of making both white and
red lead but it is probable that English production was on a small scale
and that the Dutch dominated European output in the seventeenth
century. They appear to have replaced Venetian white lead in many
markets by the simple expedient of adulterating their white lead with
chalk, a practice subsequently adopted by the English, which enabled
them to undersell their Venetian competitors. As a result Venetian
white lead retained a small, quality market (especially for the supply of
artists) while the Dutch and English fought it out for the larger market
for general painting purposes. Savary, writing in the mid-eighteenth
century, commented, 'Painters use it in great quantities; and, to afford
it them cheap, it is generally adulterated with common whiting. The
English and Dutch ceruse are very bad in this respect; the Venetian
ought always to be used by the apothecaries.'[17]

Until the mid-eighteenth century the Dutch and Venetian industries
were able to dominate the European market. The advantages of having
the technical knowledge and long experience of a troublesome process
outweighed the disadvantage of having to import their pig lead, much

of it from England and Wales. By contrast the British industry, which had the advantage of home supplies of lead, was still struggling with production problems which prevented it from reaching internationally competitive levels. The total value of English exports of red and white lead (probably composed chiefly of red lead) rose only from about £2,000 to £3,000 per annum around 1700 to about £5,000 in the 1720s and £7,000 in the 1760s, in a period when the range for total lead exports (made up chiefly of pig lead) was from about £100,000 to £175,000 per annum.[18]

Little change seems to have taken place in the white lead process over this period. Gabriel Jars, a French metallurgist, visited England in 1764 and his description of white lead manufacture differs little from earlier ones.[19] He does, however, state that the lead remained in the stacks for four to five weeks before it was withdrawn, with five layers or 'heights' of lead in each stack. The time which the process took for the conversion of lead into lead carbonate was, however, by no means standard. Greek producers had removed the lead after ten days, scraped off the corrosion and then returned the metallic lead remaining for further corrosion. This made sense for small-scale production in classical times but would be inconvenient with the more highly structured stack process, especially in a temperate climate where loss of heat as a result of opening the stack was important. It was, therefore, necessary to leave the lead in until it was felt that as high a corrosion level as possible had been achieved. At the Sheffield white lead works, visited by Jars, corrosion took six to eight weeks, although Savary, writing shortly before Jars, gives a period of 10 to 14 days for the conversion process, which suggests that some manufacturers were settling for lower corrosion levels in return for more rapid production.

Jars' account also gives us useful evidence on the major centres of lead manufacturing in Britain, since he visited Derbyshire and north Wales, two major centres of lead mining. Smelting of the lead ore (galena, lead sulphide) usually took place on or close to the ore-field in this period, since the ore would contain 70 per cent or less lead and it was obviously sensible to remove the impurities and then transport pig lead. In the seventeenth and early-eighteenth centuries it is likely that most lead left the ore-fields in pig form and that much was exported. There were, however, manufacturing works near the ore-fields and Jars visited a white lead works at Sheffield and red lead works in north Wales. Most of these works and especially those set up in conjunction with lead smelters, appear to have been on a small scale with one, or at the most two, lead products. Towards the end of the eighteenth century, however, there was a tendency for integrated works, producing a range of lead products, to develop.[20]

Jars' selection of Derbyshire and north Wales as areas for investigation are confirmed by the evidence of Richard Watson, Professor of Chemistry at Cambridge. In his *Plan of a Course of Chemical Lectures*, first published in 1771,[21] he selects Holywell in Flintshire for an account of white lead manufacture and refers to both Flintshire and Derbyshire for accounts of the manufacture of red lead. By this time red lead was being manufactured by the modern process of calcining pig lead to litharge or massicot,[22] which was cooled and ground, and then oxidising the resultant powder until it took on a bright red colour. The process was undertaken in a reverberatory furnace in which coal was used to heat but not allowed to come into contact with the metal (to avoid impregnating it with sulphur impurities).[23] The reverberatory furnace came into wide-scale use for the smelting of many non-ferrous metals in the first half of the eighteenth century, including lead and was most rapidly innovated in Flintshire for lead smelting. For a long time in the eighteenth century it was known as the Flintshire furnace as contrasted with the Scotch or northern ore-hearth — a furnace in which ore and fuel were mixed and the blast provided by bellows.

Both Jars and Watson describe the manufacture of red lead in reverberatory furnaces in Derbyshire and the latter states that there were no fewer than nine red lead works in that county.[24] While it is certain that the consumption of red lead in the eighteenth century was considerably less than that of white lead, it was expanding, being used not only as a pigment but also as a pottery glaze and in flint glass manufacture, where about 20 per cent of the final weight of the glass was composed of red lead. The use of red lead in the manufacture of flint or crystal glass dates from the 1670s when George Ravenscroft developed the process, which enabled the glass manufacturers to use the reverberatory furnace.[25] As an added bonus the inclusion of lead increased the refractive index of the new glass, which focused light more effectively from a lens of a given size and this development had a considerable impact on the quality of microscopes and telescopes. It also led to the pre-eminence of English crystal glass for tableware. While this development caused an increase in the demand for red lead it was not until the nineteenth century that it experienced a major demand as a protective pigment as iron and then steel became major constructional materials. Unlike white lead, however, it is probable that red lead did not experience much foreign competition in the eighteenth century. Manufacturers in this country had the advantage that the technology, the reverberatory furnace, had been invented here, while continental producers had no technical advantages and had the disadvantage of having to import their raw material.

While white and red lead were the chief chemical products there was an increasing number of uses for lead as a metal. Chief of these was in building work where the metal's high level of resistance to corrosion by the weather made it an excellent, if expensive, material especially for roofs. Because of its high cost[26] it was used chiefly on the more prestigious buildings with churches particularly prominent, but a lead clad roof would last for a couple of hundred years without attention. Sheet lead for roofing purposes was manufactured by a process of casting molten lead on a bed of sand. This was a relatively simple and easy technique and in the eighteenth century was certainly done by the plumbers[27] who were going to lay the roofs. The melting point of lead is low which enabled pig lead to be melted easily in an iron pot, which was then lifted and poured on to the prepared bed of sand. The workmen then smoothed the surface with an iron bar in order to ensure an even thickness, the metal was allowed to cool and the sheet lifted from its sand mould and then rolled to be stored until needed.[28]

By the later eighteenth century cast sheet was facing a challenge from rolled or milled lead. The origins of the rolling mill appear to be obscure but hand-rolling certainly existed by the later Middle Ages and the use of horse and then water power to turn the rollers followed not much later. Although Singer stated that 'Rolled lead rapidly replaced cast sheets because of its cheapness and freedom from defects',[29] it is clear that cast lead did not give up easily. In the first place it is not clear that the production costs of milled lead (which required considerable capital expenditure) were markedly lower than those of the more simple casting process. Secondly, contemporaries were far from certain that milled was superior in quality to cast sheet. Rees, for instance, wrote:

> [lead] is either cast into sheets in a mould, or milled, which last, some have pretended, is the least serviceable, not only on account of its thinness, but also because it is so exceedingly stretched in milling, and rendered so porous and spongy, that when it comes to lie in the hot sun, it is apt to shrink and crack, and consequently will not keep out the water. Others have preferred the milled lead, or flatted metal, to the cast, because it is more equal, smooth and solid.[30]

Such a controversy was not to be settled easily and it is likely, at least for the nineteenth century, that cast sheet continued to be preferred for quality roofing work while milled (in rapidly increasing quantities) was used for cheaper work and in the chemical industry. Although it subsequently became more concentrated, in the eighteenth century lead rolling was a relatively small-scale, localised industry, with a considerable number of firms.

Probably next in quantity to its use in sheet came the use of lead in pipe. With an extension of prestige building from the Middle Ages and the slow beginnings of a return to Roman levels of cleanliness, the demand for pipe was rising. At first it was produced by the Roman method of taking narrow sheets of lead, bending them and welding the edges by applying molten lead to them.[31] Joints between two lengths of pipe were made by tapering the end of one, fitting it inside the other and welding with hot metal. It is clear that the low melting point of lead and its malleability made it an ideal metal for such purposes. These ensured its use in circumstances where it was physically impossible to use iron for instance, which required much more expensive equipment which was not available at the site of building operations nor within the financial range of the small-scale plumbers.

The Roman method of manufacturing pipe was already somewhat archaic by the mid-eighteenth century and had largely been replaced by casting, of which there were at least two versions. That illustrated by Diderot in his *Encyclopaedia* shows a mould in which holes were left at strategic intervals into which the molten lead was poured, while a second method was to have a mould split in halves. In both instances the mould was fitted with a central core which was pulled out when the metal had solidified sufficiently. A subsequent process was the drawing out of a thick lead pipe cast around a central iron core or mandrel. The pipe was either pulled through successively smaller holes in a steel plate, or passed through grooved rollers of successively smaller size until the correct diameter was obtained. Again it is clear that it was lead's malleability which made it suitable for such a process. Rees states that casting was the common method in the early nineteenth century (although some of the material in his *Cyclopaedia* is undoubtedly out of date), but that 'A very great improvement has been made in the manufacture of lead pipes, by drawing them in a manner similar to wire.' It is probable that pipe drawing was becoming common at the end of the eighteenth century at the very time when experiments were being made with the process which was to replace it, extrusion. The first patent for a pipe extrusion process was taken out in 1797 by Joseph Bramah but it was not until 1820, when Thomas Burr, a Shrewsbury plumber, took out a patent (no. 4445), that an effective extrusion press was available.[32] To an even greater extent than lead sheet, the manufacture of pipe was suited to small-scale production and the output was heavily localised in the hands of a large number of small firms.

There was a number of other lead products but their consumption of metal was not great. Lead shot was used for both sporting and military purposes. Large shot was cast in moulds while small shot was dropped through a sieve into water,[33] a process which was to be considerably improved in the late eighteenth century. Both of these processes had considerable disadvantages. Hand casting of large shot for bullets was tedious and time consuming, although by the eighteenth century bullets were cast in multiple moulds attached to a bar of lead from which they were snipped as required. The pouring of small shot suffered the disadvantage that the shot experienced distortion from the impact on the water before solidifying, which caused it to form uneven shapes. The dropping of small shot appears to date from the mid-seventeenth century. Robert Hooke, in 1665, referred to a report to the Royal Society on this process and described it as 'a very pretty curiosity, and known but to a very few'. 'Auripigmentum' (arsenic tri-sulphide, As_2S_3) was added to molten lead, which was poured through holes in a copper plate into a bucket of water four inches below the plate – a process in which it was claimed 'the Lead will constantly drop into very round Shot, without so much as one with a tail in many pounds'.[34] That claim for the process appears to have been optimistic, although it remained in use until the late eighteenth century but had certainly been replaced by 1819 when Rees described it as the only method in use (which is one of the reasons for regarding his *Cyclopaedia* as somewhat unreliable).

Lead was also in use in the eighteenth century for a variety of decorative purposes and the big expansion of country house building of the period undoubtedly led to an increase in demand for lead statuary and ornaments, again from the ease with which the metal could be worked. The first major use of lead in this way came at Louis XIV's Chateau de Versailles and many accounts of large country houses refer to this use of lead.[35] Other uses included type-metal and pewter, where the lead content had sunk considerably from the level of 50 per cent in the Roman era, while pewter was also experiencing a decline in demand in the face of pottery and china for mugs and plates. Lead was also used as a material for containers for storage purposes. A new use in this area came with the invention in 1746, by John Roebuck, of the lead chamber process for the manufacture of sulphuric acid,[36] which chemists regard as one of the most significant developments during the Industrial Revolution. Because of its widespread use in nineteenth-century industry, output of sulphuric acid was a useful proxy for industrial growth and its production certainly led to a considerable demand for sheet lead, which, under cold conditions, is hardly acted on by sulphuric acid – a thin layer of lead sulphate being built up which protects the lead from further attack. While other metals such as tin and nickel have a higher level of corrosion resistance than lead, they are and were much more expensive.

Lead was also used for the manufacture of water cisterns and in the West Country for tanks to contain the juice from apple pressings for cider manufacture. In both instances, and especially as a result of the reaction of the acid in the juice with the metal in the latter case, lead poisoning could result and Rees noted that Devonshire colic was a well-known illness resulting from the drinking of cider which had been stored in lead tanks. There were also other areas where the use of lead was dangerous, even if the quantities involved were small. Possibly the most common in the eighteenth century was the use of litharge to sweeten sour wine. The acid in the wine was converted to lead acetate (which tastes sweet, hence its common name of sugar of lead) which is highly soluble and dangerous to the human body.[37] That similar uses continued is made clear by a number of studies of the adulteration of foodstuffs in the first half of the nineteenth century, with the colouring of the rind of cheese with red lead, in order to make it look more attractive, being probably the most common.[38] Another dangerous, if small-scale, use of lead was in medicaments. This use, both internal and external, has a long history which is lost in the mists of antiquity but continued in the eighteenth century despite considerable criticism. Thus Richard Watson commented dryly, 'Paracelsus made no vain boast, in saying that he could cure two hundred diseases by preparations of lead; but he does not tell us of the many hundred persons he, probably, sent to their graves by his attempt.'[39] The external application of lead was commonly in the form of lead plasters of which one delightful account runs as follows:

> Take two pounds and fore ounces of the best and greenest sallet oyl: with a pound good red lead and a pound of white lead. Beat them well into dust, then take twelve ounces of Castle sope, incorporate all thees well together in a well glassed and great earthen pot that the sope may com upwards: set it on a small fire of Coales ... The

vertues of the leaden plater 1 If it be lade to the Stomack it provoketh appitite: and taketh away any grief in the same. 2 If lade to
the belly it is a present remedy for the ache. 3 If lade to the reines of
the back it cureth and healeth the bloudy flux, the runing of the
reans, heat in the liver or weakness in the back. 4 It healeth all
bruises and swelling it taketh away aches, it breake the felons,
pushes and other impostumes, and healeth them. 5 It draweth out
any running humer without breaking the Skin and being applied to
the fundament it healeth any disease there growing. 6 The same laid
to the head is good for the eyes. 7 The same laid to the belly of a
woman provoketh the tearms and maketh apt for conception.[40]

During the eighteenth century, however, the use of lead as a medica-
ment was on the decline and in 1819 Rees could write that it 'is apt,
however, to bring on colics of so violent a kind, that the remedy often
proves worse than the disease'.

While these were the uses for lead there were lead manufacturing
centres based on the ports as well as those in the major ore-producing
areas such as north Wales and Derbyshire, which have already been
mentioned. In the seventeenth and early eighteenth centuries, with
British lead output exceeding home demand and manufacturing capacity, lead was exported, particularly to Holland. A considerable proportion of this was in the form of flake litharge, the by-product of desilverising of pig lead by cupellation. Many British lead ores had a sufficiently
high silver content to justify its recovery and this was done, usually in
the smelting works in a cupola furnace whereby the pig lead was
oxidised to litharge, leaving a small 'button' of silver.[41] The litharge
could be converted back to pig lead in a reverberatory furnace, but this
was expensive, and it was often sold to continental manufacturers who
were short of supplies. It was, therefore, natural that lead manufacturing was to develop in the ports from which the lead was exported.
Bristol, although of little significance by the end of the eighteenth
century for lead shipments because of the decline in output of lead ore
from the Mendip mines, had a considerable manufacturing industry
which was now buying its lead from more distant parts in Cornwall and
Wales. Having developed from the earlier output of Mendip ores the
town had several works in the late-eighteenth century smelting lead ores
and producing red and white lead, cast sheet lead, shot, etc. London, of
course, as a major centre of population, had a large demand for lead
products and several works were established there in the eighteenth
century, receiving their supplies of pig lead from the Pennines via Hull,

Stockton and Newcastle. Among these was a white lead works, in existence by 1752 at Limehouse and worked by the London Lead Company, probably with lead from its north Pennine mines.[42] Hull was another port which was significant within the industry, conveniently situated as it was to ship the products of the Yorkshire and Derbyshire lead mines. For most of the eighteenth century Hull would appear to have been not only the major outlet for pig lead but also for manufactured lead.

Table 1.1: Outward Shipments of Lead from North-eastern Ports, 1705-70 (tons)[43]

Year			Hull	Newcastle	Stockton-on-Tees
1705/6		Lead	6,379	2,530	3,251
		Manufactured lead	428	47	0
1725/6		Lead	4,057	1,928	2,560
		Manufactured lead	742	133	0
1755/6		Lead	5,813	3,249	4,485
		Manufactured lead	1,609	306	0
1769/70		Lead	7,513	7,542	3,478
		Manufactured lead	2,997	432	0

It is not possible to tell the extent to which the manufactured lead being shipped from Hull was actually made in the town rather than being produced on the ore-fields or in towns such as Sheffield and then distributed via Hull. It is, however, certain that there were at least two white lead works, Kirkby's and Bielby's, in the town.[44] It is also clear that, wherever it was produced, the town's shipments of manufactured lead were expanding rapidly in the mid-eighteenth century, while shipments of pig lead show a much less well-defined trend, suggesting industrial development in this country in place of the previous export of raw materials for other countries to process. By comparison it is clear that the northern Pennine area remained for most of the eighteenth century a supplier of raw materials which would be worked up in other parts of the country or abroad. Throughout the period covered by the table the shipment of Teesdale lead through Stockton led to no manufacturing in that town at all, while little of the north-west Durham and Northumberland lead which came to Tyneside remained there for manufacture. In commenting on the figures presented in Table 1.1 Burt suggests that there was in fact a large manufacture of lead in Newcastle but that it was used locally in the flint glass and paint industries.[45] While it is certainly true that some lead was manufactured for these purposes it is unlikely that this was of great significance. Burt, himself, suggests a figure for 1794 of only 600 tons of red lead used in flint glass manufacture throughout the country and it is unlikely that the consumption on Tyneside in 1770 was much more than 100 tons. Similarly, although paint was certainly made on Tyneside, it is almost certain that the consumption of lead for this purpose was small. In the first place the North-east's population was small and growing more slowly than the national average in the eighteenth century, and secondly, it cannot even be argued that paint was made but shipped elsewhere since at this period most paint was mixed by painters themselves from white lead (which would, of course, show up in the shipments from Newcastle given in Table 1.1). It does, therefore, seem clear that the Tyne, which was to become a major centre of manufacturing in the nineteenth century, was of little significance in the eighteenth.

At the end of the eighteenth century, when the first companies were formed which were eventually to merge into Associated Lead Manufacturers Ltd (ALM), the demand for many of the products mentioned earlier was expanding and this was to bring a change to the locational structure of the industry outlined above. A rapid increase in the number of firms in the lead industry in the last quarter of the eighteenth century, and the fact that they were frequently integrated firms making a wide range of products, confirm this as the take-off period for the lead industry. It was not, however, the result of the Industrial Revolution creating a major new demand for a particular lead product but rather the general pressure of population and industrial growth which created a broad demand across the wide range of lead products which have already been discussed.

Before we look at the ways in which the new firms met this demand, it is well to see how the supply of lead was developing. From a position of shortage of lead in the Middle Ages, England and Wales had reached a position by the early-seventeenth century where output of lead exceeded home demand and they had become important suppliers of pig lead and litharge to European manufacturers. Lead was second only to iron in the volume of production among the metals and by the late-eighteenth century Britain was probably the leading lead producer in the world. During the eighteenth century exports of lead did not rise very much[46] which may imply that rising home demand was restricting the surplus available for export, although the price of lead shows no secular tendency to rise during the century, except in the last two decades. Although there are no reliable figures it is probable that British output of lead ore was stagnant in the late-eighteenth century and that rising demand was slowly bringing about a price increase which was to be made more rapid during the inflation of the French wars. While lead output may not have increased in the country as a whole it did so in the northern Pennines, where Weardale lead output rose from about 2,500 tons per annum in the 1770s to about 4,500 in the late 1780s and about 6,500 in the 1790s, while at Coalcleugh in West Allendale there was a rise from about 1,000 tons per annum in the 1770s to about 2,500 tons in the 1780s.[47] It is therefore clear that home supplies of ore were capable of coping with the initial phases of rising demand, and since there was an increasing supply of raw material channelled through the Tyne for shipment, it is perhaps more than just fortuitous that the first new lead manufacturing development came on that river.

Notes

1. Quoted in W.H. Pulsifer, *Notes for a History of Lead* (New York, 1888), p.v.
2. John Read, *Prelude to Chemistry* (1936), p. 88.
3. John Gay, *The Beggars Opera* (1928 edn), p. 37.
4. Pulsifer, *History of Lead*, contains considerable detail on the uses of lead in the classical period. See especially Chapter VII. There is also considerable detail on the uses of lead in the ancient world in W.W. Krysko, *Lead in History and Art* (Stuttgart, 1979), pp. 47-63. See also J. Needham, *Science and Civilisation in China*, vol. 5, part 3 (Cambridge, 1976), pp. 15-17. At a number of points the argument and supporting evidence in this book has had to be reduced in order to meet publishing limits. A fuller account may be found in D.J. Rowe, A History of the Lead Manufacturing Industry in Great Britain 1778-1980, with Special Reference to the Constituent Companies of Associated Lead Manufacturers Ltd (University of Newcastle upon Tyne, PhD, 1982).
5. Krysko, *Lead in History*, pp. 13-18.
6. Ibid., pp. 51-2. See also R.T. Falconer, Lead into Gold. A Short Account of the Lead Industry in the North (Ts., 1971). This article has drawn my attention to a number of quotations and references.
7. See R.F. Tylecote, *A History of Metallurgy* (1976), pp. 48-9 and 61.
8. See, for instance, J.A. Smythe, 'Roman Pigs of Lead from Brough', *Transactions of the Newcomen Society*, 20 (1939-40), pp. 139-45. In common with iron, bars of lead were known as 'pigs'. The derivation is supposedly that as molten metal came from the furnace it was run into bar moulds, which, because they were arranged in rows to a channel along which the metal ran, looked like piglets feeding from a sow.
9. D. Greninger *et al.*, *Lead Chemicals* (New York, 197–).
10. The most useful sources on this matter are Pulsifer, *History of Lead*, and C.D. Holley, *The Lead and Zinc Pigments* (New York, 1909).
11. Quoted in R.M. Clay, 'Blue Lead to White Lead', *Country Life* (23 June 1950), p. 1903.
12. See, for instance, Holley, *Lead and Zinc Pigments*, pp. 2-6.
13. Quoted in Clay, 'Blue Lead to White'.

14. The stack was the, usually brick, windowless building in which the conversion took place.

15. Quoted in Holley, *Lead and Zinc Pigments*, pp. 9-11.

16. Pulsifer, *History of Lead*, p. 276. A. & N.L. Clow, *The Chemical Revolution* (1952), p. 382, comment 'From the time of Queen Elizabeth, lead was converted into white lead at Newcastle.' Their comment was probably drawn from S. Miall, *A History of the British Chemical Industry* (1931), p. 210, in which the same unsubstantiated statement is made. This may be true but the scale of production was certainly small for the next two centuries.

17. Jacques Savary des Bruslons, *The Universal Dictionary*, vol. II (2nd edn, 1757), p. 23.

18. Mrs E.B. Schumpeter (ed.), *English Overseas Trade Statistics 1697-1808* (Oxford, 1960), table VII.

19. L. Willies, 'Gabriel Jars, (1732-1769) and the Derbyshire Lead Industry', *Bulletin of the Peak District Mines Historical Society*, 5, 1 (April 1972), pp. 31-9.

20. For this development in Derbyshire, see L. Willies, Technical and Organisational Development of the Lead Mining Industry in the Eighteenth and Nineteenth Centuries (University of Leicester, PhD, 1980), p. 258.

21. R. Watson, *Chemical Essays*, vol. V (1787), pp. 247-8.

22. There is considerable confusion over the historic use of the common names, litharge and massicot, used to describe different forms of lead monoxide. Lead monoxide, represented by the chemical formula PbO, exists in two crystal forms, tetragonal and orthorhombic, may take the form of flakes or powder and range in colour from reddish-brown to yellow. There appears, from reference to a number of historical texts, never to have been any standardisation of common name either for the crystal or the physical forms.

In the eighteenth century litharge and massicot were used interchangeably, although litharge had originally been used to describe the end product of the cupellation of lead to obtain its silver content. In this form litharge was described as consisting of white or semi-transparent flakes resembling mica (see A. Rees, *The Cyclopaedia* (1819)). It is clear that it was in this form that litharge obtained its name, from the Greek, *lithos argyros*, silver stone. Gradually, however, litharge had come to be used to describe the monoxide produced by drossing lead in a reverberatory furnace, a powder which was also called massicot. Eighteenth-century observers sometimes named the monoxides according to colour, with massicot describing the brown and litharge the yellow forms. With the development of modern crystal analysis, however, it is now known that colour alone is insufficient to distinguish the two crystal forms. The factors determining which of the crystal forms will predominate in the final monoxide are the temperature to which it is heated and the rate at which cooling occurs. For a recent discussion of the subject see E.J. Ritchie, *Lead Oxides, Chemistry – Technology – Battery Manufacturing Uses – History* (Largo, Florida, 1974). I am most grateful to Mr L. Williams, ALM's chief chemist at Chester, for going to a great deal of trouble to assist me with this problematic topic.

23. A reverberatory furnace had two hearths, one for the coal fuel and the other for the metal to be smelted, divided by a low wall called the fire-bridge which prevented immediate contact between fuel and metal but did not reach to the top of the furnace. The heat from the fuel was thus allowed to reverberate from the roof on to the metal. Such furnaces are described in J. Percy, *The Metallurgy of Lead* (1870), pp. 222-34, and in A. Raistrick and B. Jennings, *A History of Lead Mining in the Pennines* (1965), pp. 121-2. The initial use of the reverberatory furnace in lead smelting is attributed to Viscount Grandison who took a patent (no. 206) in 1678 and had works at Bristol. He was, however, heavily dependent upon the earlier work of Samuel Hutchinson and his

experimentation was not totally satisfactory. About 1698 the reverberatory furnace was more widely and successfully used in lead smelting as a result of the work of Dr Edward Wright, Chairman of the Royal Mines Copper Company and a major figure in the formation of the London (Quaker) Lead Company. For these developments see J.N. Rhodes, The London Lead Company in North Wales 1693-1792 (University of Leicester, PhD, 1970), vol. I, pp. 19-29 and vol. II, pp. 358-61, R. Jenkins, 'The Reverberatory Furnace with Coal Fuel, 1612-1712', *Transactions of the Newcomen Society*, XIV (1933-1934), pp. 67-81, A. Raistrick, 'The London Lead Company 1692-1905', ibid., pp. 119-62 and idem, *Quakers in Science and Industry* (1968 edn, Newton Abbot), pp. 163-72.

24. Willies, Technical and Organisational Development of Lead Mining, pp. 258-9, and Watson, *Chemical Essays*, vol. III (5th edn, 1789), p. 339.

25. See R.J. Charleston, 'George Ravenscroft: New Light on the Development of his "Christalline Glasses" ', *J. of Glass Studies*, X (1968), pp. 156-67.

26. See Appendix I for the course of pig lead prices.

27. Plumbers were originally entirely workers with lead, from the Latin name for which, *plumbum*, their name and also the chemical symbol for lead, Pb, are derived.

28. The process was described in a number of contemporary publications and detailed accounts are given in R. Neve, *The City and Country Purchaser and Builder's Dictionary* (2nd edn 1726, reprinted Newton Abbot, 1969), pp. 183-6 and in A. Rees, *The Cyclopaedia* (1819). It is illustrated in Diderot's *Encyclopaedia*.

29. C. Singer et al., *A History of Technology*, vol. III, *From the Renaissance to the Industrial Revolution* (Oxford, 1957), p. 45.

30. Rees, *The Cyclopaedia*. There is further discussion on the subject in L. Hebert, *The Engineer's and Mechanic's Encyclopaedia* (1837), pp. 52-4.

31. Tylecote, *History of Metallurgy*, p. 62.

32. C.E. Pearson, 'The History of the Hydraulic Extrusion Process', *Transactions of the Newcomen Society*, 21 (1940-1), pp. 109-21. For the general development of the processes of lead pipe manufacture see Hebert, *Engineer's and Mechanic's Encyclopaedia*, pp. 54-9.

33. Singer, *History of Technology*, vol. III, p. 45.

34. R. Hooke, *Micrographia* (1665), pp. 22-4, where the process is attributed to Prince Rupert of the Rhine [1619-1682], grandson of James I of England and nephew of Charles I.

35. See, for instance, the National Trust guides, *Wallington Gardens* (1975), p. 9 and *Charlecote Park* (1979), p. 14, C. Hussey, *English Country Houses: late Georgian 1800-1840* (1958), p. 40 and Krysko, *Lead in History*, pp. 105-27.

36. Clow, *Chemical Revolution*, p. 133, and Singer, *History of Technology*, vol. IV, *The Industrial Revolution c.1750-c.1850* (Oxford, 1958), pp. 242-6.

37. J.C. Drummond and A. Wilbraham, *The Englishman's Food* (1939), p. 240.

38. See, for instance, F. Accum, *Treatise on the Adulteration of Food and Culinary Poisons* (1820) for several references to the use of lead.

39. Watson, *Chemical Essays*, vol. III, p. 372. Paracelsus was a Swiss physician of the sixteenth century, infamous for his use of a variety of dangerous remedies. See also Savary, *Universal Dictionary*, vol. II, p. 23.

40. *Jane Mosley's Derbyshire Remedies 1669-1712* (Derbyshire Museum Service, 1979), pp. v-w.

41. For an account of the process see Singer, *History of Technology*, vol. IV, pp. 137-8.

42. Raistrick, 'London Lead Company', p. 143.

43. Figures drawn from R. Burt, 'Lead Production in England and Wales

1700-1770', *Economic History Review*, XXII, 2 (1969), Appendix 1, p. 267. This article (pp. 259-63) contains information on the uses of lead in the eighteenth century. There are also some figures for shipments of lead from Hull in G. Jackson, *Hull in the Eighteenth Century* (1971), pp. 56, 61 and 91, but some of them are unreliable. Jackson's figure for exports of red lead from Hull in 1783 of 4,000 tons is obviously incorrect since the total British export of both red and white lead in that year was less than 400 tons (Schumpeter, *English Overseas Trade Statistics*, p. 26). There is also some disagreement between the figures quoted by Jackson and Burt, although comparison is difficult since the figures are for isolated years. It is surprising, if Hull was as important a lead manufacturing centre as the figures quoted by Jackson suggest, that (1) none of the firms which were to develop rapidly in the late-eighteenth and early-nineteenth centuries decided to erect a works there; (2) none of these new firms, which individually would certainly not have made more than 500 tons of red lead per annum in the 1790s, seemed particularly aware of the significance of or competition from the Hull area.

44. Communication to the author from Mrs M.E. Bickford, 8 Sept. 1980. Jackson, *Hull in the Eighteenth Century*, pp. 192-4, has details of production at Pease's white lead works in Hull in the 1740s.

45. Burt, 'Lead Production in England and Wales', p. 258.

46. The figures are in Schumpeter, *English Overseas Trade Statistics*, p. 65.

47. M. Hughes, Lead, Land and Coal as Sources of Landlord Income in Northumberland between 1700 and 1850 (University of Durham, PhD, 1963), facing p. 60 and p. 63.

THE FORMATION OF WALKERS, FISHWICK & CO., 1778-1800

> Love with White lead cements his Wings,
> White lead was sent us to repair
> Two brightest, brittlest earthly Things
> A Lady's Face and China ware.
> > Jonathan Swift, 'The Progress of Beauty' (1719)

Walkers, Fishwick & Co. the earliest of the companies later to be absorbed in ALM: the early partners and the choice of Newcastle as a site; early history of the works from 1778, output figures and marketing policy; Archer Ward and Richard Fishwick, the active partners, their activities in the firm; extension of product range from white lead to include other lead products and colours; sales statistics 1780-1800; sources of power supply for machinery; sources of raw materials; effect of supply expansion on London market.

Into this situation of rising demand for lead products and plentiful supply of raw materials, in the last quarter of the eighteenth century, came the first of the new companies which were to dominate the lead industry for the next two centuries. This was Walkers, Fishwick & Co. (sometimes referred to as Walkers, Fishwick and Ward), a partnership set up in Newcastle and arguably the earliest founded firm in what has become ALM. Of other possible contenders for that title, the earliest is undoubtedly the firm of Cooksons on Tyneside, usually dated to the early eighteenth century and sometimes precisely to 1704. At that time Isaac Cookson, son of a Penrith brazier, came to Newcastle to seek his fortune.[1] He found it as merchant, coal-owner and manufacturer but his manufacturing interests were in iron, glass and salt and lead manufacture was only introduced to the family activities in the mid-nineteenth century by a great-great-grandson of the original Isaac. There is, however, one earlier link with lead, since Isaac had a nephew, also confusingly called Isaac Cookson, who also came to Newcastle, probably in 1720. He became one of Newcastle's most famous silversmiths at a time when the town had a considerable reputation for the quality of its silverware and about 1743 went into a partnership to work lead ore at leases in Middleton-in-Teesdale. The partners also took a lease from John Hodgson of land at Elswick for lead refining works. Here they not only desilverised the pig lead, the silver going to Isaac Cookson's work-shop, but also made sheet lead and cast shot. In 1754 however Isaac died, without a son to follow his trade, the Elswick refinery passed into other hands and although Walkers, Fishwick set up their works at Elswick, they were on a different site and the earlier refinery seems by then to have fallen into disuse.[2]

The claims of another firm, James & Co., to have been the progenitor of ALM and what little is known of the firm's early history may also be disposed of here. The firm laid claim to 1774 as the year of its founda-tion but it seems certain that John James, who started the firm, was actually an importer and retailer of tobacco and other goods in Sandhill, Newcastle. His son, William, added to the earlier family interests the activities of the chemist and druggist and began to deal in dyes and colours. Since white and red lead were the most important contem-porary pigments, he therefore had an incentive to manufacture them in order to supply his customers. In April 1799 it was said that 'Hind & Co. are going forward with their buildings',[3] while the first reference in a local directory comes in 1801 when James's, Hind & Co. were estab-lished with a white and red lead works on the Ouseburn, a tributary of the Tyne, to the east of Newcastle. This firm appears to have been a partnership between William and Thomas James and William Hind, another Newcastle druggist, who had also become involved in pigments as a lamp-black and colour manufacturer in Pipewellgate, Gateshead. The firm clearly prospered since in 1817 William James was able to purchase the estate of Deckham Hall, Gateshead, with 37 acres of land, for the sum of £7,100, while his son, Edward, was able to lease what was probably the largest house in Gateshead, Park House or Gateshead Park, where he was living in 1839. By the mid 1840s, however, an estate on the edge of Gateshead, whose only claim to fame was the description of it, attributed to Dr Johnson among others, as 'a dirty lane leading to Newcastle', was regarded by Edward James as an unsatis-factory residence and he moved to the more rural delights of Wylam Hall, leased from another lead manufacturing family, the Blacketts. Subsequently he purchased some nearby land in Wylam, on which he had Holeyn Hall erected. The places of residence of industrialists not only throw light on their rising (and sometimes falling) fortunes but also show a movement away from urban and industrial scenes which was common among the Tyneside lead manufacturers and industrialists generally. Beyond this there is little that can be said about the James family and the Ouseburn works until the latter became a limited company in the 1880s. The works remained small, employing 25 men and 33 women in 1842, and, uniquely on Tyneside, concentrated production entirely on red and white lead.

Although members of the Cookson and James families were active on Tyneside earlier, the palm for being the first of the subsequent constituents of ALM to be formed must go to Walkers, Fishwick & Co. It also seems probable that this was the first lead manufacturing company of any significance in the area. This is somewhat surprising given the considerable level of shipments of pig lead from the river and the fact that smelting and refining had been established. So early as 1696 a lead smelt mill had been established on the Tyne at Ryton, the first north-eastern venture of what was to become the London (Quaker) Lead Co.[4] Neither this company nor the Blacketts, the other leading lead mining concern in the north east, appears to have ventured into the realm of manufacturing lead on Tyneside. They remained as lead merchants selling their pig lead chiefly on the London market, but it must have been the increased supply of lead from the northern Pennines which attracted the Walkers, Fishwick and Ward to Newcastle.

Samuel Walker I [1715-82] was a Rotherham iron master, the son of a local nailor, at which activity he had made a considerable fortune.[5] Also to be involved in the lead venture were his four sons, Samuel II [1742-1792], Joshua [1749/50-1815], Joseph [1752-1801] and Thomas [1756-1828], together with Richard Fishwick and Archer Ward. The latter two were Hull residents. Ward was frequently described as a merchant and it is probable that his link with the Walkers came from the fact that Hull was the port which the Walker iron foundry at Rotherham would have used for shipments of goods coastwise and abroad and Ward may well have been the Walkers' agent in the port. Fishwick is also occasionally described as a merchant, although the entry admitting him a burgess of the town in 1767 states that he had been apprenticed to 'John Banks, grocer' while the Hull Poll Book of 1774 describes him as 'book-keeper'.[6] The only clear link between him and Ward, as we shall see later, is that they were active Baptists. In 1778 Samuel Walker I, Fishwick and Ward entered a partnership to erect a white lead works at Elswick, near Newcastle. That white lead was to be the sole product suggests its importance within total lead manufacturing at the time and that demand for it was rising.

The important question as to why these people should have entered lead manufacturing is unanswerable. As Arthur John pointed out in his edition of the minutes of the Walker enterprises,[7] the step from ferrous to non-ferrous activities was far less great in the eighteenth century than it would seem now, since manufacture was far less specialised. The

Walkers had worked in brass as well as iron and steel and it is possible, as a Hull merchant, that Ward would have shipped lead. What actually sparked off the new venture is, however, totally obscure and there is no evidence to support John's suggestion that Samuel Walker I was 'approached to finance a new venture in the making of white lead',[8] although it is perfectly feasible that he did not initiate the project, given that the activity was to be chemical rather than metallurgical. One slight clue may be that the Walkers had a pottery in Rotherham, to which the first cask of white lead made at Newcastle was sent in May 1779, where it would be used as a glaze.[9] The decision having been taken to set up a works, however, it is surprising that Newcastle was chosen rather than Hull, the established base of two of the partners. The reason may have been that there were already established white lead works in Hull (with which, it has been suggested, Fishwick may have been connected, since it was he who was to show the greatest technical knowledge of the product).

In April 1778 the partners leased two acres of land at Elswick, a mile or so west of Newcastle, from John Hodgson for 99 years at an annual rent of £20. Erection of the works does not seem to have commenced until July, the first entry in the company's letter book being dated 7 July, while the intervening period had presumably been taken up in the arrangements for Fishwick and Ward to move from Hull, since they were to supervise the erection of the works and then manage it, while the Walkers remained financial partners only. Having selected Tyneside, it is clear that the actual site had been chosen carefully. It had good communications, with a road to Newcastle and access to river transport which was the major means of moving heavy commodities such as lead because of the high cost of overland transport by horse and wagon. Indeed, as we shall see, nearly all of the lead works set up in the next 100 years were based on rivers or canals for the delivery of their raw materials and access to markets for their finished goods. The Elswick site also had a water supply which was essential for the washing process in separating the corroded lead from the metal.

Although the land may have been suitable the landowner was to prove something of a handful. After a couple of years' experience Ward wrote to Joseph Walker on 9 February 1780, 'It is a very common with Mr. H to trifle with his tenants for years, he ought to be handled smartly to bring him to his sences.' The problem appears to have been that having signed an agreement for a lease, Hodgson was trying to insert covenants into the formal lease which were unacceptable to the partners and the lease was not finally signed until 1 October 1780.

William Cuthbert, Hodgson's attorney, had written to Samuel Walker I on 26 November 1779 that Hodgson, 'had from an Enquiry into your Character a perfect Confidence in you, but if you give up the concern he would wish to limit the person who succeeds you in the undertaking and oblige him not to burthen his Estate with a numerous poor . . .' Hodgson therefore wanted the lease to include a covenant which would prevent the settlement of poor people on the land. One must have sympathy with Hodgson owning a rural estate on the edge of a large and expanding town. There was no doubt some outflow of people from Newcastle who would try to establish a settlement in Elswick which would entitle them to poor relief from the parish should they become destitute, while Hodgson, as the chief landowner, would be the major contributor to the poor rate. It seems certain, however, that Hodgson was being unreasonable, since correspondence between lawyers for the two parties during 1780 shows that there were difficulties not only over this covenant but also over the area of land covered by the lease and other matters. The partners appear on occasion to have been treated rather rudely:

> Mr. Hodgson's compliments to Mr. Fishwick and acquaints him — he has fully considered the proposals, for the additional ground to the factory, and for building a private house, and disapproves thereof; as not inclining, at present, to apply any more of his estate, in that way.

It is clear that the relationship was far from harmonious and some of the events were farcical. On 15 March 1781 Ward informed Joseph Walker that the wife of an employee who lived in a works' cottage had died, necessitating the taking of the corpse out by road. He had applied to Hodgson to give directions that the corpse should not be stopped on the road but it was stopped by Anderson, over whose land the road ran. The corpse was then taken across Hodgson's fields, which he had said they might use on payment of the sum of one penny. When stopped and asked for this payment, however, Ward had said 'they should pass without any acknow't whatsoever, or their the corps should actually stand, so the same passed'.

The death, soon afterwards, of Hodgson and inheritance of the estate by his son, also John, then still a minor, must have been greeted by the partners with some relief. Relations with the landlord, at first through an agent, certainly became easier. Further land was leased for the expansion of the works and manufacture could go on unhindered.

In the middle months of 1778 the problems of the lease lay ahead and it was the building of the factory[10] which was the priority. Fishwick and Ward were involved in much travelling backwards and forwards between Hull and Newcastle as they made arrangements for a permanent move but one or the other was always at Newcastle to supervise developments, although it is clear from the copy letter book that they were feeling their way somewhat in the dark. Fishwick wrote to Joseph Walker[11] on 11 July:

we meet with Difficulties which obstruct our getting forward so quick as we wish . . .

   Being Strangers in the place & unacquainted with the real Characters of Workmen and Builders we find we are liable to be imposed on by such as wait to take Advantage this Oblidges us to proceed very leisurely & though we have received sundry Proposals we have not yet made any Agreement with any their is only 25 p.cent difference between the Estimate or Proposal of one Mason & that of another & about 20 p.cent in the Carpenters Work proposed for — could we certainly know which of the Undertakers of Buildings are or whether any of them are real honest Men we might be able to judge of the cause of this Newcastle Moderation . . .

The work proposed included a stone wall, six or seven feet high, around the works, 100 yards by 36 or 38 yards, 'with a Stack house a Stone House Melting house Wash house Brew House & [wind] Mill' with '3 small houses for the Workmen & if you approve of it a Dwelling House for Mr. Fish^k & A.W. upon as moderate a plan as may be this we think cannot be any Disadvantage as it would let from its Situation for a great rent in case the works should be given over . . .'

Despite Fishwick's suspicion of Newcastle builders there appear to have been few problems in the basic construction work for subsequent letters ignore the subject. A chief concern seems to have been preparations for manufacture. Although many comments on Fishwick refer to the fact that he had prior knowledge of white lead manufacture (usually the reason given for his inclusion in the partnership), it is far from clear that the partners knew what they were about. The very first letter in the firm's letter book asked William Bynon, the millwright they were employing, to find out the strength of the vinegar the white lead makers used. This enquiry hardly suggests a clear knowledge of the process and subsequent letters, enquiring for supplies of pig lead, show that no arrangements for the supply of raw materials had been made before the partners' arrival in Newcastle. In other areas, especially commercial ones, the partners appear to have been efficient. Early arrangements were made to bank with the Tyne Bank of Baker, Shafto, Ormston, Cuthbert and Lamb, while Fishwick reported to Joseph Walker on the various books he had purchased in which to keep the firm's records. At the same time he showed that they were not going to waste any opportunity of making a little money, by taking a crop of hay off their land – 'The Hay is cut & made the Quantity about 2 Loads Quality very good.'

By May 1779 the first white lead had been made and the first cask sent to the Walker pottery, although they may not have thought a great deal of it since there were clearly problems. On 15 May Benjamin Bowser, of Carlisle, who had placed an order, was told that there were two types of white lead sold at 30s and 25s per cwt (on terms of six months' credit or 4 per cent off for ready money), the cheaper 'is a fine colour but by being new & hastely forwarded in the drying is not so firm a texture'. More honestly a letter of the same date showed that the whole of the first make was less than satisfactory. It was sent to John Dowson,[12] with whom Joseph Walker had made arrangements for the sale of the partners' white lead on commission in London:

> . . . Our first making has proved rather Defective in point of firmness . . . when the Second parcell is suffitiently dryed we shall be able to form a better Judgment than we possibly can do at present & unless it proves Superior we shall hardly think of venturing a Tryal into so difficult a Markitt as that of London.

Nevertheless the partners took it philosophically and the letter continued, 'Few Manufactures in their Infancy attain to that Compleatness which is known to be attainable, but find Experience a skilfull Teacher.' It seems that subsequent output was of a satisfactory quality although complaints from customers did occur occasionally. The strong response, to a complaint in 1781 from Butterworth & Collier that they had been sent adulterated lead, suggests that the partners were trying to maintain a high standard:

> It is with the utmost possible degree of Astonishment & concern that we read your Letter of the 24 Inst [October] & were we not possessed of the most rigid Consciousness of having sent you always *genuine unadulterated* Lead we must of necessity have fallen under the Charge & submitted to the opprobious Censure you are pleased

to put upon us as well as an adequate Deduction in the Price of the Article you have Suspected to be reduced by a different Substance.

We *can* and *do* most Solemnly assure you that from the first of our manufacturing White Lead we have *never in any one Single Instance* sold *so much* as a pound thereof for dry White Lead which we have *known* or *suspected* to be in whole or in part mixed or adulterated with any other Substance whatsoever. We have frequent Applications for White Paint of inferior Quality & this Article we do reduce but we always sell it as such & denominate it *reduced*.

Since the firm remained as customers we may assume that the problem was overcome.

It was especially for entry to the London market that the greatest concern about quality was shown, for here the Elswick production would come into competition not only with white lead from Sheffield and Nottingham but also from London makers who were rising in significance. The chief problem appears to have been the suitability of the pig lead used. As was to be expected the partners had begun by using local pig lead, bought chiefly from Sir Thomas Blackett, but they soon began to purchase lead from Yorkshire and Derbyshire to mix with the local pig in order 'to obtain a firmer body to our lead'. The difference in suitability appears to have been reflected in price. The market price of local pig lead at Newcastle was 5s or more per fodder[13] below that at Hull (when the price was about £13 per fodder) and the partners continually grumbled at having to buy lead at Hull at the higher price, together with a further charge of 5s per fodder for shipping it to Newcastle. On several occasions in the early years attempts were made to get information which would enable them to improve the quality, as on 1 June 1782 when the London agents (by now Lockwood, Russell & Parkinson of Southwark) were asked to find out whether the London white lead manufacturers used Derbyshire lead and how they treated it.

While entry to the large London market was of considerable importance for expansion most of the early output, estimated in 1779 to be between 200 and 250 tons p.a., went to the north-east, with Tyneside accounts, most of them small, predominating. Competition here was slight and the partners were able to make personal contacts with painters, druggists and drysalters, but the wider area of Northumberland and Durham also offered a relatively easy market, while most of the remaining accounts were in Scotland or the east coast of England where coastal shipping provided a simple transport mechanism. As with the first London agents, some of the earliest customers came through the Walkers' existing trading contacts, which certainly provided some Yorkshire customers who would not otherwise have been obtained. Occasionally these contacts were not without their embarrassments — on one occasion as a result no doubt of the fact that not all clerks in the eighteenth century yet had copper plate handwriting. On 15 May 1779 an order of white lead was consigned to John Wells of Portsmouth, with whom Joseph Walker had made arrangements for sale on commission. A week later a letter was sent to John Wilks of Portsmouth pointing out that the order and previous letter had been sent in error to John Wells. The problem was then compounded by a letter from a Mr Wells telling the London agents (who were responsible for shipping the goods on to Portsmouth) not to send him goods about which he knew nothing. By this time, however, they had already been shipped and the vessel was to fall into the hands of a French privateer, from which it had to be ransomed. The chapter of accidents was as yet incomplete. The white lead eventually reached Wilks who failed to sell it. Several letters from the partners tried to get some action from him but it is clear that he was, as Fishwick described another customer, 'a very dilatory chap'! Eventually in July 1780 he must have replied for on the 18th Fishwick wrote to him expressing horror at his suggestion that he should auction the white lead and asking him to ship it back to the London agents.

Ventures which ended so unfortunately as this were rare but there was often difficulty in obtaining payment, even from regular customers such as Hartley, Greens & Co. of Leeds. On 31 July 1781 a letter to them commented, 'When it suits your Convenience we shall be obliged to you to remit us for the Lead sent you in October last amounting to £22 17s 9d and charged on six Mo^s Cr.'. A month later, when more lead was sent to the same firm, the plea was repeated and payment was finally received on 7 September. To a new firm the giving of long credit could produce severe cash flow problems, especially in an industry such as white lead, where working capital was already large because of the stocks of lead tied up in the stacks. Walkers, Fishwick & Co. do not seem to have been troubled too much since they were fortunate in having a sound financial backing through the Walkers. They were highly credit-worthy and were, therefore, able to run up an overdraft at the bank and, in their turn, get credit from their suppliers of raw materials. This must have stood them in good stead as the practice by customers of taking more than six months' credit became frequently formalised by the end of the century in quoting twelve months' credit terms.

Despite some early problems it is clear that the company was profitable from the beginning and in the first twelve months to May 1780 a turnover of nearly £4,000 gave a profit of £712.[14] This seems to have encouraged further investment to expand capacity. It is probable that the partners had begun with a single white lead stack (or at the most two stacks) but the discontinuities of production (given a period of about six weeks from filling to emptying a stack) would soon make it clear that greater capacity was needed in order to smooth the flow of output, providing the demand was sufficiently high. By 1783 turnover had risen to about £10,000 p.a. implying an increase in the number of stacks and it is clear from the stock book that there were five in 1784, increasing to eight in 1788, when the turnover of white lead again shows a large increase. Since the stacks were merely brick buildings the increase in capacity required very little in the way of additional fixed capital and in this sense the firm bears out the standard view that fixed capital formation was not high during the early stages of industrialisation. It must, however, be remembered that the increase in capacity made considerable demands for extra working capital in the form of stocks, work in progress and, in view of what has been said above, credit sales.

These increases in capacity appear to have been financed without any serious difficulties, either in terms of provision of capital or in cash-flow. Profits were high and it is possible that they alone were able to finance the expansion of fixed capital which took place. Without any annual figures for new investment or partners' drawings against annual profits, it is, of course, impossible to say anything of use on the means by which expansion was financed, but total profits of almost £150,000 over the period 1780-1800 would leave scope for considerable internally financed expansion even after allowance for partners' drawings. Indeed, while recognising the difficulties attached to making such comparisons,[15] the consistency in profit earnings over the period is remarkable, fluctuating only a little around 20 per cent of turnover, with three very good years in the middle 1780s when it was around 30 per cent and a poor year in 1789 when it slipped to 8 per cent during a time of adjustment to a large expansion in capacity.

It has often been said that it was the Walkers who provided the major part of the capital — 'Samuel Walker appears to have found the larger portion of the capital, whilst Archer Ward placed his business connection, and Richard Fishwick his practical knowledge of the manufacture of white lead at the service of the new firm . . .'[16] The evidence of the partnership's ledgers, however, suggests that the partners had equal shares in the capital.[17] They increased their capital commitment as the accounts for building and stocking the factory were paid, each contributing £30 at the first ledger entry on 25 April 1778, and by October 1779, when the factory had been working for some months, they each had a capital of £1,487. By October 1782, when the first ledger was concluded on the death of Samuel Walker I and the commencement of a new partnership including his sons, each of the original partners had a capital of £2,687. Bearing in mind the fact that the profit figures were almost certainly struck before any payment to Fishwick and Ward for management (and are therefore overstated), the firm showed a return on capital employed of around 15 per cent in the first year of operation and 35 per cent in 1782. If we make an arbitrary allowance of £200 p.a. each to Fishwick and Ward as salaries, these figures would be reduced to about 7 per cent and 30 per cent, which show a handsome return, justifying the original decision to erect the works and amply supporting expansion.

Apart from his contribution to its capital, Samuel Walker I, already over 60 when the partnership commenced, appears to have taken little part in its affairs. He was heavily committed to the iron business at Rotherham and from the beginning it was his third son, Joseph, who looked after the Walker interest in the partnership. The Newcastle partners sent him copy letters of partnership correspondence and there was also the occasional meeting of the partners (in June 1781, for instance, Ward was at Rotherham), especially when any major project was contemplated, and at the, usually annual, stocktaking, which was a physical count of the stocks and equipment by all the partners. Archer Ward seems to have been the less important of the two active partners, although he was responsible for office management (and probably factory management, in spite of Fishwick's supposed knowledge of white lead manufacture). He made all the payments and undertook some of the correspondence. It was, however, Fishwick who appears to have been dominant (possibly by personality) in the Newcastle activities. His letters in the early letter book are unmistakeable, signed as they are 'for Walkers, Self & Co.'. In his absences, which as we shall see were frequent, important letters, for instance from a London agent, were left for him to answer on his return. The absences from Newcastle were largely a result of the fact that he appears to have had a role which corresponds partly to that of a modern sales director, although he also had a variety of interests and a wandering spirit, as Frank Beckwith noted, 'Fishwick had a habit of bobbing up unexpectedly'.[18]

At this point it may be worthwhile to step aside from the partnership to sketch what we know about the active partners. Ward was born on 28 September 1743 and died on 22 July 1800 ' "at his house near the White-Lead Works," Greenhill House, Derby, a works which, as we shall see, he set up in 1792 after he had left Newcastle'. The chief thing that we know about him was that he was an active Baptist, yet another example of the Non-Conformist entrepreneur of the early industrial revolution. He was three times married, on each occasion his wife being related to a Baptist minister. He had moved to Hull in 1774 from Bishop Burton, some ten miles to the north-west, where he had been a farmer. He spent part of the time before going to Newcastle working for Williamson and Waller, Hull merchants. Joseph Kinghorn,[19] son of Ward's minister at the Baptist Church at Bishop Burton gives us some idea of Ward's life at Newcastle. At the age of 15 Kinghorn went to Newcastle to be a clerk at the white lead works and he lodged with the Wards. The second Mrs Ward he found to be highly unpleasant and she clearly made Ward's homelife unattractive, while Kinghorn also found it difficult to listen to Ward's frequent utterances on religious matters. It was much to Kinghorn's relief that Ward was away for some weeks in June 1782 and he was able to move to temporary lodgings with Fishwick. Ward's absence also lightened Kinghorn's workload since it is clear from his letters to his father that he was normally kept very busy by Ward for long hours.

Although Kinghorn was employed in the office by Ward, the latter seems to have spent much of his time in the factory, as a result of which his health suffered, although Fishwick thought that Ward's relationship with his wife did not help. When Ward was ill in 1787 and went to rest at the newly developing resort of Brighthelmstone, Fishwick wrote sarcastically, 'Nothing will in my opinion conduce so efficatiously to Mr. Wards recovery as a relax into the country *with his wife*.' It was, however, Ward's wife who died first in 1796 and the Baptist circle, which had not been slow to comment on the shortcomings of his family life, was soon able to speculate on the likelihood of his marrying his former wife's companion. Joseph Kinghorn informed his father that 'Marrying is with him [Ward] a *safe* experiment: he cannot jump out of the frying pan into the fire.' When the marriage took place, on 25 May 1797, David Kinghorn informed his son, perhaps a little enviously, that 'The new couple set off for Matlock . . . with 2 maids and a footman . . . Thus Lead is turned into Gold . . . The Ladies Maid is now the Lady.'

It is clear that the lead industry had raised Ward considerably in fortune and social status and two years later David Kinghorn noted the sale of a ship – 'Mr. Ward had 900£ in it, for his share he received 1000

Guineas: we are told W. has made his will and settled 300£ per annum on Mrs. W. Wealth flows in like a sea, or this could not have been done; he never could have done it by farming.' But Ward's good fortune was not to last. Although David Kinghorn had heard that Ward was 'young again ... so a young wife puts spirits into the aged', he continued, 'After all, I do not think he will ever again have good health, the Lead having tainted the inside, its effect is not soon removed.' This was correct and Ward died only three years after his third marriage.[20] In his will[21] he left his estate to his wife and this included not only his share in the lead partnership but also a share in a 'cotton, callico and printing manufactory and business' at Newcastle-under-Lyme. He also left the sum of £5,000 in trust for his recently born daughter Ann and £2,000 to his sister, Ann Brown.

Ward appears as one of the unsung entrepreneurs on whom the industrial expansion of the late eighteenth century was based. Coming from a rural background he took advantage of the opportunities which came his way and successfully diversified his fortune. In the lead partnership he appears to have been happy to let the extrovert Fishwick play the dominant role but it is clear that he was a competent entrepreneur. When the partners decided that their initial venture at Elswick justified a further expansion, with concentration on the London market, it was Ward who left Elswick in 1785 to establish a white lead works at Islington. Seven years later, when further expansion seemed justified, it was Ward again who went to Derby to build the white lead works where he was to remain until his death. Ward's was not a particularly startling career, in the form of those of the classic industrial revolution entrepreneurs such as Arkwright, Wedgwood or his own partner, Samuel Walker I, but it was one which must have been repeated many times in the late-eighteenth century and without such contributions the British economy would have grown less remarkably.

Fishwick was, of course, from a similar if rather more heroic mould, an extrovert, whose interests were manifold, who could always find time to be involved with the problems and affairs of others and a very hard-working (and hard-riding) man. His letters frequently refer to long journeys on company business by the overnight mail coach and on one occasion, after a business trip along the east coast, 'I reached home on Saturday night but not without a severe push; the last day was an 80 miles job' [on horseback].

Fishwick had been born in 1745 and was, therefore, two years younger than Ward, whom he was long to outlive, dying on 17 January 1825. By his first wife he had a son, also Richard, who followed him into the lead business, being employed as a clerk in the counting house. It was typical that Fishwick senior, writing to Ward in June 1787, when his son was being sent to Islington to gain experience, should comment, 'let him go thro all the inferior Employ the same as an apprentice'. This early example of management training did not prove successful, for although Richard (junior) was subsequently employed in the Newcastle office in the mid-1790s, he soon left the firm.

Together with Ward, Fishwick (senior) played a major role in revitalising the Baptist meetings at Newcastle which had declined in number as a result of the divisive influence of Unitarian doctrines. In the early 1780s regular weekly meetings were held alternately at the houses of Fishwick and Ward and Fishwick used his connections to bring Baptist ministers to Newcastle, although his authoritarian nature seems to have driven as many away, Beckwith noting that at least three ministers left the Newcastle Baptist church after disagreements with Fishwick.[22] Nevertheless the growth of the church owed much to Fishwick who contributed generously to the cost of building a new chapel which was opened in 1798. Indeed he was a generous benefactor of Baptist institutions in general, giving £350 of the £700 needed to purchase the North Shields Baptist Church in 1804,[23] supporting a number of potential ministers through college and even, at one point, proposing the establishment of a Baptist theological college in Yorkshire. In 1798 he was made the first treasurer of the Northern Evangelical Alliance.

That Fishwick made a considerable fortune out of the lead works is clear but he also had fingers in a number of other industrial pies and especially in iron manufacture. About 1797 Fishwick entered a partnership with George Gibson (senior) and his son, also George, London architects, together with Aubone Surtees of Surtees Bank in Newcastle, in order to commence the first modern ironworks in the north-east, at Lemington on the Tyne, a few miles west of Newcastle. There was a local market for iron, growing rapidly as a result of the demands of the Revolutionary and Napoleonic Wars, which the new firm could tap and the beginnings appeared to be as auspicious as those for the lead works had been. As befitted a man who in his other interests had partners of the significance of the Walkers, Fishwick was to erect an integrated ironworks on a considerable scale, with two furnaces, with a potential output of 5,000 tons p.a., forges and finishing shops.[24] Fishwick was thus making the reverse diversification from that which the Walkers had made some 20 years earlier and he was doing it on some scale, which may account for his withdrawal from the lead industry when the partnership with Ward and the Walkers terminated at the end of April

1800.[25] With Fishwick's drive the ironworks might have been successful but for the failure on 30 June 1803 of Surtees Bank in Newcastle with whom the Lemington company banked. Lack of bank finance for working capital compelled the stopping and eventual sale of the ironworks and Fishwick lost a lot of money. The original partnership in the ironworks was dissolved on 4 October 1803, the business sold and, much reduced in circumstances, Fishwick left Newcastle in 1806 for London, where he died at Islington, although it is probably coincidence that the lead partnership had a works in that place.

It is clear that both Fishwick and Ward were heavily involved in the practical aspects of lead manufacture and they both took an interest in and made improvements to the white lead manufacturing process, although only Fishwick's was of any great significance. His name has gone down in history as the only person to make any improvement to the stack process in several centuries. It took the form, in patent no. 1581 dated 15 January 1787, of the use of spent tan bark in place of horse-dung to provide the heat necessary for the stack process. Most of the patent is taken up with a detailed description of the existing stack process and the only section of importance states:

> that my said invention is the substituting in such process for preparing or making white-lead, as aforesaid, in the room or place of horse-litter, the use of the common tanners bark, or of the bark of the oak-tree, after the same has been previously applied or used by tanners, for making the liquor or tan which is used in tanning, and preparing hides for leather: and that I have found, by repeated experiments, that such tanners bark will communicate a more certain and equal degree of heat to the vinegar and lead, placed over or upon it, than the horse-litter has hitherto been found to do.[26]

To some people such a change seemed too insignificant to justify a patent and one Newcastle historian, with a considerable interest in industrial affairs, wrote, 'We believe a patent has been obtained ... merely for mixing oak bark with the dung.'[27] The sarcasm implied in 'merely' serves to show the gap between the interested amateur and the manufacturer. If tan bark, in its fermentation, gave a more regular heat than dung, as seems likely, then it would have done something towards the elimination of the problem of the burning and blackening of parts of the corrosion as a result of too high a temperature. Fishwick had presumably been working empirically to try to overcome the problem of poor corrosion, although there is no evidence as to how he hit upon the use of spent bark, which was, of course, widely available as a waste product in any large eighteenth-century town. It is not, however, clear as to whether Fishwick was aware of what was certainly to become an important effect of the use of tan bark − an improvement in the colour of the white lead. One result of the fermentation of horse-dung, in addition to heat and carbon dioxide which were essential to the process, was the release of hydrogen sulphide which was not, since it caused a discolouration of the corrosion and the production of off-white (or even grey) white lead. This problem was much reduced with the use of bark, although complaints continue into the early-twentieth century of poor colour white lead as a result of pieces of bark contaminating the corrosion and causing it to yellow. Another result of the introduction of bark was a further lengthening of the time taken for the corrosion process and in his patent Fishwick gives the usual period as 'two calendar months'. As a corollary to producing 'a more certain and equal degree of heat', bark produced its heat more slowly since it had a lower rate of fermentation. As a result it took longer for a stack to build up to the temperature at which the acetic acid would evaporate and, therefore, longer for a satisfactory level of corrosion to build up. By the end of the nineteenth century the time-scale for corrosion seems to have standardised at about 13 weeks, 50 per cent longer than the figure given in 1787 by Fishwick, which was itself several weeks longer than the norm in the mid-eighteenth century.

One result of this lengthening of the process was that the working capital tied up increased proportionately, giving advantages to large firms and making entry to the industry difficult for those without considerable resources. As in most industries in the period it was access to adequate working capital, rather than the fixed capital required to erect a plant, which was the barrier to both entry and subsequent expansion. The higher costs involved limited the use of Fishwick's process and although his patent referred to the substitution of tan bark for dung it was a long time before this took place and white lead makers tended to try to make a balance between too much bark (with rising costs) and too much dung (and poor quality white lead). At Elswick, Fishwick continued to use dung (mixed with bark) into the 1790s, and although there are no figures to suggest the proportions in which they were used, the monetary figures for stocks of dung in the annual accounts imply that its use was continued on a considerable scale.

By contrast with Fishwick's concern about the manufacturing process, Ward's improvement was more concerned with the environment and health of the employees. In 1794 the Society of Arts had offered a prize of £50 for a 'method of preparing white lead which shall not be prejudicial' to the health of workmen. Ward responded to this and was awarded the Society's gold medal for the improvement which he communicated to the Society by a letter dated 2 January 1795.[28] Referring to the fact that corroded lead, after being removed from the stack, was passed through iron rollers in order to separate the white lead from any remaining metal, Ward noted that the result was that 'a very considerable quantity of fine dusty white lead is raised, which almost covers the workmen thus employed, and is very pernicious to them'. He went on to propose the use of rollers partly submerged in water in order to obviate the raising of dust but the principle of wet rather than dry grinding was not new, nor was wet grinding to become the norm. It perhaps interfered with the effective separation of white from blue lead, and on several occasions up to the end of the nineteenth century new processes for wet grinding were heralded as breakthroughs in dust prevention. It does, however, seem that the process was used more widely than merely at the Derby White Lead Works, where Ward was managing partner, since a testimonial from Samuel Walker Parker to the Society of Arts, dated 19 January 1795, stated:

> I do hereby certify that the White Lead made at the manufactory at Islington belonging to Walker, Ward, and Co. is, and has been for some time past, manufactured in the manner stated by Mr. Ward, in his letter, . . . and that many tons have been manufactured; and that, since Mr. Ward's plan was adopted, no other method has been used.

The firm had much cause for concern to reduce dust problems since lead poisoning was a common result of employment in a white lead works. Nevertheless there does seem to have been a difference in the severity of lead poisoning in various works and at Newcastle Fishwick gives the impression that it was not common, and that the firm went to surprising lengths and expense to look after employees who suffered the disease. So early as 2 June 1779, within a month of having produced the first white lead, he wrote to his London agents:

> one of our men having fallen ill of the Lead Distemper which is either through obstanacy or foolhardness he subjected himself to he has been exceeding careless of taking the necessary precautions & we fear it will go very hardley with him — a skilfull Surgeon is ordered to attend him & everything afforded that seem likely to promote his recovery.

Thereafter references to lead poisoning at Elswick are few and that this is a result of infrequent occurrence rather than indifference is suggested by some correspondence in 1787, between Fishwick and Ward at Islington. Ward had asked for some Newcastle girls to be sent to Islington to work in the stacks there and in August Fishwick noted that five girls had been sent.[29] On 2 November Fishwick commented on the return to Newcastle of three girls, 'two of them in a very distressing Condition, Grearson I think will not be long in mending but Reveley will not be stout of several Weeks, they are both under the Surgeons Care'. Fishwick's evident concern and his subsequent account of the efforts to restore the latter girl's health imply that working conditions at Islington were more deleterious to health than those at Newcastle.

The correspondence between the partners also gives some insight into the partners' characters and especially that of Fishwick, who wrote all of the outward correspondence from Elswick (most of it to a new London partner, Thomas Maltby), after Ward's departure for Islington in 1785. We have seen that Fishwick was something of a mercurial and domineering character, although Joseph Kinghorn found him very sympathetic and helpful. His letters, however, tend to bring out the former side of his character and Maltby must have done well to tolerate some of the remarks, although, not having the other side of the correspondence, we can only conjecture as to how he replied. On a matter on which they disagreed Fishwick wrote on 7 March 1789, 'you are *wrong* in your *premises* — false in your *conclusions* — and *erronious* in your conjectures'. A couple of weeks later Fishwick picked up a word which Maltby had used to describe Alderman J.E. Blackett; 'There is a word in your letter which affects my risibles, I mean the word you apply to the Alderman "Ter-gi-ver-sa-ti-on" which were it not very long legged I might make use of sometimes . . .' Inevitably this meant that he would use it, and against Maltby, and later in the same letter he replied to Maltby's further comments on the subject about which they had disagreed earlier in the month, 'You are yet I think incouridgeable [incorrigible] about Terrys litharge matter and I wish you may not be guilty of "Tergiversation" respecting what has passed about it' and, after a further explanation of his own actions in the matter, 'Therefore let us have less *Tergiversation*.' Two weeks later on 3 April there was more, rather pointed banter. Fishwick had asked Maltby to send letters on whole sheets in order to reduce the cost of postage — now he wrote:

I *half* thank you for *half* attending to my recommendation by sending *half* of your papers in *whole* sheets to save postage as we have only had 18d to pay when we used to pay 2/- had the other *half* been upon a *whole* sheet it would have saved 6d more, when you manage *that* you will have my *whole* acknowledgement.

Three days later Fishwick wrote, 'I did not say much about the extra postage the last time, nor do I say much about the same now, but beware the 3rd time!', whereafter the subject was dropped and we can assume that Maltby had been bludgeoned into doing what Fishwick wanted.

Certainly for a man who was making a lot of money Fishwick appears to have been very careful in money matters and not extravagant, either in personal or business affairs. The house which he lived in at the Elswick works, although more grand than the 'dwelling house . . . upon as moderate a plan as may be', which had been proposed to Joseph Walker in 1778, was a modest Georgian house without the luxury trimmings which it was to take on in the nineteenth century. It was built by local men at an expenditure of £570 in 1780-1, perhaps ten times as much as each of the cottages, built at the works to house some of the essential employees, would have cost. There is occasional evidence of presents passing between the partners but these were hardly luxurious. In December 1786, for example, Fishwick wrote to Maltby, 'I thank you for the barrell of Porter, I shall send you some nice Newcastle Ale in lieu of it.' On 8 October 1787 after thanking Maltby for the 'scraps of politics you favour me with', Fishwick asked, 'Pray what will it cost annually to have a newspaper sent hither? I could like the London Packet and I could like to make *the Company* willing to pay for it?' There were, of course, local newspapers[30] but they contained little of the up-to-date political news which Fishwick required in order to manage his export sales. He obtained his London paper and the Company paid for it.

Fishwick was equally anxious to economise on the Company's expenditure as on the demands made on his own pocket and he clearly found it impossible to understand the Edinburgh merchant to whom he had offered £18 per ton for pig lead but who insisted on selling it at 17s 4d per cwt to a Dutch buyer. 'This is the first instance I have ever known of a *Scotsman* not seeing, or seeing not immediately resolving to make use of an advantage so self evident! but he has "lang exported a' his *Leed himsalf* and choizes to do so".' It is in the internal dealings between partners that Fishwick's carefulness in money matters is most apparent. There are numerous references to the making of unreasonable claims and charging too high a price in inter-company trading. On 27 December 1786 Fishwick wrote to Maltby:

You say you suppose I mean to make you some remittances soon, to be sure I do if any are due or if you want to *borrow* a small sum, but before I can judge of the former I must have a few more *Acc'ts Sales* than those you have sent me. As I do not know but you may be like the Islington House calling for money before the proper time of payment & forgetting that we were in advance for you sometime ago?

Careful though he may have been in money matters, Fishwick was profligate with his time in the Company's interest and the expansion of the 1780s owed much to his trips around the country to drum up sales. Once a sales base had been established on Tyneside and along the east coast and the quality of the white lead had been improved sufficiently to tackle the London market, Fishwick turned his activities to wider markets. He was away for nearly the whole of October 1782 on a tour of Yorkshire, Nottinghamshire, Lincolnshire, Staffordshire and Lancashire, 'to find sale for our articles'. If we take another year, 1788, for which Fishwick's letter book was complete, we find him visiting customers in Glasgow at the end of August, York in mid-September and then from late September away for a month of business calls covering Leeds, Manchester, Warrington, Liverpool, Chester, Newcastle-under-Lyme, Derby, Sheffield and York, the major northern towns in which white lead would be used on any scale. On 1 December he wrote to Maltby:

I am meditating a journey on the East Coast of the Kingdom & think of setting from Norfolk, thro Stockton, Whitby, Scarbro', Hull, Lincoln, Boston, Lynn, Norwich & Yarmouth — can you furnish me with any hints likely to be useful in my peregrination *pro bono Co.*?

He left Newcastle in mid-December, spent Christmas with friends at Hull and after Yarmouth went on via Cambridge to London, arriving back in Newcastle in mid-February, during which period Ward had returned from Islington to run the Elswick works.

From the beginning the partners had clearly no intention of running a small white lead works to provide for a local and perhaps regional demand, which had been the role undertaken by most such provincial works of the period. A determined attempt to enter the large London market had been followed by efforts to build up national outlets. In doing this the partners were breaking into local monopolies — especially in towns like Derby, Sheffield, Hull, Nottingham and Chester, where local works were strongly based on the nearness of supplies of pig lead.

In order to achieve sales in such places the partners had to overcome the disadvantage of distance from markets and transport costs which, being of necessity by road for part of the journeys, would be high. Their success in achieving such sales points to efficient, low-cost production and throws some doubt on the frequently expressed opinion that poor transport facilities inland (even with some canal development by the 1780s) made for regional monopolies in the sale of heavy or bulky commodities.

The extension of sales was not left entirely to Fishwick's 'peregrinations', since in addition to personal visits the partners endeavoured to appoint agents in the major areas of demand. We have seen the early appointment of an agent for the London market and a similar step was taken for Staffordshire, where the potters were increasingly using white lead as a glaze. Agents were, however, inevitably more interested in their own than their supplier's affairs, were prone to be bribed to take someone else's wares, would lose interest in particular products and so on. Walkers, Fishwick and Co. went through two London agents before, in 1785, deciding that a London partner was a more satisfactory way of developing that potentially large market. The Staffordshire agent was little more satisfactory until Fishwick hit on the plan of giving him sole rights to sell for Walkers, Fishwick in the area. This worked for some time until on a visit in 1788 Fishwick discovered that the agent had recently died and that his son had transferred his allegiance to a rival firm of white lead manufacturers. It is an interesting pointer to the power which the Elswick firm had established after only a decade in the industry that Fishwick's threat of direct sales to the potters brought the young man to heel. A further method of obtaining sales was shown at Derby during Fishwick's same trip. Here he obtained an order from Richardsons who 'sell a great deal of Lead ... on conditions of our occasionally taking Lead from them in Pigs' and a similar agreement to supply manufactured lead in exchange for pigs was made with another Derby firm. Since most of the partners' pig lead needs were obtained from the northern Pennine mines in Northumberland and Durham, this method of tying agents by obtaining raw materials from them could not be widely used. In any event in most areas of the country there was no pig lead available and agents had to be trusted to stock and sell the firm's goods. A final but very insignificant way in which sales were increased was by exports. Before 1800 there appears to have been no serious attempt to expand such sales which came only fortuitously, as with shipments to Thomas Fishwick (presumably a relative) at St Petersburg to be sold on commission.

If these are the areas in which demand for the firm's products increased it is well to look at the ways in which supply was expanded. After the initial development of white lead the earliest expansion was into the related field of painters' colours. This development was under consideration so early as September 1779 and in the following November the first colours were made and Fishwick was endeavouring to obtain agents to sell them. It was a sensible extension of interest since it was not only closely related to the manufacture of white lead but raw materials could easily be obtained locally, molasses from the Gateshead and Newcastle sugar houses, copperas from local collieries, while chalk and sand came as ballast in returning colliers and tar, pitch, turpentine and rosin were readily available from Tyneside's considerable trade with the Baltic. The range of colours made was considerable and included Spanish brown, purple brown, Venetian red, yellow ochre, colcother (red iron oxide used by jewellers as a rouge), chocolate, ivory black and lamp black. It may be that colour production was initially commenced at Elswick but by 1780 the partners had certainly taken the lease of a building known as Bussell's Factory House at Paradise in nearby Benwell.[31] Initially the project appears to have been successful, although with an employment of only four or five workers in the early 1780s it was much smaller than the white lead works, which had about 20 employees at that time. Around 1785/6 the early success was followed by an extension of the colour branch with the annual stock-taking figure for the capital invested in the colour works' buildings, mill and utensils rising from about £550 at the end of 1782 to about £1,100 in August 1786. This satisfactory development was, however, short-lived, for on 14 January 1788 Fishwick wrote to Maltby, 'I intend throwing off all the colour business except white and black at Paradise'. It is difficult to explain this sudden change of mind but it seems likely that the colour trade was a highly competitive one in which little capital was required, there were many small firms and margins were low. The Benwell works was kept on, probably until its lease ran out around 1799 when it was taken on by John Gibson, son and brother of Fishwick's partners in the Lemington Ironworks.[32]

If diversification into colour manufacture was unsuccessful as a means of expanding the firm's activities, the extension of the range of lead products it manufactured was just the opposite. We have seen that there was gradual growth of white lead output in the first few years of the partnership but it was not until 1785 that any diversification of production away from white lead occurred. Table 2.1 shows that in the years to August 1785 and 1786 some red lead was sold but it hardly seems likely that the firm had installed a reverberatory furnace to produce such small amounts and it is probable that these sums represented the resale of bought-in red lead before production commenced in the late autumn of 1786.

Up to the stocktaking of August 1786 it is clear that the partners were consolidating their early experience in the manufacture and sale of white lead, but in the following two years came a massive expansion which more than doubled total sales. In this expansion the production and sales of red lead were clearly of considerable significance, usually accounting for between one-third and two-thirds of the value of the sales of white lead in each of the remaining years of the partnership up to 1800 and in the particularly poor period to March 1794 actually exceeding the value of white lead sales. The letter book for the autumn of 1786 is missing but by the time the next book commences, in December, red lead was being shipped to London from the Elswick works and it was stated that there were four red lead furnaces, capable of making about 2½ tons daily, but that it could not be ground that fast because of a lack of power. Unlike the gradual expansion of capacity in white lead to adjust to market demands, it seems clear that the partners had recognised the scope for red lead sales and begun production on a considerable scale. At this time the production of white lead was about two tons per day (although it was soon to increase as a result of the erection of new stacks) and it is certain that red lead output was below the capacity of 2½ tons daily – the value of sales to August 1788 would suggest an average weekly sale of about six tons over the previous two years. Sales were mainly to potters and glass-works and the network of contacts which the firm had already made in selling its white lead now made it easier to sell the new product on a national market – in August 1787, for instance, a dozen barrels of red lead were despatched to the Bristol Glass Works.

1786-8 also brought significant expansion in the sales of what were described as white lead paint.[33] Since the normal practice was for painters to buy dry white lead and mix their own paint, it seems likely that the expansion of paint sales was a way of dealing with either surplus or poor quality white lead, since it was less easy to detect poor quality or the adulteration with chalk when the white lead was ground in oil. Table 2.1 supports this interpretation, showing larger sales of paint in 1781 to 1782 when there was difficulty in getting the quality of the white lead right, and after August 1786, with the big increase in output. One new product which made some contribution to total output at this same time of expansion was lead shot (to be dealt with in

Table 2.1: Walkers, Fishwick & Co. — Sales of Lead Products[a], 1780-1800 (£)

Date of stocktaking	Period covered (months)	White paint	Milled lead	White lead	Red lead	Shot	Total sales value	Total sales on an annualised basis[b]
May 1780	12	146		3,559			3,834	3,834
November 1782	30	1,744		12,261			14,307	5,723
November 1783	12	124		8,829			9,396	9,396
August 1784	9	65		6,788			7,489	9,985
August 1785	12	42		9,392	253		10,127	10,127
August 1786	12	162		9,773	172		10,424	10,424
August 1788	24	2,803	3,170	21,133	17,038	10,082	55,070	27,535
August 1789	12	1,638	394	15,013	7,468	181	25,475	25,475
August 1790	12	1,522	294	14,649	10,481	249	28,202	28,202
August 1791	12	2,251	14,895	16,973	9,333	3,122	47,720	47,720
August 1792	12	1,969	14,814	26,387	10,416	1,886	56,560	56,560
August 1793	12	1,604	9,593	23,580	7,339	1,411	44,602	44,602
March 1794	7	1,218	4,513	5,603	6,337	447	18,417	31,572
March 1795	12	2,203	9,769	20,456	10,476	2,225	46,541	46,541
August 1796	17	4,465	19,562	41,918	16,129	3,141	87,190	61,548
April 1797	8	1,801	7,989	19,500	7,341	859	38,549	57,823
April 1798	12	3,065	12,536	17,753	12,836	1,475	48,946	48,946
April 1799	12	3,465	11,071	19,815	12,572	3,600	51,209	51,209
May 1800	13	3,845	18,604	29,815	14,595	5,266	73,218	67,117

Notes: a. The sales of a few insignificant commodities are not shown here — the sum of the individual sales shown therefore falls short of the total for each period. b. Monthly average sales for each period times twelve.

Chapter 3), but the other new product, which was to be of greatest significance, in many years exceeding in total value the sales of red lead, was sheet or milled lead.

The decision to manufacture sheet metal brought to a head the problem of power supply which had bedevilled the Elswick works from the beginning. The initial power source for the grinding of the white lead had been a brick windmill, 45 ft high, and costing the considerable sum of £570.[34] Wind was, however, an uncertain power source and lack or excess of it, both illustrated on occasion at Elswick, could be very frustrating when pressing orders had to be met. Comments such as 'a deficiency of wind has thrown us back in the grinding part', tell their own story, while on another occasion Fishwick commented 'on my reaching the factory found all well, only our wind mill has by a bois- terous gail of wind been bereft of two of her sails, providentially no lives were lost nor any hurt, neither much mischief to the buildings has been occationed'. Delays caused by unsuitable weather conditions could not be long tolerated if the firm was to establish a satisfactory clientele by delivering orders promptly. On 20 July 1781 Fishwick replied to a complaint from Butterworth & Collier by stating that lack of wind was delaying production and went on to inform them that:

> we have other orders by us besides yours unfulfilled which have lain by us from 3 to 5 months but as we are erecting a Horse Mill . . . we hope soon to be able to supply the demand of our friends in a more expeditious manner than heretofore.

A horse mill would have the advantages of being cheaper in capital cost and more regular in working than a windmill but the disadvantage of having higher running costs, in feeding, stabling and replacing horses. Between the two, adequate power seems to have been available until the introduction of red lead manufacture placed additional demands for power for grinding. This was at first provided by a third power source, with the erection of a water mill[35] but, like wind, water often comes in inadequate or more than adequate amounts. Fishwick was soon writing:

> We require so much water to grind red lead that we have been unable to grind any of this article for some time and have very little on hand so that we have to my no small mortification been obliged to refuse an order for 12 bar[rel]s of 8 cwt each. But the steam engine will I hope alter the case in *June* next.

Thus a fourth power source was to be tried in less than ten years. It has recently been argued that the impact of the steam engine on industry was much less significant than is usually claimed.[36] Without denying that other power sources remained important well into the nineteenth century, nor that the steam engine only provided a considerable power source from the nineteenth century, the Elswick example does add a little evidence that certain industries could not have expanded as they did without steam power.

The decision to erect a steam engine must have been taken in the early winter of 1786, for on 20 December Fishwick communicated estimates to his London partner of £1,200 from Cameron & Co. for erecting an engine 'proper for rolling or grinding; the Yorkshire Mechanic for 500£ *less!*' It is probable that the 'Yorkshire Mechanic', a man called Oakes (and sometimes Oaks), had been introduced to Fishwick by the Walkers and their Rotherham foundries certainly provided the castings for the engine which Oakes eventually built. Fishwick clearly had some doubts about the cheapness of Oakes' quotation but did not wish to offend the Walkers. He wrote to his London partner, Thomas Maltby, on 21 December to tell him that he had informed Joseph Walker of Cameron's estimate of £1,200 for an 8 hp engine but that he (Fishwick) wanted a 10 hp engine which would cost about £1,500. He went on to point out that Cameron had recently erected on Tyneside a 'Coal drawing engine', which 'is now coming very fast into repute, her motion pleases me exceedingly and I think we had better employ them & have a complete affair than get one for 500£ less & have much trouble and vexation therewith'. Although Joseph Walker left the final decision to Fishwick, the latter was under considerable financial pressure (as a result of the fact that Oakes had told Walker that he could build an engine for £500 while Cameron was holding out for £1,600) as he told Maltby, 'I find myself a good deal embarassed thereby for the Difference of £1,000 is a prodigious sum.' Despite his suspicions of Oakes, Fishwick eventually decided to place the contract with him, although not until he had been to York to inspect an engine which Oakes had erected there.

By the early summer of 1787 erection of the engine was underway and Fishwick could comment that Oakes' work is 'very substantial and neat & withal exact' but it is clear that he had already found cause to regret his decision to employ him. Consecutive letters to Maltby in June comment on Oakes' drunkenness — 'what a pity such a workman should be given to Intoxication, which with respect to him is too often the case & he will sometimes be absent 2 or 3 days nay even four days drinking together drinking with – anybody'. This was to be a common cry during the next few months with ensuing delay in the completion of the engine. To some extent the delay may not have been entirely Oakes' fault, since it appears that he was also made responsible for the construction of a rolling mill which the steam engine was to power (which had not been mentioned in the early negotiations with Oakes, when the steam engine was intended to power grinding machinery only). At the beginning of October Fishwick was hoping the engine would be ready in November but as late as 9 January 1788 he wrote to Maltby, 'I thought to have rolled some lead today [for the first time], but Oakes has been so buy rolling or trolling Newcastle beer down his throat that he has not shewn his face at the Factory and when I shall see him again I know not!' A few days later, however, lead was success-fully rolled for the first time.

While lead rolling had existed for well over a century, it was prob-ably a new undertaking for Tyneside and clearly was uncommon in the rest of the country. The idea to erect a rolling mill at Elswick and sell sheet lead in London had originated with Maltby, who was aware that a proposal to erect a new rolling mill in London had been given up because of the cost. It was Maltby who dictated the measurements for the mill, 60 ft in length by 6 ft 3½ ins in width and he who had the job of obtaining, what was obviously a rare commodity, an employee with experience of lead rolling. From October 1787 onwards Fishwick asked him on several occasions to find a lead roller and on 5 November noted:

> I am glad you are likely to get a Man to roll Lead or to teach us so to do, I think the latter would be a more eligable plan as such a man as you may send would expect great wages and this is rather out of our Line especially as I conceive the Business is somewhat like shot making & of course contains no great degree of witchcraft in it?

Nevertheless when Maltby finally found a man in January 1788, Fishwick still had to pay 'great wages' – 18s a week, plus house and firing, when he reckoned that one of his own men would only require 9s and would do other work when he was not rolling lead.

The production of sheet lead was not as successful as Fishwick hoped at the end of January 1788, when he was expecting to roll 500 tons a year – working a twelve-hour day. Such an output would have produced sales of more than £10,000 p.a. but, as Table 2.1 shows, this figure was not reached until 1791 and sales in the years to August 1789 and 1790 were very low, after a promising start in 1788. There is no evidence to explain this failure, since the letter books for the period are no longer in existence but it is possible that failure of the steam engine was a contributory factor. In 1795 and again in 1798 Walkers, Fishwick & Co. ordered steam engines for Elswick from Boulton & Watt, which it is unlikely they would have done had they been satisfied with the Oakes' engine.[37] The addition of a further two steam engines in the 1790s appears to have provided adequate power for the Elswick works and the production and sales of sheet lead were well established. In addition to supplying sheet lead for the London market a profitable outlet was found in the supply of chemical lead to the sulphuric acid manufacturers. In the period 1796 to 1798 milled lead to the value of well over £1,000 was supplied to the Prestonpans Vitriol Co., the company set up by John Roebuck to manufacture sulphuric acid by the lead chamber process, and in 1797 Losh & Co. of Tyneside, another sulphuric acid manufacturer, included milled lead in its purchases of no less than £4,000 of lead goods from Walkers, Fishwick & Co.

It is difficult to assess the capital cost of the expansion of the later 1780s since there are, unfortunately, no figures for the cost of specific capital projects but there are valuations of stock at the periodic stock-takings. These give figures for buildings, plant and machinery and, although they need to be treated with some caution, give a rough order of magnitude for fixed capital. The value of buildings[38] rose from about £2,000 in the early 1780s to c.£4,500 in 1788; wind- and horse-mill valuations rose from c.£600 to c.£1,200 from 1786 as a result of an additional mill and the steam engine valuation appears at £1,655 in 1788.[39] The expansion of 1786 and 1787 also required an increase in other machinery and equipment and the value of 'utensils', which had been about £1,500, doubled by the 1788 stocktaking, presumably to include the value of the rolling mill among other equipment. Utensils would also have included the 'bellows and a smelting furnace' (pre-sumably a slag hearth[40]), which were in operation by December 1787, having been necessitated by the extension of the range of production, enabling the firm to recover lead from the waste products of the various manufacturing processes. During this period of expansion in the mid-1780s, although it was not directly connected with manufacturing processes, the partners also invested in a share in a ship to carry their products to London.[41] It would seem, therefore, that the expansion involved capital expenditure of not much more than £6,000 between 1782 and 1788, which could have been financed out of profits (which totalled about £10,000 between 1782 and 1786, with an additional £9,500 in the two years 1786 to 1788).

One would have expected that the biggest impact of the expansion would have been in the level of variable capital required to finance the increased stocks needed. In fact the stock book valuations do not show this to be true, although they do not, of course, take into consideration credit sales. Total value of stocks (including the fixed capital assets mentioned above) rose from an average of £11,000 in the period 1782 to 1785 to about £26,000 between 1788 and 1791, with fixed capital rising from about 40 per cent to 45 per cent of the total. This is not a true reflection of the proportions of fixed to working capital because it does not include the cost of financing credit sales. There are no figures for the normal level of credit which had to be financed, although one might guess that most sales were on six months' credit (sometimes unofficially extended) and, therefore, that this form of working capital required approximately one-half of the annual value of sales. At the end of the 1780s this would mean an addition to working capital of about £12,000 — less the usual credit value of purchases of pig lead.[42]

Between 1782 and 1788 an increase in fixed capital investment of about £6,000, together with an increase in working capital which might have been of the order of £20,000, enabled sales to increase from less than £10,000 to more than £25,000 p.a. and, with margins held fairly well, annual profits to rise from a little over £2,000 to £5,000 to £10,000. The expansion had been justified and indeed provided the profit on which further expansion at Newcastle and the opening of new factories could be based.

Expansion was not, however, without its problems. Possibly the most important of these was obtaining an adequate supply of pig lead. We have seen that the partners thought that Derbyshire lead was essential to the production of good white lead and for many years the obtaining of pig lead suitable for this purpose was the major aspect of the problem. A shortage often meant that they had to make do with common, unrefined pig lead which was unsatisfactory. On one occasion Fishwick told Maltby that they were emptying a stack in which common lead had been used and that the corrosion was so poor that the lead 'keeps almost as blue as it went in'. The quality problem led to difficulty with customers. On 21 March 1787 Fishwick noted, 'Some of our first customers are displeased with our Lead and I must get you or Mr. Ward to send them some [of Islington manufacture] instead of us.' By May 1787 the supply of lead had become much more free and Fishwick decided to empty his stacks and fill them with Derbyshire lead in order to 'retrieve our caracter in white lead'. Over the next couple of years the difficulties in using north-eastern lead were overcome and in March 1789 Fishwick informed Joseph Walker that the Elswick works had not used Derbyshire lead for several months.[43]

Dependence on local, north-eastern supplies, chiefly from the Blacketts' and London Lead Company's mines in Allendale and Weardale, still, however, faced problems of shortages. Natural causes, especially weather conditions, were significant. The river Tyne sometimes froze above Newcastle bridge and delayed the movement of lead, while road conditions became so treacherous in the lead dales that it was impossible to move lead by pack-horse during the winter months. Fine weather at other times could delay the supply of lead. Fishwick told Maltby in February 1787, 'it is not the weather being fine that will facilitate the coming down of lead to this river but rather retard it for the farmers are the carriers and they must and will attend the *seed time* in preferrence so that very little will be done these two months'. The following year it was fine weather in the summer which led Fishwick to complain to Maltby:

> we have long had very dry weather here & in the neighbourhood & the lead hearths where the bellows are blown by water must feel the effects of it — Blackett's wheels turn very slow & the Lead Company's are nearly at a stand which will throw them back very much & perhaps throw the carriage so far into the winter that a quantity of it will not get here until next spring . . .

One of the side effects of natural delays in the supply of lead was that the lead producers took advantage of the opportunity to push up the price and Fishwick was certainly convinced that the producers were in league. His main object for criticism was Alderman Blackett, agent for the Blackett lead interests, whom he described as, 'certainly an *irresolute indecisive* caracter and to make the most of him requires a more than ordinary degree of patience and command of temper'. The letters to Maltby give evidence of the cat-and-mouse tactics between Blackett and Fishwick — the former avoiding the latter if he knew that Fishwick needed lead, promising him lead and then failing to deliver or increasing the price, while Fishwick would occasionally dine the Alderman or do him a good turn in an attempt to get the first supply of available lead, and at other times he would buy from other agents, both local and in other lead areas, in order to stress his lack of dependence on the Alderman. It may be that the collusion of lead producers in order to force up pig lead prices was merely a figment of Fishwick's imagination, since seen from the other side it could be argued that

Fishwick and other purchasers' refusal to buy when lead was plentiful amounted to collusion to force prices down. Drawing his evidence from the Blackett estate papers Mark Hughes claimed that:

> In 1789 the London lead houses, under the leadership of a Mr. Fishwick, entered into 'a kind of combination' for the purpose of reducing the price, and J.E. Blackett as the agent for the largest single producer took it upon himself to break it by underselling to non-members and refusing sale to members.[44]

The evidence is insufficiently clear to establish whether or not either producers or consumers were operating some form of collusive agreement but it seems likely that community of interest and common sense led them to react in ways which might have suggested the existence of informal cartels. It is certainly unlikely that Fishwick was in collusion with London lead producers (other than his London partner, Maltby), since they felt threatened by the competition of an increased supply of lead goods from Elswick after the expansion of the mid-1780s.

With the successful establishment of sales of white lead through agents in the London market in the early 1780s, it had become evident that the market was sufficiently large to justify a more effective presence and in 1785 Maltby was taken as London partner and the Islington works opened to supply white lead to the London market. This established the base from which Elswick could be expanded, with effective methods for a large London sale and by the end of the 1790s the annual level of sales between the Newcastle and London partnerships was from £50,000 to £60,000 (including supplies of pig lead). The London lead companies, of which Lancaster & Co. was certainly the largest, remained the price setters but they were aware of the competition which was developing as a result of the fact that Walkers, Fishwick sold at 'from 30/- to 40/- p. ton less than the London houses'. Towards the end of 1786 Joseph Walker had suggested the possibility of coming to some form of merger with Lancaster & Co. and Fishwick passed the idea on to Maltby, who, to Fishwick's evident relief, opposed it. Fishwick was confident that they could fight the London makers:

> If they [Lancaster & Co.] mean to contend the matter and make a fighting trade of it I think we at present stand upon even ground with them & should we get an Engine & it answers I am persuaded if they could cope with us in pig lead we should be better situated to sell White Lead, Red Lead, Sheet Lead and Shott at a lower price

than they could & an accommodation must of course be the consequence.

Two years later, however, Fishwick raised the matter again but it was obviously rejected by Maltby, since Fishwick subsequently wrote:

> The hints I furnished you with about a coalition were only of a desultory nature − the impulse of the moment and I by no means wish to press them upon you being fully persuaded of your being quite competent to judge of the propriety of any such proceeding, and as such fully resting the management thereof with you.

It is evident from this that Maltby was an important member of the partnership to whom even Fishwick deferred on matters relating to the London market[45] and Maltby's rejection of an agreement with the London makers seems to have scotched the idea. It was not until the end of the nineteenth century that cartelisation with the London manufacturers was to become a serious proposition and by then the Elswick company was the only survivor of those which were in competition in the 1780s.

Certainly that competition does not appear to have had any adverse effects on Walkers, Fishwick & Co. Although there is little detailed information on developments in the 1790s, it is clear from the sales and profit figures that successful expansion of the existing activities continued. By the time that the partnership agreement between the Walkers, Fishwick and Ward expired, on 30 April 1800, a very large and successful concern was firmly established. Moreover, an era had come to an end, since Fishwick left the partnership and Ward was soon dead. A Newcastle historian, writing at the time, provided a fitting epitaph for this first stage of the development at Elswick:

> West from Skinner-bourn is a manufactory for making white-lead, one of the most extensive north of London; and where there was not a single hut before the commencement of this rich and advantageous manufactory, by the constant accretion of new buildings, it has now the appearance of a considerable village. Here, by chemical process, vast quantities of lead are converted into white and red lead, so useful in forming the basis of paints. These are mostly sent to London, to the great emolument of the proprietors, Messrs. Walker, Ward & Co.[46]

Notes

1. W.P. Hedley and C.R. Hudleston, *Cookson of Penrith, Cumberland and Newcastle upon Tyne* (Kendal, n.d.).

2. Ibid., pp. 15-16 and *Newcastle Journal*, 1293, 25 Feb. 1764. A plan of the Elswick estate, drawn in 1778, shows that the land leased by Walkers, Fishwick had no buildings on it. Newcastle Central Library, Seymour Bell Collection, portfolio 9, Elswick Estate (1).

3. Northumberland Record Office (in future NRO), Blackett (Wylam) Mss, Letter book 1798-1800, Christopher Blackett to John Locke, 24 April 1799. An effective date for the commencement of lead products of 1801 is suggested by the fact that the books of the company relating to lead manufacture commenced in that year. There is a plan of the Ouseburn Lead Works, dated c.1810, in Newcastle Central Library, Bell plans, portfolio 6 (28).

4. Raistrick, *Quakers in Science*, pp. 168-72 and 176-7.

5. A.H. John (ed.), *The Walker Family, Ironfounders and Lead Manufacturers 1741-1893* (1951).

6. Hull City Library, Hull Poll Book 1774; Fishwick also appears in the 1802 Poll Book for the town, where he is described as 'ironmaster of Newcastle', as does his son Richard described as 'gentleman of Whitby'. Hull Record Office, Register of Freemen, BRG 4, fo. 94. I owe these references to Mrs M.E. Bickford.

7. John, *Walker Family*, pp. iii-iv.

8. Ibid., p. iv.

9. Much of the detailed evidence which follows is taken from the records of the companies which were drawn into ALM. These have been deposited in Tyne and Wear Archive Department, accession number TWAD 1512. Since they are uncatalogued it is often impossible to give a useful reference, and to avoid overloading this book with notes it has been decided not to reference evidence drawn from this source other than to indicate in the text that it is from minute books, annual accounts, correspondence, etc.

10. Factory is the word which contemporaries used to describe the works at Elswick, not the more common 'manufactory', from which it is obviously a corruption, but one which implies something different from hand power.

11. Although the sons of Samuel Walker I do not appear in the partnership deeds of 1778 it is clear that they, and especially Joseph, who was the Walker subsequently responsible for the lead side of the family interests, were very much involved on a consultative basis from the beginning.

12. Of the firm of Dowson and Atkinson, Wharfingers of Chamberlain's Wharf, 12 Tooley Street, Southwark.

13. The fodder varied in weight from 19½ cwt at Hull and London, through 20 cwt at Chester and Grassington, 21 cwt at Newcastle, to 22½ cwt at Derby and Wirksworth. See Burt, 'Lead Production in England and Wales', Appendix II, p. 268. Dr L. Willies has pointed out to me the fact that in the High Peak district of Derbyshire the fodder equalled 23½ cwt, John Rylands Library, Bag 8/3/89. See also Raistrick and Jennings, *History of Lead Mining*, pp. 42-4.

14. See Appendix II for statistical details on the partnership for the period 1780-1800. 'Profit' in eighteenth-century accounts is a difficult concept. A very useful treatment of the problems involved is G.A. Lee, 'The Concept of Profit in British Accounting 1760-1900', *Business History Review*, XLIX, 1 (1975), pp. 6-36. 'Profit' in the sense used by Walkers, Fishwick & Co. was the difference between total revenue and total expenses, with the latter including repairs and renewals. The profit figure therefore represented the total return to the partners both for their capital and (in the case of Fishwick and Ward) entrepreneurial and

management investments.

15. See, for instance, ibid., p. 17, it is 'almost impossible to compare the profitability of different firms, even in the same industry at about the same time, or to assess the progress of individual businesses over a short series of years'.

16. John, *Walker Family*, p. 33.

17. It is, of course, possible that Ward and Fishwick's shares in the capital were nominal and that the original cash was supplied by Walker who was to be repaid out of their share in the profits, although there is no hint of this in any of the partnership records. It was, however, a very common way of giving those who were in effect managers a partnership and was, as we shall see later, frequently used in the lead partnerships.

18. F. Beckwith, 'Fishwick and Ward', *The Baptist Quarterly* (April 1954), p. 264. Most of the following material on Fishwick and Ward is drawn from this article, which in turn is heavily based on the Kinghorn correspondence. Some of the letters which Beckwith used are no longer available, some are in Norwich Public Library and a bundle covering the period 1770-83 is in the hands of C.B. Jewson, Horsfold Hall, Norwich, to whom I am most grateful for access to and permission to quote the correspondence.

19. For Joseph Kinghorn (1766-1832), who was to become a famous minister of the Norwich Baptist Church, see M.H. Wilkin, *Joseph Kinghorn of Norwich* (Norwich, 1855) and *DNB*.

20. The minutes relating to the lead partnership state that Ward retired from the partnership in 1809, John, *Walker Family*, p. 35, but this seems to be a misreading for 1800.

21. The will was preserved in a collection of documents and partnership deeds relating to the Walker family by M.L. Walker, a great-great-great grandson of Samuel Walker, original partner in the Elswick lead works. I am deeply grateful for his permission to consult and quote from these records, which it is hoped will be placed with the remainder of the company records in Tyne & Wear Archives.

22. Beckwith, 'Fishwick and Ward', p. 268. See also D. Douglas, *History of the Baptist Churches in the North of England from 1648 to 1845* (1846), pp. 205-7 and 218-19.

23. Rev. F.G. Little and Rev. T.F. Walker, *The Story of the Northern Baptists* (Newcastle, 1945), p. 20.

24. One local historian stated of the Lemington ironworks, 'The company are of the first respectability, and, it is said, have embarked in the enterprise above an hundred thousand pounds', Rev. J. Baillie, *An Impartial History of the Town and County of Newcastle upon Tyne* (Newcastle, 1801), pp. 504-6. Even in the inflationary years at the turn of the eighteenth century such a figure seems extreme considering the size of the works built. It may of course have included the working capital, stocks, work in progress and credit sales but it is inconceivable that it could refer to fixed capital alone, in the way in which some subsequent writers have interpreted it. Nevertheless financing the works did press hard on Fishwick's resources. He noted that his 'altered circumstances would make his own contribution [towards the cost of the new Baptist chapel in Newcastle in 1798] smaller than he could otherwise have wished', Little and Walker, *Story of Northern Baptists*, p. 21.

25. The termination of the partnership, at the end of the 14-year term from 1 May 1786 for which it had been renewed, appears to have been done amicably. A legal document dated 29 July 1800, in M.L. Walker's collection, gives the details. Fishwick was to receive his share of the capital less £7,283 1s 7d which he owed to Walker, Eyre and Stanley, bankers of Rotherham. (For this firm see 'Ironmasters and Duke's Agent, the Sheffield and Rotherham Bank – the Early Days', *Three Banks Review*, 73 (March 1967), pp. 36-50). Fishwick was to receive

£1,000 on 1 May and 1 November, plus the interest on the outstanding capital, until his full share in the partnership had been paid. The Walkers were also to pay him an annuity of £1,200 p.a. (also payable half-yearly on 1 May and 1 November) so long as he did not re-enter the lead business. Whatever prior knowledge of white lead manufacture Fishwick had in 1778, it is clear that he had accumulated enough in the next 20 years to make it worth while for his ex-partners to pay him handsomely not to compete with them.

26. The patent specification was reproduced in full in *The Repertory of Arts* ... vol. III (1795), pp. 225-30, in itself evidence that the process patented, though a simple change, was an important one. It also brought a useful royalty income from other white lead manufacturers, totalling more than £3,000 between 1791 and 1800.

27. Baillie, *History of Newcastle upon Tyne*, p. 509.

28. *Transactions of the Society . . . of Arts . . .*, XIII (1795), pp. 221-30.

29. Female labour was used to a considerable extent in white lead manufacture for unskilled work since it was cheaper than male. Although it may seem surprising, there were not infrequent movements of labour between the Newcastle and London areas, as evidenced by the Elswick records. Travel was almost invariably by sea, which although slower, was much cheaper than by road.

30. The *Newcastle Courant*, first published in 1711, was the first newspaper issued north of the Trent.

31. Paradise was on the Benwell Estate, west of the Elswick works, close to the Tyne and to Benwell Colliery. It is possible that Bussell is an incorrect derivation from a carelessly written word. Russel is more common and 'Russel's factory' is given as an address by Joseph Oxley for his letters sent from Newcastle to Sir John Hussey Delaval between 1763 and 1765. NRO, 2 DE 4/10/1-6. The building is shown on a plan in Newcastle Central Library, Seymour Bell Collection, Portfolio 3, Benwell Estate (House & Tower), plan 2A, plan of Benwell 1808.

32. See Benwell Community Project, *The Making of a Ruling Class* (Benwell, 1978), a fascinating account of the development of the industries and families of west Newcastle, if one in which evidence is hammered to fit into a preconceived model of class relationships. Here it is stated (p. 118) that the colour works was sold to Gibson. I have no evidence that Walkers, Fishwick & Co. ever purchased the works and certainly in 1780 they leased it. The works became part of the Tyneside paint company, Hoyle, Robson, Barnett & Co. (which eventually became a constituent of British Paints Ltd), but was closed by 1888.

33. It is probable that this was what later became known as ground white lead (as opposed to dry white lead) where basic lead carbonate was ground with linseed oil. For use as a paint, thinners would be needed.

34. That this included the wages of a London millwright to supervise the construction of the working parts, is an interesting reflection on the partners' lack of confidence in local builders. Windmills were a common source of power on Tyneside and there were many competent local builders including, in John Raistrick of Morpeth, who bought white lead from the Elswick works, at least one of some eminence. See S. Macdonald, 'John Raistrick, Civil Engineer of Morpeth', *Durham University Journal*, XXXVIII, 1 (1976), pp. 67-75.

35. There is no documentary evidence for the siting of a water mill at Elswick and it is, therefore, possible that the mill was on another site, possibly at the Benwell colour works, where there is evidence of the existence of a mill. See Newcastle Central Library, Seymour Bell Collection, Portfolio 3, Benwell Estate (House & Tower), plans 2A and 2B.

36. G.N. von Tunzelmann, *Steam Power and British Industrialisation to 1860* (Oxford, 1978).

37. Birmingham Public Library, Boulton and Watt Collection, 'Catalogue of Old Engines', pp. 44-5 (portfolio of drawings 119), 16 hp (nominal), sun-and-planet, double acting, cylinder 21¼ x 5 ft 0 ins, ordered July 1795, price without erection costs £705; and pp. 100-1 (portfolio 262) double acting beam engine, 23¾ x 5 ft 0 ins, July 1798, £698. It is interesting to compare these prices with those quoted ten years earlier. In the absence of a massive fall in production costs it seems likely that Cameron's figure of £1,600 was a gross exaggeration of the likely cost, perhaps made possible by a shortage of suppliers of rotative engines in the mid-1780s, while Oakes' estimate was not as low as it had previously seemed. There are some figures for the cost of steam engines in the late-eighteenth century in von Tunzelmann, *Steam Power and British Industrialisation*, pp. 47-61.

A small (4 hp) Boulton & Watt engine was ordered for the Lambeth shot works of the firm by Thomas Maltby in January 1799 and a 16 hp engine for the Islington works in April 1799. Since the Newcastle-under-Lyme calico and cotton works, in which Joseph Walker and Archer Ward were partners, also ordered a large engine in 1796, Boulton and Watt did very well out of the firm. They also built in 1800 a 12 hp engine for the 'Newcastle Lead Co.', which, as will be seen in Chapter 5, was almost certainly for the Gallowgate works of Locke, Blackett & Co.

38. Presumably this included Fishwick's house and the cottages built for the workmen, since it is difficult to believe that the initial factory buildings could have cost £2,000.

39. This is, of course, well above Fishwick's statement that Oakes would build an engine for £500. The figures usually quoted for steam engine prices merely included the cost of the basic engine materials, sometimes excluding even the boiler and always excluding the base and framing for the engine and costs of erection. Even allowing for these, which could put several hundred pounds on to a basic price, it is clear that Oakes' initial figure was much exceeded.

40. For details of the various types of furnace used in lead smelting see R.F. Tylecote, 'Lead Smelting and Refining during the Industrial Revolution, 1720-1850', *Industrial Archaeology*, vol. 12, 2 (1977), pp. 102-10.

41. It also became necessary to purchase a wherry to carry lead products down the Tyne from Elswick to a point below the Newcastle bridge which could be reached by sea-going vessels. The ship, the *Britannia*, cost £1,690 and Walkers, Fishwick & Co. appear to have had perhaps a one-third share in it – since the value first appears in the accounts at £499. Control of a means of shipment of goods was very important, since obtaining cargo space in the ordinary shipping market was time consuming and led to delays in delivery. The purchase of the ship proved satisfactory and a profit of £400 (presumably this was total profit and not just Walkers, Fishwick & Co.'s share) was made in the first year of operation and Fishwick proposed taking a larger share in a second vessel.

42. There is, unfortunately, no evidence in the company's records of the terms on which the partners were able to purchase pig lead. Blackett/Beaumont lead appears to have been sold on six months' credit, NRO, Blackett (Wylam) Mss, Letter Books.

43. By the end of the 1780s the Elswick works was mainly dependent on lead from the northern Pennines in Northumberland and Durham, although small amounts of pig lead were purchased from other lead mining areas. In the 1790s and 1800s the Elswick works purchased occasional parcels of North Yorkshire lead through Matthew Wadeson of Stockton, who acted as the firm's agent for the sale of lead products in that area. Wadeson purchased large amounts of North Yorkshire lead for Walkers, Maltby in London, however. Wadeson also purchased for Walkers, Fishwick, in 1793, some old sheet lead, the covering of a local chapel. This is the only evidence for this period that I have found of the use of

secondary lead. That Wadeson wrote to Walkers, Fishwick to ask whether they would be interested in purchasing this material and that the price he eventually quoted for it was £14 per ton against a current price of about £20 per ton for pig lead, suggests that secondary lead was little used. I owe these references to Miss M. Hartley and Miss J. Ingilby, Coleshouse, Askrigg, Leyburn, North Yorkshire, DL 8 3 HH, who have Matthew Wadeson's letter books for the period 1792-1808.

44. M. Hughes, Sources of Landlord Income in Northumberland, p. 99. The marketing of lead is discussed on pp. 97-115. Since the Walkers partnership rapidly became a major consumer of lead it is not surprising that it tried to influence the price of pig lead nor that it was accused of trying to do so. This occurred on at least one other occasion – see F. Hall, *An Appeal to the Poor Miner* . . . (1818) (an account of the proceedings at a public meeting in the London Tavern, 24 Feb. 1818).

45. From the fact that there was no reference to the other partners being consulted it would seem that Fishwick and Maltby were very much in charge – the former of production and country markets and the latter of London markets. To some extent this conclusion is the inevitable result of the fact that only the Elswick letter books (and only a few of those) are available but the regularity of the correspondence, usually two letters a week and sometimes more, from Fishwick to Maltby, while months pass without letters between Fishwick and either Walker or Ward, suggests that it is reasonably accurate.

46. Baillie, *History of Newcastle upon Tyne*, p. 508.

3 THE MANUFACTURING OF LEAD SHOT

> Here, too, they would tell old legends of what the Thames was in ancient times, when the Patent Shot Manufactory wasn't built, and Waterloo Bridge had never been thought of . . .
> Charles Dickens, 'Scotland Yard' (1836).[1]

> We went big St. Paul's and Westminster to see,
> And aw warn't ye aw thought they luck'd pretty,
> And then we'd a keek at the Monument te,
> Whilk ma friend ca'd the pearl o' the city:
> Wey hinny, says aw, we've a Shot Tower se hee,
> That biv it ye might scraffle to heaven,
> And if on St. Nicholas ye once cus' an e'e,
> Ye'd crack on't as lang as ye're livin.

> 'Bout Lunnun then div'nt ye mak' sic a rout,
> There's nouse there ma winkers to dazzle:
> For a' the fine things ye are gobbin about,
> We can marra iv canny Newcassel.
> Thomas Thompson, 'Canny Newcassel'[2]

Shot manufacture from the seventeenth century to the present: early methods of production; invention of the shot tower in 1782 by William Watts; innovation of the tower by various firms and history of their development; closure of many of the towers in the twentieth century.

As an instrument of war lead goes back certainly as far as the classical Greek period, since lead sling bullets could be fired further, because of their greater density, than stones. As we have seen by the middle of the eighteenth century there were two distinct methods of making lead shot.[3] While large shot were cast in moulds, small ones were manufactured by dropping them into water and it was this latter process which was to be improved in the later-eighteenth century. Into a perforated vessel was poured molten lead to which had been added a small percentage of arsenic, which caused the lead to pass through the perforations in the bottom of the vessel in droplets rather than as the continuous stream which molten lead would otherwise do. The short fall, apparently only a matter of inches, from the vessel into a tank of water in which the shot was collected, led to a considerable proportion of the shot being of uneven shape. It was necessary for the shot then to be placed in a churn and rotated until friction wore away the worst of the imperfections and more or less round shot were produced. Clearly this was an unsatisfactory method of production as demand for shot increased and one might well expect that experimentation would occur to replace it. As it happened, however, it was neither scientific investigation nor empirical efforts but apparently extra-sensory perception which led to the new process.

One version of the story goes that William Watts, a Bristol plumber, after imbibing rather too heavily one evening, was unable to make it home to his bed and fell asleep at the foot of the tower of St Mary Redcliffe. He then had a dream that the church caught fire and that as the lead on the roof melted it dropped to the ground, where it landed in pools of water and solidified as perfectly spherical shot. An alternative version, which more appropriately catches the likely wanderings of a drunken Watts' sub-conscious, states that he dreamed that his wife was pouring molten lead on him from the tower of the church through the holes in a rusty frying pan.[4] The story continues that Watts and his wife, determined to capitalise on the dream, obtained permission to use the tower of the church to experiment to find out whether the process would actually work. From the roof of the tower they dropped molten lead into a bucket of water on the ground below, where it was found to have formed good shot.

We do not know very much about Watts. Since he was apprenticed on 27 February 1765 to Philip and Elizabeth Rose (by which time his father, John Watts, a hooper, was dead),[5] we may assume that he was in his very early thirties when the dream took place. He became a freeman of Bristol on 4 July 1772 and set up in business as a plumber at 126 Redcliff Street and local directories record him as being in business there until 1784. His training and business experience would give him a sound knowledge of the properties and uses of lead, especially in building work, but there is no evidence to suggest that Watts had any specific connection with lead shot. However, since Bristol was a centre of lead manufacturing it may be that Watts was aware of the problems involved in the manufacture of small shot from acquaintances in the city. Unsatisfactory though the dream stories are, it seems unlikely that it is possible to replace them with a more rational explanation.

The results of his experiments having proved satisfactory, whatever it was that originally led to them, we move on to firmer historical ground, since Watts took out a patent for his process (no. 1347) on 10

December 1782. Much of the patent is taken up with a description of the method of producing 'poisoned lead', that is to say, arsenical lead, and with preparing the lead to be dropped.[6] There was nothing new in this and Watts' 'said invention of making small shot solid throughout, perfectly globular in form, and without the dimples, scratches, and imperfections, which other shot, heretofore manufactured, usually have on their surface', came in the last sentence of his patent: 'If for the smallest shot, the frame must be at least ten feet above the water, and for the largest shot about one hundred and fifty feet, or more, above the water, and so in proportion, according to the size of the shot intended to be made.' It is noticeable that the patent did not mention a shot tower but only covered the process of dropping shot a considerable distance into water, a very simple development of the existing process. That Watts' patent could define the distances required for the dropping of certain types of shot adds verisimilitude to the stories of his experiments.

At the time that he took out the patent Watts was described as a plumber and shot-maker but there is no evidence as to how he carried on the latter activity. On 1 January 1785, however, Watts took the lease of an old house in Redcliff Hill,[7] which was a continuation of the street in which Watts was already in business. That these premises, intended to provide a specific shot manufactory, were not leased until two years after the patent was taken out suggests that little interest was shown in the process until it was worked on a production scale and that this could not be done until Watts had raised the capital. By 1785 Watts was in a position to proceed and set about preparing the Redcliff Hill premises. A hole was cut through the floors of the house, an existing well in the basement deepened and a small tower erected on the roof, which gave a total drop of about 90 feet.[8] This was less than the patent had suggested but it was presumably found to be satisfactory, since this Heath-Robinson construction, true to the traditions of British industry, remained in production for not far short of two centuries until it was demolished for road-widening in 1968.

The first evidence of the effect of Watts' production at Redcliff Hill appears to have been a complaint by neighbours about its nuisance value. In December 1786 he replied to the complainants as follows:

WILLIAM WATTS

A CARD

William Watts presents his compliments to the GENTLEMEN who united for the purpose of taking *legal measures* to procure *the removal* of his SMELTING and SHOT-WORKS, and begs leave to ask them, whether it is not as unreasonable, to expect that he should knock down his Shot-Works, because *some people* are offended with the smell occasion'd by that particular process, which may be conducted (with very little additional expence) as well on the heights of Mendip, as on Redcliff-hill – as it would be to require *Mr. Cross* to demolish his great Distillery, merely because those nauseous PIG-STIES, offend the *delicate Noses* of a *few Individuals*? However, to obviate every cause of complaint, as well as to disappoint that malignity which would be gratified by involving him in an expensive suit, William Watts will as soon as possible cause that process, to be discontinued at Redclift-Backs, which alone can furnish the least complaint.[9]

If Watts actually did anything about the complaints it must have been to cease the smelting process at Redcliff, for the shot tower continued in production. Watts was soon, however, to establish his position by going into partnership with some powerful Bristolians.

The traditional account is that, after experiencing opposition to the process at Bristol, Watts left the industry and sold his patent to the Walkers' lead partnerships for £10,000 in 1787.[10] This is now largely untenable as the following evidence will show. On 17 September 1788 Fishwick, having heard that Watts had taken partners, wrote to Maltby:

I think there must certainly have been a Cypher too much in the premium given Watts to share in his Shot Patent & that it has been only 1000, instead of 10,000£? I suppose the former Sum is more than you would have advised our Company to give for it.

It is evident that if anyone paid £10,000 to Watts for a share in the patent it must have been his new Bristol partners and that Fishwick at this time clearly did not regard the patent as particularly important.[11] Watts' new partners were Philip George, of the Bristol distilling and porter-brewing family and Samuel Worrall, the somewhat flamboyant Bristol banker and, from 1787 to 1819, Bristol town clerk. Its minute book[12] shows that the new partnership, called P. George & Co. and sometimes Watts, George & Co., commenced in January 1788. With the aim of making as much use as possible of the patent, which was due to expire in 1796, the partners made a rapid decision that the largest source of demand for shot was in London. As a result they had built for them a shot tower, opened in 1789, on the bank of the Thames, just to the east of the subsequently erected Waterloo Bridge, at Cuper's Bridge, Narrow Wall, Lambeth. Two years later it was described as 'a new structure ... [which] cost near six thousand pounds, but cannot be considered an object ornamental to the river Thames'.[13]

Watts, George & Co. were undoubtedly more concerned about obtaining an adequate return on their capital investment than the visual impact of the tower on the Thames river-line. The new process does not, however, appear to have enabled them to produce shot sufficiently cheaply to undercut traditional producers. Perhaps the fixed capital investment in the Lambeth tower was too great a burden (£6,000, if that figure was accurate, would place the tower in the same kind of bracket as a large water-powered cotton mill of the period in costs of construction and equipment). Perhaps also the partnership suffered from lack of technical expertise, for William Watts appears to have played little part in the business. With his new-found wealth, from the sale to his partners of a share in his patent rights, he began to speculate in building activity.

After 1787 Watts no longer appears in Bristol directories as a plumber and shot-maker but as 'Watts, Wm., gent., Clifton', and it was at Clifton, high above the Avon gorge, that he purchased land about 1790 and began to build Windsor Terrace.[14] It was a brave but doomed attempt to build on a dramatically imposing site but the nature of the site is supposed to have taken up the whole of Watts' capital in building secure foundations and by 1794 he was declared bankrupt.[15] Watts by this time may have left the partnership with Philip George and certainly from 1795 the partnership appears in directories as Philip George and Patent Shot Co. both at Bristol and London. After his bankruptcy Watts reappeared in his old role as a plumber of Redcliff Street, Bristol, although in different premises from those he had previously occupied. He had however left his mark on history:

> Mr. Watts very soon a patent got
> So that only himself could make Patent Shot;
> And King George and his son declar'd that they'd not
> Shoot with any thing else — and they ordered a lot.
> The Regent swore that the smallest spot
> In a small bird's eye he'd surely dot;
> And every sportsman, both sober and sot,
> From the peer in his hall, to the hind in his cot,
> Vowed that they cared not a single jot
> When the game was strong and the chase was hot,
> For any thing else than the Patent Shot.[16]

It was not, however, the withdrawal of Watts from the partnership but financial difficulties which caused Philip George & Co. to give up the Lambeth works. In the early 1790s the partners had borrowed from a number of sources in order to provide the working capital necessary not only for the two lead works but also the Bristol brewery in which Philip George was a partner. They had, however, been caught up in the credit shortage and financial panic of the middle 1790s, partly a result of a poor harvest and high food prices in 1795 and partly of the French war, which was to lead to the crisis of 1797 and the Bank Restriction Act of that year. In the circumstances the partners had to raise money and decided to sell the Lambeth works, which was peripheral to their main interests in Bristol. It is noteworthy that the sale occurred about the time that Watts' patent, which the partners held, ran out in December 1796. Obviously the partners expected competition from other firms which would now build shot towers and decided that the Lambeth works was not viable. It will come as little surprise that the purchasers were Walkers, Maltby & Co., the London branch of the Walkers' lead partnership, who appear to have paid about £8,000 for the works as a going concern.[17]

Under the title of Philip George & Patent Shot Co., the Redcliff Hill shot works remained under the old partners who added new lead products. In 1848 the Georges were replaced by new partners in control of the firm and by 1863 the present owners, Sheldon Bush and Patent Shot Co. Ltd, had purchased the business.[18] The firm gradually came into the hands of the Thomas family where it is today with S.W. and C.W. Thomas as directors, very much in the style of the managing partners of the early concerns. In its continuing field of the production of sheet, shot and pipe, by diligent concern for the needs of its customers, the firm has managed to show that there is still room for a small producer in the lead industry. It remains the only extant firm, with an eighteenth-century background, not to have been drawn into the embrace of ALM.

Unfortunately the Redcliff Hill shot tower no longer exists. It had a long and useful existence and at least in its latter days the tower had two dropping heights. This was quite common in other shot towers in order to produce different sizes of shot,[19] but whereas in most towers it was the point from which the shot was dropped that changed, at Redcliff Hill it was the point to which the shot was dropped which was altered, while the dropping point at the top of the tower was unchanged. The original shot tower was as different from its successors in almost every way possible while remaining a shot tower. For shot up to and including No. 5 the shot was dropped about 50 feet into a tub of water at ground floor level but for larger shot that tub was moved aside and the shot dropped into a tank of water, at the bottom of the pit, which was then hauled up by pulleys. When the Redcliff Hill premises were purchased for road widening, Sheldon, Bush received compensation which enabled them to build their splendid modern shot tower at their Cheese Lane premises. The building, which has a Y-shape cross section, won a design award and its three limbs contain lift, fire escape stairs and shot drop.

While the initial years of the Redcliff Hill tower do not seem to have caused serious problems for the existing makers of common shot, a letter from Fishwick to Maltby on 31 May 1787 shows evidence of some concern. It mentions the visit of a man from Watts, to check that the Elswick firm was not infringing his patent, who had told Fishwick that 'the expence of making Patent Shot is not above 5/- p. ton above the common shot & that they think of making shot in London & *knocking up the making of common shot*'. It is possible that at this stage Watts was merely trailing his coat in order to see what the response would be from existing shot manufacturers and from whom he might get the best terms, since his partnership agreement with Philip George lay more than six months ahead. Fishwick, in his letter, suggested that Maltby might consider becoming Watts' London partner, which would usefully safeguard the position of Walkers, Fishwick. Nothing came of this and the subject was not raised again until 17 September 1788 when Fishwick commented on the supposed premium of £10,000 paid to Watts. At this stage Fishwick was clearly concerned about the development of patent shot but was trying to put on a brave face: 'Before I went to Scotland I visited a shot maker who lives not 100 miles from York who makes shot equal if not superior to Watts & spent a good while in his shot house, wondering at his *ingenuity*!!!' This was clearly an attempt to imply that Watts' process was nothing special but the laconic aside in the same letter to 'acquaint me at the same time what Phillips reports about the supposed defect in Watts' Patent', betrays concern. Three days later Fishwick had moved from concern almost to panic:

I think it will be highly necessary to take every legal method to *circumvent* the sale of *shot* going intirely in the hands of the successors of Watts the patentee — Inclosed I send you a copy of his specification agreeable to your request & it will be right I think to loose no time in laying the same before one or two Counsel together with our mode of dropping on a woolsack, a hair cloth or straw instead of water? I have no doubt at all of our being able to accomplish it [dropping shot] provided we can do it legally ... I think it very right to take the *alarm* respecting the *shot* trade going intirely into the hands of a set of *opulent* men ...

Fishwick's concern is understandable. If patent shot were successful it would break a production link for his firm and it would provide the basis on which the newcomers might extend their range of production of lead goods, thus competing further with existing firms. Three days later Fishwick wrote again to Maltby, having in the meantime been to York and back! The shotmaker there had a tower '65 or 70 feet high 14 foot by 10 square, has two or more departments at different heights, for different sizes I suppose – he drops it into the water'. Fishwick had not been allowed into the building but although the man claimed that the process had been his own idea, nearly five years had elapsed since the publication of Watts' patent, from which it might have been derived. Fishwick's evident hope that the York process would be a safeguard against a monopoly by Watts and his partners had to be discarded. Nothing seems to have resulted from the proposal to put to counsel a method of dropping on to a material other than water[20] and after the frantic concern shown in September the topic of shot drops from Fishwick's correspondence. It seems likely that he had decided to live with the situation and was gradually to discover that the patent shot was not going to deprive him of his market completely.

Towards the termination of Watts' patent at the end of 1796 the situation was obviously different. Philip George informed one of his correspondents on 26 March 1798 that 'Since the expiration of the Patent we have reduced the price of Pat Shot to that of Common which has prevented any other House going to the expence of Erecting a Building for the same purpose ...', apparently in ignorance of the fact that Fishwick had commenced the building of a tower at Elswick in 1796, before the end of the patent, which was completed and in operation in 1797.[21] There is a tradition that when completed the tower was found to be out of true – 'On completion (in 1797) the tower was found to be two feet out of the perpendicular. This was rectified by digging down to the foundation and removing a certain amount of soil on one side, when it gradually righted itself.'[22] In 1908 the tower was thought to be 4 ft 3 ins out of plumb and the company's minutes record that estimates were obtained for straightening it. A more careful survey showed that it was in fact only 2 ft 1 in out of perpendicular and it was decided that nothing should be done. It seems probable that the rather horrifying method of righting the tower, supposedly adopted in 1797, is nothing more than an enjoyable myth.

Like the other shot towers, that at Elswick was connected with some singular events because of its conspicuousness. It was illuminated to celebrate the defeat of Napoleon and with the arrival of electricity shot towers became attractive for advertising purposes. In the autumn of 1907 Lipton Ltd, the grocery chain, was in negotiation for the use of the Lambeth tower for an electric sign. Originally offering to pay £150 p.a., the minutes show they were pushed up to a payment of £550 p.a.[23] This experiment must have been successful since on 1 November 1910 an electric sign, advertising Liptons, was erected for the first time on the Elswick tower. This was less successful and was removed in 1913 while the Lambeth sign continued to flourish and after the interval of the war it was noted in 1930 that Liptons had a new sign, especially ordered from the United States. The shot towers also attracted more esoteric proposals. On 11 September 1912 the board of Walkers, Parker considered a request from a Capt. Penfold for 'permission to make a descent from the top of the Lambeth Shot Tower by means of a special parachute, but ... the Directors could not see their way to affording the necessary facilities'.

Many accounts of the Elswick tower state that it was 190 ft in height and that the distance from the shot-running floor to the ground was 175 ft. In fact a survey of the building before demolition showed it to be 174 ft high with the dropping point 150 ft from the ground, exactly the height recommended in Watts' patent for the largest shot.[24] Manufacture of shot in the tower ceased in 1951 and despite its being listed as a building of historical interest in 1954 it was not maintained effectively, its tilt increased and it had to be demolished in 1969, before it collapsed, only a year after the original Bristol tower was demolished.[25]

From about the same time that the Elswick tower came into operation the Walkers partnership also took over the Lambeth tower, said to have been 140 ft high with a shot drop of 123 ft. This was destroyed by fire on 5 January 1826[26] and its replacement was not completed until 1830. There is some confusion since there were two shot towers at Lambeth at this time. The 1830 tower, a rebuild on the site of the original tower of 1789, east of Waterloo Bridge, belonged to Walkers, Parker, while in 1826 Thomas Maltby, who had withdrawn in 1824 at the expiry of his partnership agreement with the Walkers, opened a rival tower only a few yards from the first, but on the west side of Waterloo Bridge.[27] Maltby's operation does not seem to have been very successful and in 1839 the works, which was at 63 Belvedere Road, Lambeth, was taken over by Walkers, Parker, who now possessed two shot towers in close proximity (and, as we shall see, four in all round the country). The decision was taken to give up the original Commercial Road premises,[28] east of Waterloo Bridge and these do not appear to have been worked for some time, although by 1856 they were certainly occupied by Burr Bros. This firm worked the tower until 1870 when it was taken over by Lane & Nesham who worked it for some years. In the late 1880s, however, the partnership was in the Court of Chancery and was eventually dissolved. In order to continue the shot works Harry Lane formed Lane Sons & Co. Ltd, a company which had some financial backing from Walkers, Parker, but which was wound up in 1892. It is probable that the tower subsequently remained unused for manufacturing shot until its demolition in 1935.

Meanwhile the second Lambeth tower, west of Waterloo Bridge, which Walkers, Parker had taken over from Maltby in 1839, continued to produce shot for the same owners. According to Dickens who visited it in the 1850s, it had cost £30,000,[29] a figure which seems very high. Nevertheless Dickens gave an accurate and very readable account of the manufacturing process from being 'Led by a steady, rushing noise, like the sound of a great waterfall' (the shot falling into the barrel of water) to the finish 'in the quiet granary of death' (the sacks of finished shot looking much like sacks of grain). Had Dickens been able to be present a century later, he would have found little difference in the process which he described. He would still have found the Lambeth shot tower he had visited in 1858, but a century later it was no longer operational. It had closed for the manufacture of shot in 1949 and two years later was incorporated in the Festival of Britain Exhibition, complete with radio transmitter, from which messages were beamed to the moon. When the Exhibition was closed the tower was retained, but failure to find a suitable use led to its demolition in 1962.

Discussion of the second tower at Lambeth has led us out of the chronological order of the construction of shot towers in Britain, but while in south London we may mention another short-lived venture in Southwark. About 1808 a tower was built near St Olave's Church, Tooley Street, London Bridge. It was built for T. Preston & Co. but by the time it was destroyed by fire on 19 August 1843 this use had been given up and it was the London end of a chain of telegraph stations to the south coast. This was run by a London firm of merchants to give information of the arrival of ships in the Channel and depended on a series of manually operated telegraphs, erected on tall buildings along the route, of which the shot tower was a suitable example.[30]

To return to the chronology, the next tower to be built after Elswick came at Chester, commenced probably in 1799 and completed in 1800, a works which was opened at that time by the Walkers' partnership.[31] The 168 ft high tower has remained in production ever since, with a brief break in 1899 when fire destroyed the plant, and is now the senior shot tower in the country, by more than a century. The only change, and one which has rather spoiled its appearance, was the addition of an external lift shaft in 1968, which saved the shot runner the climb to the top of the tower, an internal electric hoist having been earlier installed to lift lead to the dropping point. At its peak the works had a capacity of about 2,000 tons p.a. but in 1950 it produced 1,550 tons, from which it fell to 1,022 tons in 1960. At that level it has stabilised and in 1980 it accounted for the whole of ALM's shot production of a little over 1,000 tons p.a. About 450 tons of this was ordinary shot for sporting guns and fishing weights (of which the latter took about 10 per cent), while some 600 tons was of steelmakers' shot. The latter is the only important modern development in the use of lead shot and consists of the addition of very small shot to molten steel, an addition which improves the steel's subsequent machineability.

Walkers, Parker also came to own one other tower and to purchase a company which had erected a shot tower on a site which it had subsequently sold. The latter was the Morledge works, Derby, begun by Thomas Cox & Co. in 1806, at which the shot tower was probably completed in 1809. It was demolished in the city's central improvements scheme in 1931. The details given in a paper in the Cox records suggest that this was an interesting tower. The paper gives the overall height as 149 ft 0½ in with no fewer than eight landings. Since these were measured precisely, the first being 8 ft 10 ins from the ground, the second 18 ft 1½ ins and the next four between 21 ft 7 ins and 22 ft 10½ ins apart, it would seem to be accurate and to imply that the tower had multiple dropping points for different shot sizes. An account of the tower at the time of its demolition also refers to the existence of nine floors. The other tower which Walkers, Parker owned was at Greenfield, near Bagillt, Flintshire. It was taken over in 1866 when its previous owners, Newton, Keates & Co., left the lead trade and sold this and two other works to Walkers, Parker. It is probable that this tower never worked under its new owners and it was resold to Newton, Keates & Co. in 1872 and demolished soon afterwards.[32]

From the number of shot towers built in the decade or so after the expiration of Watts' patent at the end of 1796, it seems that there was a considerable expansion in the trade and a high level of competition,[33] which was to continue throughout the nineteenth century and make several of the ventures short-lived and not very profitable. Walkers, Parker controlled a considerable proportion of the total trade and, since all three of the firm's main towers remained operational until the mid-twentieth century, it seems likely that the firm were price leaders and remained profitable in shot manufacture. As with other lead products their geographical dispersal, with works in Chester, London and Newcastle, made for considerable advantage in the supply of customers in most parts of the UK.

Philip George's expectation, mentioned earlier, that no other firm would venture the capital in building a shot tower, since his Redcliff firm had reduced the price of patent shot to the level of common shot, was therefore soon falsified. Walkers, Parker were obviously to feature as the chief competitor to the Bristol shot works, but they also experienced competition from other sources. We have seen that two towers were built in London and provided intermittent competition there, but the most interesting development was on Tyneside. Baillie, writing in 1801, contrasted the Elswick shot tower with the shot manufacture of a new firm which, 'at the twentieth part of the expense, found a mode of casting shot equally good, if not by a similar erection, yet, which answers equally well, by dropping the melted shot down an old coal pit, fitted up for the purpose'.[34] The firm was Locke, Blackett & Co., whose works were in Gallowgate, Newcastle upon Tyne, just outside the city wall, but the shaft used for dropping shot was at Wylam, on the Tyne, ten miles to the west. Here, Christopher Blackett, one of the partners, had a colliery and it seems almost certain that later references to the firm's use of a mine ventilator shaft for dropping shot must relate to the Wylam period. By 1826 at the latest a shaft had been sunk at the Gallowgate works, although it may not have immediately replaced the Wylam works which was still listed in local directories until the 1830s. In an account of the Gallowgate works in August 1844 the *Penny Magazine* illustrated the process and stated that the shot dropped into a pit 200 ft deep (presumably an exaggeration) by 6 ft diameter and that 'When two tons weight of shot have thus fallen into the pit, operations are temporarily suspended; the aperture of the pit is opened, a man is lowered to the bottom, and he sends up all the shot from the water into which they had fallen.' Rather surprisingly this seemingly inefficient process lasted until the early 1900s. By this time, however, the firm's output of shot had been inconsiderable for some years and a board minute of Walkers, Parker & Co. Ltd of 10 January 1906 notes that

Locke, Blackett had offered to give up shot manufacture if the Elswick firm would take over their trade. The offer was rejected on the grounds that the terms were unattractive but it seems likely that Locke, Blackett gave up their shot pit soon afterwards.

By the late-nineteenth century all the traditional shot manufacturers were under pressure from a new product, 'chilled shot', and it is almost certain that this had caused the decline in Locke, Blackett's output. Chilled shot was merely shot hardened by the addition of about 3 per cent antimony but it had a considerable impact on the shot trade and on sportsmen.[35] In May 1899 the Elswick manager reported to the Walkers, Parker board a falling off in his shot trade as a result of competition from the Abbey Improved and Newcastle Chilled Shot Companies. He obtained permission to produce new shot sizes and to use the word 'chilled' and this increase in the level of competition was apparently too much for the Newcastle Chilled Shot Co. Ltd, which was in liquidation by September 1902, and the following month Walkers, Parker purchased its goodwill and trade marks for the sum of £400. Although chilled shot continued to be produced it seems likely that this was in name alone. On 16 August 1946 an internal ALM memorandum noted:

> The mention of Chilled Shot in your notes is rather amusing, as the only difference at Lambeth between Ordinary Shot and Chilled Shot is a 'C' stamped on it. In other words, the contents of the bag of either shot is exactly the same. This seems to have been the practice carried out for very many years, and could really only be described as 'Sales spoof'.[36]

In the late-nineteenth century shot manufacturers and purchasers were involved in controversy not only over the merits and demerits of chilled shot but about the variation in the numbers of shot in different manufacturers' cartridges of the same nominal size. Late in 1888, *The Field*, through its letter columns, provided the ground for this particular debate. Cox & Co., especially, was criticised for considerable variation in the number of shot in its cartridges. The editor of *The Field* wrote to Walkers, Parker, as the leading producer, to try to obtain standardisation and this was gradually achieved by agreement between the manufacturers, who also were gradually to adopt a common standard for the numbering and lettering of shot size.

The late-Victorian and Edwardian eras were the period of the huge, organised shoot and the market for shot was sufficiently large and growing so rapidly that it probably repaid considerable attention. Having moved towards the standardisation of shot numbers and sizes and beaten off the competition from chilled shot, Walkers, Parker probably felt that their troubles were over. They were, however, to be faced with a revolt by one of their chief customers, the cartridge manufacturers, Eley Bros. In 1907 Eley erected a shot tower at their Edmonton, north London, works, having previously bought from the lead manufacturers all the shot they packed. Considerable concern was expressed by the Walkers, Parker board, which estimated that this would mean a loss of trade of about 1,000 tons of shot p.a. It was also noted that one of the Lambeth shot-runners had left and accepted a job at Eley, which had no doubt recognised that the technique of shot-running was a skilled job and decided to pay enough to attract an expert from an existing firm. Within a few months the Walkers, Parker board faced a further worry with the discovery that Kynoch Ltd of Birmingham, another major cartridge manufacturer, was considering the building of a shot tower. Following so closely on the heels of the Eley decision this may merely have been a ploy aimed at obtaining improved terms from suppliers who had already lost an important market. If this was the case it worked, for on 20 May 1908 Walkers, Parker agreed (jointly with the Bristol firm of Sheldon, Bush) to supply Kynoch's estimated demand of 1,000 tons of shot p.a. for five years at a delivered price of 42s 6d per ton over the pig lead price.[37] For some reason renewal terms for the contract were discussed in 1911, after three rather than five years, and Kynoch endeavoured to drive a wedge between its suppliers. Claiming not to be satisfied with the renewal terms Kynoch approached Sheldon, Bush with the offer of the whole contract but it is clear from the Walkers, Parker board minutes that Sheldon, Bush was not prepared to indulge in double-dealing. It was suggested between the two firms that Sheldon, Bush might take the contract and then buy half the shot involved from Walkers, Parker at Chester. At this point Kynoch, unaware of these negotiations, approached Walkers, Parker to supply the whole contract independent of Sheldon, Bush. No doubt this offer was received with some cynical amusement and it was promptly turned down. There is no further evidence on the subject in the Walkers, Parker minute books but it is likely that the firms eventually came to an understanding. Certainly after the First World War the shot trade was carefully arranged by agreement between shot manufacturers and cartridge makers in order to ensure that competition was restricted to that from foreign producers.

On 31 January 1924 the contract to supply Nobel Industries Ltd, which had taken over Kynoch Ltd and was shortly to become part of Imperial Chemical Industries Ltd, came up for renewal.[38] The contract had been based on the total trade of 1913 with Walkers, Parker supplying 58 per cent, Eley 26 per cent and Sheldon, Bush 16 per cent of Nobel's requirements. Since the pre-war golden age of game-shooting the demand for shot had fallen off and deliveries for 1922 amounted to only 3,838 tons, against an unspecified but obviously higher pre-war level. As a result Eley were pressing for a minimum share of 1,500 tons p.a. in order to make the working of their tower economic. In part their claim was met, with the inevitable squeeze on the shares of the other two suppliers, as the final agreement signed on 9 April shows:

It is hereby agreed that the total orders for Shot including those of the Associated Companies, the Gun Makers and other buyers in the United Kingdom and for export shall be divided between the Shot Manufacturers as nearly as possible in the following proportions namely: Walkers, Parker & Co. Ltd. $52^1/_3\%$, Eley Bros. Ltd. $33^1/_3\%$, Sheldon, Bush & Patent Shot Co. Ltd. $14^1/_3\%$.

The agreement also set prices, with an increase of 7s 6d per ton in the shot manufacturers' margin and an additional 12s 6d per ton on sales to UK buyers other than Nobel.

With this restriction of home competition the shot manufacturers could face the decline in consumption with some ease. During and after the Second World War, however, the decline in game shooting was considerable, leading to the works' closures we have seen. It is possible that they were premature, since the 1960s and 1970s have seen an enormous resurgence in the demand for shot as social and economic changes have brought game shooting to a much wider social strata and clay-pigeon shooting has become a significant sport. Precise figures for lead shot alone are not available but at the end of the 1970s the annual consumption of shot (including bullet rod) was about 8,000 tons p.a. In 1980 this figure fell to about 6,500 tons and imports of bullets and lead shot (coming chiefly from Italy and Russia) were about 1,200 tons, which was greater than the output of ALM in that year.

Notes

1. Charles Dickens, 'Scenes', Ch. IV 'Scotland Yard', in *Sketches by Boz* (London, Chapman & Hall, 'Autograph Edition', n.d.), p. 37. See also *Punch*, 2

March 1878, for a debate between the Lambeth shot tower and Cleopatra's Needle.

2. *Allan's Illustrated Edition of Tyneside Songs . . .* (Newcastle, 1891), p. 48.

3. To devote a whole chapter to shot manufacture may require some defence! Within a book which is itself wide ranging and complex, the introduction of the manufacture of shot in shot towers possesses these attributes (or defects) in considerable degree. It seemed sensible to isolate the problem and discuss the development of shot manufacture up to the present in a unified section. Further justification may be given by the fact that the shot tower is the only really original development in lead manufacturing in a very long period. A technical account of the shot tower process is contained in W. Johnson *et al.*, 'Small Spherical Lead Shot Forming from the Liquid, Using a Shot Tower', *Metallurgia and Metal Forming*, March 1976, pp. 68-72.

4. There is a number of variants on the basic theme of a dream of drops of molten lead falling a considerable distance and solidifying in perfect spheres. See R.A. Buchanan and N. Cossons, *Industrial Archaeology of the Bristol Region* (Newton Abbot, 1969), and Bristol Record Office, R.A. Steedman, A History of Lead Smelting in Bristol (typescript, October 1966), p. 13.

5. Bristol Record Office, apprentice register 04356(14).

6. The patent is reproduced in full in *The Repertory of Arts and Manufactures*, vol. III (1795), pp. 313-15.

7. Bristol Record Office, lease 52061. Redcliff(e) appears to be spelled alternately in eighteenth-century references. The modern street names are, however, Redcliff, while confusingly the parish is St Mary Redcliffe. In modern road-numbering the premises were numbers 28-30 Redcliff Hill, on the corner with Redcliff Parade and between that street and Ship Court. The premises are described in a number of sources including A. Gomme *et al.*, *Bristol, an Architectural History* (1979), pp. 94-5; *Illustrated Bristol News* 2, 10 (1960), pp. 36-7; *Industrial Archaeology*, 5 (1968), pp. 409-10; and J. Mosse, 'Redcliff Shot Tower', *Bristol Industrial Archaeology Society Journal*, 2 (1969), pp. 4-5.

8. There is much confusion over the height of all the shot towers built in this country. Partly this is a result of confusion between the actual height of the tower and the distance the shot was dropped. Several sources state that the Redcliff Hill tower gave a drop of 110 feet but the drawings taken by Mosse (ibid.), show that the actual drop from the top floor to the water level of the tank in the basement was 89 ft 3 ins. As a result of the fact that the first shot tower was built at Bristol, there have been some unwise deductions – for example, the statement that Bristol was noted 'for its near monopoly of the production of lead shot', B.W.E. Alford, *W.D. & H.O. Wills and the Development of the U.K. Tobacco Industry 1786-1965* (1973), p. 14. This statement cannot be substantiated.

9. Felix Farley's *Bristol Journal*, 2 Dec. 1786.

10. See, for instance, John, *Walker Family*, p. 34.

11. Nevertheless there is a mystery figure of £10,000 in the accounts of Walkers, Fishwick & Co., as may be seen in Table 2.1. For the two years to August 1788 sale of shot is given as an incredible £10,082 while in the following year it fell to only £181. This may have given rise to the idea that the firm purchased Watts' patent but the sum appears on the revenue not the expenditure side of the firm's accounts and does so two months before Fishwick, who signed the accounts, expressed disbelief that anyone should pay £10,000 for a share in the patent.

12. The minute book is in the archives of Courage (Western) Ltd, Bristol, to whom I am grateful for permission to consult and quote from their records.

13. London County Council, *Survey of London*, vol. XXIII (1951), p. 16, quoting Samuel Ireland *Picturesque Views on the River Thames* (1791). The

Survey goes on to say that it was not the first tower built in London – 'In 1758 Henry Raminger, of Christ Church, Southwark, had taken out a patent for the manufacture of lead shot, and a tower was built in that parish some years before the one in Prince's Meadows [i.e., Cuper's Bridge, Lambeth].' If this were true it would throw doubt on the originality of the Watts patent. There is, however, no evidence to support the statement made in the *Survey*, which appears to be based on a misreading of its source – H.E. Malden (ed.), *Victoria County History of Surrey*, vol. 2 (1905), p. 413, which in its turn is based on O. Manning and W. Bray, *History and Antiquities of the County of Surrey*, vol. III (1814), p. 536. After mentioning the Raminger patent, Manning and Bray, without connecting the two, quote an unidentified source as stating that there was 'a slender tall Brick Tower for the manufacture of lead shot under a patent', which undoubtedly refers to the Lambeth tower of 1789. There is no independent reference to a tower in Christ Church, Southwark, and, moreover, Raminger's patent, no. 725 of 29 June 1758 has nothing to do with a shot tower: 'RAMINGER, HENRY – Shot and bullets made exactly round and solid. The lead is purified by melting red lead with one-third of its weight of calcined flints beaten with whites of eggs and eggshells. The shot or balls are cast and made round by pressing them through a cutting engine with steel rollers.' *Abridgments of Specifications, Firearms and other Weapons, parts 1-2 1588-1866* (1869), p. 28.

14. For Windsor Terrace see W. Ison, *The Georgian Buildings of Bristol* (1952, reprinted Bath, 1978), p. 226. On the speculative building boom at Bristol generally, although Watts is not specifically mentioned, see J.R. Ward, 'Speculative Building at Bristol and Clifton, 1788-1793', *Business History*, XX, 1 (1978), pp. 3-18. See also John Latimer, *Annals of Bristol in the Eighteenth Century* (Bristol, 1893), p. 454.

15. The notice of the Commission of Bankruptcy is contained in Felix Farley's, *Bristol Journal*, 1 March 1794. See also *London Gazette* (1794), pp. 166, 554 and 1,257.

16. This is an extract from a long poem about Watts and the discovery of the process, etc. in J. Dix, *Local Legends and Rambling Rhymes* (Bristol, 1839), pp. 8-13.

17. P. George to Walkers, Maltby & Co., 26 March 1798. The letter book is held by Sheldon, Bush and Patent Shot Co. Ltd, Bristol, successors to the lead business of Philip George & Co. I am grateful to Mr Christopher Thomas for the opportunity to consult and permission to quote from the letter book, a microfilm copy of which is now in Bristol Record Office.

18. Some sources state 1868 but *Matthews Bristol Directory* for 1863 gives Sheldon, Bush & Patent Shot Co. established at Redcliff Hill. See also *Work in Bristol* (1883) (articles reprinted from the *Bristol Times and Mirror*).

19. The size of the shot depended upon the gauge of the shotmaker's card (a wire-mesh grid) through which the molten metal was run, but the larger the shot the longer the drop necessary to form round shot. Since there is no evidence to suggest that small shot suffered as a result of being dropped unnecessarily long distances, it is difficult to see why more than one dropping point was necessary. A single dropping point, at the height necessary for the largest shot which could be produced in a tower, seems to be all that was needed and would save the expense of an extra floor, lead melting pot, etc. The only advantages of having more than one dropping point were (1) larger quantities of small than large shot were needed and a lower floor saved the shotmaker from climbing and hauling the pig lead an unnecessary distance and (2) shot of different sizes could be run at the same time – although that could have been possible for a single floor with two dropping holes.

20. It may have been thought that water was essential for cooling the shot,

whereas in fact it was merely there to break the fall, which any soft material would have done. In October 1939 T.G. Robson of the Lambeth works of Walkers, Parker & Co. Ltd, wrote to R. Mitchinson of the firm's Elswick works to inform him that at Lambeth they were dropping shot from 30 ft or less on to a canvas catch-sheet but that this 'should not on any account be made public, because any Jack, Tom or Harry could commence to manufacture'.

21. A number of different dates has been given for the erection of the Elswick shot tower, including one of 1786 (N. Pevsner and I.A. Richmind, *The Buildings of England: Northumberland* (1957), p. 260), and another of 1787 (Falconer, Lead into Gold, p. 7). These were presumably based on the assumption that the tower was built when the patent was, supposedly, purchased from Watts. The dates of 1796 for commencement and 1797 for completion are drawn from an analysis, of local newspaper and other accounts, in John Sykes, *Local Records*, vol. 1, part II (Newcastle, 1833) and are therefore likely to be accurate.

22. John, *Walker Family*, p. 34. The earliest reference of which I am aware to the tower being out-of-true comes in E. Mackenzie, *View of the County of Northumberland*, vol. II (Newcastle, 1825), p. 408, who quotes the 'extraordinary story' as related by another, unacknowledged, writer. It is, however, noticeable that that very thorough local historian, Baillie (*History of Newcastle*), writing in 1801, refers to the tower but makes no mention of the story.

23. In 1926 W.D. & H.O. Wills paid £750 plus £650 p.a. for a site at Hammersmith Broadway and £565 plus £520 p.a. for one in Holloway Road, Alford, *Development of the U.K. Tobacco Industry*, p. 339. Dewars, the whisky manufacturers, also used a shot tower for an electric advertisement of a kilted Scot with a glass of whisky, D. Daiches, *Scotch Whisky: its Past and Present* (Fontana, 1976), p. 95.

24. There are two useful accounts of the Elswick tower. I. Glendenning, 'Shot Making and the Shot Tower at Elswick', *Proceedings of the Society of Antiquaries of Newcastle upon Tyne*, 5th Series, 1 (1951-1956), pp. 351-60, plus illustrations and R.M. Higgins, 'Lead Shot Tower at Elswick, Newcastle upon Tyne', *Bulletin of the North East Industrial Archaeology Society*, 10 (1970), pp. 9-16. In part the latter is merely a word-for-word copy of the former, but it does have accurate drawings of the tower.

25. Without wishing to be associated with those who wish to preserve everything that is old, I regard it as a matter of considerable regret that within the decade of the 1960s the three oldest shot towers in the country were demolished.

26. *Gentleman's Magazine* (1826), under 'Domestic Occurrences', p. 77.

27. Details of the site leased to Maltby are contained in documents in M.L. Walker's collection.

28. It is difficult to see what the reasoning behind this shift of premises was. It may be that the lease position with regard to the Commercial Road premises was less satisfactory than that for Belvedere Road, where the lease was for 98 years from 1824. It is also possible that the latter premises offered greater room for expansion.

29. Dickens' account of his visit is in *Household Words*, 26 June 1858, pp. 34-7.

30. See A.G. Hardy, 'The Old Telegraph from London to the Coast of Kent', *Archaeologia Cantiana*, XLIV (1932), pp. 211-17; *Illustrated London News*, 16 July 1842, pp. 148-9, which contains an illustration of the Tooley St. tower complete with its telegraph arms; and ibid., 26 Aug. 1843, pp. 137-8. There is a useful collection of material on the subject in Southwark Local Studies Library, PC654, Telegraph and Shot Towers.

31. D.A. Nicholls, 'Chester's Shot Tower', *Cheshire Life*, vol. 24 (May 1958), pp. 63-5.

32. John, *Walker Family*, pp. 44-5 and 47. The reasons for this unusual double transaction are made clear in the conveyance of 12 Nov. 1872, in M.L. Walker's collection, by which Walkers, Parker sold the land back to Newton, Keates. While the latter firm had given up lead manufacture in 1866 they had continued their neighbouring copper works and by 1872 discovered the need for land for further expansion. The land, including the shot tower, sold to Walkers, Parker, was now reconveyed to Newton, Keates who covenanted to demolish the shot tower before the end of 1872, and not to use the works for lead smelting or manufacture. I have not been able to establish the date of construction of this tower but it was certainly in existence in 1842.

33. It is difficult to isolate the sources of demand for the increasing output of lead shot which the shot tower process presumably brought. I am assured that there was no military demand for small shot in this period and the French wars were not, therefore, a factor. Although game shooting was of increasing significance it appears to have become of major importance only from the nineteenth century; see F.M.L. Thompson, *English Landed Society in the Nineteenth Century* (1963), pp. 136-44.

34. Baillie, *History of Newcastle upon Tyne*, p. 511. It is also said that there was a shot pit at Benwell Colliery, T.H. Hair & M. Ross, *A Series of Views of the Collieries in the Counties of Northumberland and Durham* (1844), p. 26. It is probable that many disused coal pits in various parts of the country were used for shot manufacture for short periods. Glendenning ('Shot Tower at Elswick', p. 357) mentions the belief that there were in 1830 three pit shafts in the Newcastle district being used for running shot. He also mentions (pp. 357-9) several other towers (including a second one at Chester) and shafts – although it is likely that they were all ephemeral. Prof. W.E. Minchinton, however, informs me that there was a 60 ft tower at Crane Park, Twickenham, Middlesex, which remained in production until 1927. See his article to be published during the summer of 1983 in *History Today* to celebrate the centenary of Watts' process.

35. See correspondence in *The Field*, 23 Nov. 1882 and subsequent editions. Much of this correspondence is reproduced in the ALM house journal, *Southern Area News Bulletin*, nos. 6-8 (June to August 1964). For the Newcastle Chilled Shot Co. Ltd, see 'Tyneside Industries', *The Tyneside, Newcastle and District. An Epitome of Results and Manual of Commerce* (1889), p. 164. This firm had commenced about 1870, under the title Roberts, Lampen & Co., to manufacture shot using the old shot pit at Wylam. From about 1875 it was established at Oakwellgate and West Street, Gateshead, under the title of the Newcastle Chilled Shot Co. Ltd. The Wylam works was then abandoned although there is no evidence on the method of manufacture at Gateshead. There was also, in Newcastle in the 1890s, the Abbey Improved Chilled Shot Co. Ltd, whose shot drop was said to be 240 ft – see *Rivers of the North* (1894), p. 163.

36. H.G. Lancaster, Lambeth works manager, to T.H. Deane, Advertising Dept.

37. It will be clear from Appendix I that the fluctuations in pig lead price were considerable and were the major factor in determining the price of all lead products.

38. In November 1918 Explosive Trades Ltd was formed to control the whole of the UK explosives industry. Its major constituent was Nobel's Explosives Co. Ltd, which dominated the new company, but it included Kynoch Ltd. See W.J. Reader, *Imperial Chemical Industries, A History vol. I The Forerunners 1870-1926* (1970), p. 313.

4 EXPANSION OF WALKERS, PARKER & CO. FROM 1800 TO 1889

Thou are a soul in bliss; but I am bound
Upon a wheel of fire, that mine own tears
Do scald like molten lead.
                    Shakespeare, *King Lear*, Act IV, Sc. 7, 1. 46

*Extension of the firm of Walkers, Parker & Co. from its beginnings at
Elswick — the establishment of additional manufacturing works and
their development; the overall development of the firm — changes in
partnership structure, profitability and output; the partnership dispute
of 1881, Court of Chancery action and sale of the partnership to a
limited company in 1889; financial success of the partnership and the
social position of the partners.*

We may now return to the chronological development of the lead companies, following the successful establishment of Walkers, Fishwick at
Elswick, firstly to consider the extension of that partnership by the
opening of other works and secondly to look at the creation of other
partnerships around the turn of the eighteenth century.

When Samuel Walker I died in 1782 his share in the lead partnership
passed to his sons Samuel II, Joshua, Joseph and Thomas. Of these it
was only Joseph who took any part in the affairs of the lead partnership and Fishwick and Ward continued to have day-to-day control. With
the Elswick works established and its product successfully marketed in
London, the idea of opening a London works in order more expedi-
tiously to supply the metropolitan market emerged.

Islington

There is no evidence as to who first proposed this development but in
1785 the partners took a lease of land and buildings in Hoxton Fields,
St Mary Islington 'formerly called or known by the name or sign of the
Rosemary Branch House'.[1] One of the buildings had previously been
used as a public house, known as the Rosemary Branch, which had been
a popular resort on the edge of the countryside.

We know that Ward went from Elswick to erect and manage the new
works, which, like its predecessor, was to manufacture only white lead.
This involved the necessity for power in order to grind the corroded

75

lead and, as at Elswick, this was to be supplied by a windmill. It was
this which most interested contemporaries about the Islington works.
John Nelson commented, for instance, on the building in 1786 of

> a curious windmill for the purpose of grinding lead, differing in two
> remarkable particulars from common windmills; viz. *first*, that the
> brick tower of it is crowned with a great wooden cap, to which is
> affixed, on one side, the flyers; and, on the opposite side, a project-
> ing gallery, terminated by a very small machine of four flyers, by
> means of which the whole cap is turned around at pleasure, so as to
> bring the sails into that direction which is most convenient with
> respect to the wind; and *secondly*, that instead of four, the usual
> number of flyers, this was furnished with five.[2]

The continuing use of windpower at Islington implies that introduction
of the steam engine was largely an inverse response to the price of coal
in different parts of the country. After their difficulties at Elswick with
power supply one might be surprised that the partners installed a windmill at Islington in 1786, even more surprised that they should install
another in 1792, four years after their Elswick steam engine commenced
working, and amazed that the mills at Islington were still in operation
when Nelson wrote in 1811. The difference in the price of coal between
Newcastle and London was the reason. The extra running costs of a
steam engine in London were too great to be worth the extra efficiency
in production over a windmill.[3] Fishwick had recognised this when he
wrote, on 6 January 1787, to Maltby, who clearly feared that steam
engines were going to make the newly-erected Islington windmill
obsolete: 'coals are very low here [Newcastle] so that you need not
fear similar constructions [steam engines] being fixed in your neighbourhood, a London Chalder [25½ cwt] does not cost us 3/- laid down
upon the ground and I suppose they are 30/- with you'. By the mid-
1830s, however, the situation was different, the Islington windmills still
existed, but bereft of their sails they had been replaced by a 20 hp
steam engine, since the running costs had fallen.

There is no detailed evidence of the early running of the Islington
works and there are no statistics which would give an idea of its productive capacity. That Elswick continued to supply white lead for the
London market does, however, show that Islington was incapable of
meeting the demand. Both works continued to supply the metropolitan
market and there seems to have been some friendly rivalry between
them. Fishwick wrote to Maltby on 24 September 1787, 'compare the

samples of the 8 casks [of white lead] last sent you with theirs and you will see there is room for improvement at Islington'. When Nelson wrote in 1811 he thought that 'Between 40 and 50 persons, chiefly women, are employed ...' and this number had expanded by 1842 to about 70, 'two-thirds of whom are women, whose constitutions are supposed to be less injuriously affected by the unwholesome processes of the manufacture than those of the more robust sex'.[4] Since Islington remained entirely a white lead works, these figures suggest that so far as that product was concerned Islington was much the same size as Elswick, which had 45 female employees in 1842. According to some statistics relating to the partnership,[5] the average capital employed at Islington from 1815 to 1823 was £53,000 and the average annual profit £2,196, giving a return of 4.1 per cent on capital employed.

Red Bull Wharf, Upper Thames Street

It is not clear whether Walkers, Fishwick and Ward had decided to take a London partner before they commenced the Islington works or whether the London partner was deemed necessary as a result of that decision. In either case the link was very close, since Thomas Maltby was taken into the lead partnership in 1785, the year in which Ward went to Islington. Thomas Maltby & Son were established as merchants, with an office at 6 Queen Street, Cheapside, at least as far back as the 1750s. Maltby had, therefore, established trading connections in London which would be useful to Fishwick and Ward for the marketing of Elswick and Islington products. It is not, however, clear whether Maltby was established at Red Bull Wharf before he entered the partnership or whether those premises were taken subsequently.[6] It is obvious, though, that a Thames wharf was essential if the partners were to take direct control of their London marketing operation. It was a step in the total integration of the process from manufacture to sale, which was logically followed in 1786 by the purchase of a share in a ship involved in the Newcastle to London trade.

The Thames premises were used only for wharfage and warehousing and involved no manufacturing activities. Nevertheless Maltby cannot be seen just as a merchant, since his correspondence with Fishwick shows that he took a very active and important part in the lead partnerships and was keenly involved in major decisions on the manufacturing side. His connection with the Walkers' partnership was to last for 40 years, and he was certainly the longest surviving partner who was unconnected with the Walker family. At the time of negotiations for the renewal of the partnership agreement,[7] which expired at the end of

April 1824, there was disagreement between Maltby and the other partners. Although negotiations continued into the following year, he then left the partnership with a settlement which gave him the ownership of the Islington works in exchange for the five shares (out of a total of 28) which he had held in the partnership. As we have seen he also started a rival shot manufactory at Lambeth, which was carried on after his death in 1830 by his sons. In 1839 the Lambeth works was sold to Walkers, Parker at the time of the bankruptcy of one of Maltby's sons, while the Islington works was also sold in the same year, eventually to come into the hands of Champion, Druce & Co.[8] It is much less clear what happened to the premises at Red Bull Wharf, which became the freehold property of the Walkers' partnership in 1825. In a property valuation in that year it was recorded at a value of £3,900 out of the total of £62,000 for the partnership's freehold and leasehold property. Some other statistics were also given for the works for the period 1815 to 1831, unfortunately amalgamated with those for Lambeth. The average capital employed was £91,000, with an average annual loss over the period of £2,218, giving a return on capital of −2.4 per cent. This was a much worse return than that experienced at other partnership works and it seems likely that provision for bad debts was a major cause. Running at an average of about £3,000 p.a., bad debts were much higher than for other partnership works.[9] Although one would expect Lambeth to make a positive return it may be that Red Bull Wharf, which was not a manufacturing works but only a trading centre, was in fact carrying overheads which ought strictly to have been charged against the profits of the works for which it was selling goods. After these figures end in 1831 there is no further mention of Red Bull Wharf and it is to be assumed that the freehold was sold, at a time when the partnership was experiencing some financial difficulty.

Lambeth

That Red Bull Wharf and Lambeth figure conjointly in the profit figures suggests that they were run as a single unit. As we have seen Lambeth was taken over about 1797 and was soon extended to the manufacture of products other than lead shot. Expansion of the product range necessitated an extension of the site and in 1810 a new lease[10] was taken. The original shot works of 1789 had a river frontage of 76 ft but the new lease (of which the old works formed part) gave a frontage of 160 ft to the Thames and a depth of almost 400 ft to Commercial Road. The new premises were more than six times in area the old one and this was reflected (together with war-time inflation) in an enormous rent, compared to anything so far paid by the partnership, of £770 p.a. Lambeth now appeared to be well placed for development and soon afterwards 'considerable alterations improvements and additional buildings were paid out of the general funds of the Co-partnership'.[11] The developments must have related to the manufacture of sheet and pipe since Lambeth remained a blue lead works and did not undertake chemical production.

With this limited range of activities, Lambeth was a large works but despite this and the poor profit figures (admittedly in conjunction with Red Bull Wharf), the partners were sufficiently satisfied to pay a premium of £6,750 for a new lease in 1824, seven years before the old one expired.[12] While in the cyclical upswing of 1824 an extension of the lease seemed worthwhile, in the less prosperous trade of the later 1820s and early 1830s it seemed less attractive. In August 1831 the lessor replied to an initiative of the partners 'declining to take Lambeth lease off our hands' and attempts were subsequently made to obtain sub-tenants for part of the site.[13] Subsequently the works was given up, probably before the lease expired in 1845, and the partnership's activities transferred to the old Maltby works on the west side of Waterloo Bridge. Why this should have been preferred is unclear, although the fact that it was a smaller site than the works vacated and was occupied entirely by Walkers, Parker may have been sufficient reason. It continued the tradition of being only a blue lead works, producing pipe, sheet and shot. When it was put up for sale in 1884 it contained five hydraulic pipe presses, an 86 ins rolling mill by Bramah, and a 36 hp beam engine by J. & E. Hall to provide power. By this time the Lambeth works, at 63 Belvedere Road, had become the centre of the partnership's London activities, it having been noted that in 1856 'In consequence of great robberies of lead from the Lambeth Works, the office was removed to them from the City.'[14]

Orange Street, Southwark

In 1824, following Maltby's withdrawal from the partnership and his purchase of the Islington works, the Walkers' partnership found itself without a lead chemical works in London and an existing works was taken over in Southwark.[15] Orange Street (since 1936 renamed Copperfield Street), was off Gravel Lane in the parish of St Saviour and part of it, known as Loman's (sometimes Leoman's) Street or Pond, had early attracted lead manufacturers. It was an area of poor quality terrace housing which before the end of the eighteenth century was already being infiltrated by manufacturing premises. Given the high incidence of lead poisoning among employees in the manufacture of white lead, it was conveniently near to St Saviour's workhouse. The works taken over by Walkers, Parker had originally belonged to Enderby & Clark, who appear in directories as white lead manufacturers of Gravel Lane from 1786. There was also a white lead works in Loman's Pond belonging to George Lane in the late-eighteenth century and subsequently to Thomas Johnson.

It is difficult to produce a convincing reason for Walkers, Parker's decision to take the Orange Street works, other than one of expediency. It had the advantage that it was only perhaps half-a-mile from their Lambeth shot works, from which it was possible to oversee its running, but it had several disadvantages. It was in a heavily built-up area which would make for congestion affecting road transport, it had no immediate access to water transport and it must have been a fairly small works.[16] In 1831, by which time part of the premises had already been sublet and at the same time that they were endeavouring to rid themselves of the Lambeth lease, Walkers, Parker had decided that Orange Street was not an attractive proposition any longer and were trying to find an alternative tenant.[17] They appear to have been unsuccessful and have soldiered on with the works for a few more years but in 1837 the rate books show the premises to have been empty. In future Walkers, Parker would supply lead chemicals for the London market by sea from their Elswick works.[18] The only other evidence on Orange Street is the statistical return which shows the average capital employed from 1825 to 1831 to have been almost £30,000, and average annual profit £1,211, giving a return of 4.1 per cent on capital employed, identical to the figure for the Islington works which it had replaced.

Derby

In 1792 Ward moved from Islington to supervise the building of the Derby white lead works at Mill Hill (or Windmill Hill) and remained there as managing partner until his death in 1800. We have already seen that Derbyshire was an important centre of lead mining, although the region's output was in decline before the end of the eighteenth century, and that Derbyshire lead was highly favoured for conversion to white lead. These seem to have been the basic reasons for the decision to set up a works in the area. Given the availability of local pig lead, a Derby works made considerable sense as the next area of expansion from Tyneside and London, since it was well placed to supply an expanding market in the Midlands. With hindsight, however, it is easy to see that it was not a wise move and at least part of the justification for that conclusion should have been evident at the time. The move was to an area where there was already evidence of declining local raw material supply, and although the works was close to the River Derwent and had important canal links with the rest of the Midlands and with Lancashire and Yorkshire, the transport facilities were clearly inferior to a site with rapid access to coastal shipping. Finally it may be said that Derby was a town which to some extent was bypassed by industrialisation and, as a result, it did not offer the growth in local demand which a works in Birmingham or Manchester might have obtained.

There is little detailed evidence on the Derby works which remained smaller than the other works and may have continued to produce only white lead. For the one year for which there is output evidence, 1830, the minutes record that 592 tons of white lead were made at Derby (while Orange Street made 536, Chester 986 and Elswick 1,231 tons). In addition to the works there was also an office at Full Street in the city itself.

There are profit figures for the Derby works which commence with the year ending 15 April 1794, which presumably sets the beginning of production around March/April 1793. The profit rose from £1,500 in the first year to rather over £5,000 in each of the years to April 1798 and 1799, when the figures cease. As with the Elswick figures for the same period they appear to be struck before deduction of interest on capital and management charges but after charging capital improvements to revenue. As with the other works there are financial statistics for the period 1815 to 1831, although it is unlikely that they are directly comparable with those for the earlier period. With an average capital employed of almost £30,000 but an average annual profit of only £819 (giving a return of 2.8 per cent on capital employed), they do, however, suggest that profitability was lower in the years after the Napoleonic War. In 1839 the works were sold to Cox & Co., the local lead manufacturers.[19]

Chester

The next stage in the expansion of the Walkers' lead partnership came at Chester in 1799 when they began to build works on a site which is still occupied by ALM. It may now appear to have been a strange choice since we do not regard Chester as an industrial city, but at the turn of the eighteenth century it made some sense and the fact that it still operates today suggests that the original location decision was sound. Chester had been the traditional port for the shipment of north

Wales lead, which was brought via the Dee in small ships and barges from the little harbours and creeks of the north Wales coast. In the 1760s and 1770s some 4,000 tons of pig lead were exported from Chester per annum and even the dramatic rise of Liverpool had failed to disrupt this trade. In addition Chester, astride what was to become part of the Shropshire Union Canal, had good transport connections with the Midlands and northern industrial districts. Local raw materials and good transport facilities clearly attracted the Walkers' partnership to a site which they purchased on the bank of the canal, which was, for a long time, to be the major means by which pig lead was moved in and manufactured goods out.

Tradition has it that the partnership experienced difficulty in obtaining permission to trade in the city, as a result of the fact that none of the partners was a freeman of the city, whose guilds retained some of their medieval powers to regulate industry and trade. It seems unlikely, even in a city with such strong traditions as Chester, that the erection of the works could have been prevented, but it is said that the problem was finally overcome by the fact that Maltby, a freeman of the City of London, was able to claim reciprocal privileges in Chester.[20] An area of probably ten acres was purchased and from the beginning Chester was planned, unlike any of the other partnership sites, as an integrated works. It was to make white lead and began with a complement of six stacks, in addition to red lead furnaces, a paint mill and the shot tower. Within little more than a decade pipe drawing machinery[21] and a rolling mill for the production of sheet lead had been added. It seems clear that the Chester works was intended to supply the full range of lead products to customers in the western half of the country.

As with the other works there are profit figures for the period 1815 to 1831 for Chester, but they are amalgamated with those for the Liverpool and Newcastle-under-Lyme branches which came under the administration of Chester. The Liverpool branch was almost certainly only a warehouse and office, leased from about 1801 to provide the partnership with a convenient base for the export trade and also for coastwise shipment. As a centre for making white lead as a glaze for the potteries, a works at Newcastle-under-Lyme would obviously make sense, but no documentary evidence has survived and there is no mention of the works except in the centenary minutes, which state that 'In 1807 the White-lead works at Newcastle-upon-Lyne first appear in the books and are valued at £21,191; but there is a rough stock book of them as early as 1802.'[22] It is said that the works probably contained five or six white lead stacks and they are included with Chester and

Liverpool in capital valuations of 1817 and 1826 but receive no subsequent mention. For all three establishments the average capital employed for the period 1815 to 1831 was £131,000 (but it fell from £145,000 in 1825 to £99,000 in 1831 and it is possible that the Newcastle-under-Lyme works was disposed of during this period). The average annual profit was £6,258, giving a return on capital of 4.8 per cent, little better than those of Derby or the London works.

Unlike those works, however, Chester did not come under the threat of sale in the 1830s, perhaps because its geographical position made it important for the firm's sales coverage of the whole country and Chester came to be regarded as a major base for the partnership. As at Elswick a house was built for the managing partner, on an even more lavish scale with extensive grounds of five acres, which came to include a lawn tennis court, an ice house, aviary, vinery, fernery and greenhouses for peaches, cucumbers, etc. The house, as described in 1884, was 'approached by 2 Carriage Drives' and had 'a private way into Chester Station'. There is no accurate date for its erection but a cast-lead rain water head, still extant on the building, bears the date 1819. It was not, however, merely the scale on which the Chester works and its accessories were established which encouraged its continuation, but also the involvement of the partners with the works and city. The Chester works was set up under the title Thomas Walker & Co. and while there is no evidence to suggest that the founder's fourth son actually resided at the works as managing partner, this was a policy which was later to become established. In particular E.S. (later Sir Edward), third surviving son of Joseph Walker, became managing partner at Chester and had a considerable influence, not only on the works but also in the town, of which he became mayor in 1837 and again in 1848.

No doubt this was a useful way of cementing relations with the city which must often have looked askance at a works which hardly added to the city's beauty and must often have added very obvious pollution to its atmosphere. It may also have facilitated the doubling of the site, as a result of further purchases of land in the 1840s and 1850s.

In 1839 to 1840 Chester employed between 150 and 170 employees, of whom about 25 to 30 were women in the white lead stacks. Also included in the total were the crews of five barges employed in bringing lead from north Wales. The works was considerably extended in the second half of the nineteenth century with new white lead stacks being built in 1851, 1858, 1872 and 1875, giving a total of 61 by 1908. It also had railway sidings, running into the works from the London and North Western Railway, which to some extent had superseded the canal, and these connected with tramways laid down throughout the works for the easier handling of materials.

Bagillt and Dee Bank

North Wales was one of the more important lead mining areas in Britain, an early base for the London Lead Company, and the area in which the reverberatory furnace became most strongly established for smelting lead.[23] There also developed on the Flintshire coast a number of lead manufacturing works, producing red lead especially. In setting up at Chester in 1799, rather than on the north Wales coast, Walkers, Parker had made a deliberate break with the policy of manufacturing close to the ore-field. It is interesting, therefore, to find that they subsequently reverted to the old policy, buying one of the north Wales works.

Since the output of ore from the north Wales mines had begun to decline in the late-eighteenth century and several of the local manufacturing works were sold as unprofitable, it might be thought that Walkers, Parker were rather late in entering the area. The partners may, however, have felt that the purchase would help to safeguard lead supplies at a time when demand was rising and home supply was not yet being supplemented by imports on a significant scale. In addition, while smelting works may not have been sufficiently profitable to a firm chiefly involved in smelting, they may still have been attractive to lead manufacturers to whom they offered the promise of lower raw material costs. Walkers, Parker was apparently not the only firm to show an interest in the Flintshire industry and there was a considerable inflow of English capital in the second quarter of the nineteenth century.[24]

In or about 1834 Walkers, Parker took over the lease of the Dee Bank smelting works from the Halkin Adventurers.[25] This was a relatively small works, although probably with a considerable amount of land, which 'consisted only of a single row of smelting furnaces, with slag hearth and office'.[26] In 1838 the freehold of the works was purchased following the purchase in the previous year of the land between the smelting works and the River Dee. In total Walkers, Parker now had some ten acres of land straddling the main road from Holywell to Flint, with access to the Dee for shipment.

On the land purchased to the north of the main road Walkers, Parker began to build a new works, which was in operation by 1841. Apart from the installation at other works of hearths for the working of waste metal from their manufacturing processes, Dee Bank is the first unequivocal evidence of the firm undertaking smelting operations and they appear to have done so on a fairly large scale. In 1878, for instance, A.O. Walker, the partner in charge, told the Royal Commission on Noxious Vapours that there were 16 reverberatory smelting furnaces at Dee Bank.[27] While the smelting of ore was the most important single activity, the scale of manufacturing was considerably extended. Since much of the ore received had a high silver content there was a desilverising or refining works, new red lead furnaces were built and in 1842 a rolling mill installed (in 1855 replaced by a new 100 ins mill made by the Coalbrookdale Company). As a result of this expansion the partnership's assessment of the total investment in the Dee Bank works rose from £21,000 in 1841 to £81,000 in 1856, which exceeded the valuation of Chester, and to a peak of £120,000 in 1866.[28] Nevertheless its importance never gave it more than a local manager — oversight being given by the managing partner at Chester.

Possibly the most problematical result of taking a smelting works was the necessity of dealing with pollution from the waste gases of the process, which led most such works into difficulties with their neighbours and especially farmers. In 1846 the partnership installed a Stagg's Condenser at Dee Bank, in which the furnace waste gases were pumped through water in which the lead fume was supposed to be collected. The Condenser appears to have been unsuccessful and at Elswick in 1852 Stokoe's Condenser, on the same principle, was installed. A.O. Walker subsequently claimed that the partnership expended £20,000 in putting up the Stagg's apparatus together with an 80 hp steam engine to work it but that 'the result was very unsatisfactory'.[29] Subsequently in 1867 41 acres of land to the south of the works were purchased for £2,250 with the aim of building a long flue in which to catch the fumes. The flues and chimney stack were completed in 1870, the former with a length of nearly two miles and the latter with a height of 88 yards, at a stated cost of £10,000.[30] Not even this expenditure solved the problem, since A.O. Walker told the 1878 Commission that 'Every horse or cow that has anything the matter with it within two miles of a lead chimney, even if it breaks its leg, it is always the lead that has done it, somehow.'

In 1884 Dee Bank was described as a ten-acre site, together with an additional 13 acres containing the flues and chimney. The main site was divided into (1) upper works, with engine and boiler houses and chimneys, but no manufacturing process; (2) middle works, with smelting furnaces and disused desilverising works and (3) lower works, which contained the rolling mill, red lead furnaces and a new desilverising works by the zinc process which had been introduced the previous year.

The latter development shows that investment in the works was going ahead and this is confirmed by the erection of a blast furnace in 1885. This was a significant reaction against almost two centuries' dominance of the Flintshire smelting industry by the reverberatory furnace. It did, however, reflect the much greater efficiency of modern blast furnaces, the minutes commenting, 'Blast Furnace erected at Dee Bank on the most modern principles thereby ensuing a great saving in the smelting of ores and facility in treating richer ores one had hitherto been unable to work.'[31]

Success in running the Dee Bank works had led to offers of other works to the partnership, but these had been declined. In 1854, for instance, Mather & Co. offered their Dee Bank zinc works, which was adjacent to the lead works, and in 1859 Newton, Keates & Co. made what must have been a tempting offer to join them in working the Bettisfield Colliery, which was also adjacent to the Dee Bank lead works. To have its own supply of coal would have been valuable to the partnership but the offer was rejected, 'partly in consequence of some memoranda, showing [the colliery's] unprofitable working in former years, that accidentally came into A.O. Walker's hands ... This was fortunate, as the undertaking ended disastrously.'[32] In 1866, however, Walkers, Parker accepted an offer made by Newton, Keates to purchase all their three lead works, at Greenfield, Bagillt and Glasgow. It seems likely that this decision was taken less from any desire to expand their activities than from an anxiety to restrict competition by ensuring that no one else took on the works, and that at Greenfield was closed. Like Dee Bank, the Bagillt works was primarily a smelting works but it also had desilverising equipment and a rolling mill and Walkers, Parker added a pipe press. After 1866 the smelting capacity at Bagillt was rarely used, Dee Bank offering sufficient capacity, and it was kept on as a manufacturing works. The works would appear to have cost little, the capital value of Dee Bank, under which Bagillt was presumably included, rising only from £112,000 in 1863 to £120,000 in 1866. It also appears to have been run as a poor relation to Dee Bank, which received all the new investment, and the Bagillt works was disused by 1884 and did not work again, being subsequently demolished and the land sold. Although an attempt was made to work Bagillt it was probably purchased in order to prevent competition and the subsequent fall in the price of lead and the depression of the late 1870s and early 1880s made it uneconomic. This was symptomatic of the conditions in the whole of Flintshire and in 1891 Dee Bank was the only works in production, employing 144 as compared with 461 employed in the county's lead works in 1851.[33]

Glasgow

This works, at 110-120 Cheapside Street in the Anderston District of Glasgow and close to Queen's Dock, was taken over from Newton, Keates & Co. in 1866. From the point of view of Walkers, Parker this would seem to have been a logical extension to their geographical coverage. From the late-eighteenth century the Elswick works had built up a considerable Scottish trade, although chiefly on the east coast, and there were few Scottish-based competitors. There is little evidence on the Glasgow works but it appears to have been entirely a blue lead works, producing lead pipe and sheet. It also acted as a warehouse for the supply to Scottish customers of other Walkers, Parker products. As such it seems to have been seen as an outpost, with a local manager but controlled from Liverpool. There was no attempt to extend the range of products and investment in the works was limited. This lack of interest was compounded by a fire in July 1882 which destroyed the rolling mill and left much of the works in ruins, in which state they still remained in 1884. The venture into Scotland had not proved successful, perhaps because of the difficulties in competing effectively with Alexander, Fergusson & Co., a Glasgow lead manufacturing company, set up in 1854 (whose works had actually been offered for sale to Walkers, Parker in 1857, but refused).

Elswick

One of the first pieces of evidence which we have about this works implies that some changes occurred as Fishwick's presence diminished. A works sick club was commenced on 1 January 1799, perhaps at the instance of Fishwick's subordinate, Samuel Walker Parker, who was to replace him as managing partner. Members were to contribute at the rate of 1d per week and after one month's contributions were entitled to eight shillings a week benefit from the eighth day of sickness which laid them off work. While this sum would be only perhaps one-half to two-thirds of the average wage at the time, it would make a valuable contribution when illness led to inability to work. The scheme attracted 64 initial members,[34] of whom 42 were still paying at the end of the year, and, although there was some decline from the initial level, membership of the scheme stabilised at about 40 up to 1812 and then rose slowly to about 55 by 1820, the last year for which the records are preserved. That the society remained in existence for more than 20 years in the early-nineteenth century, when many similar organisations were short-lived, is a tribute to the efficiency with which it was run. In the first years there was only a small balance of income over expenditure but an increase in contributions to sixpence a month from 1805 and a contribution of 15s per month from the company ensured financial security. In later years the annual surplus usually exceeded £10 and from 1809 entries, such as 'Paid for beef beer bread etc. £5 8s 0d', tell their own story.

The other areas in which there is some detail about the Elswick works are the customers and profits. As before 1800 the trade remained heavily east coast, from Kings Lynn to Aberdeen and with a large turnover with Walkers, Maltby & Co., (about £90,000 p.a., 1810 to 1814), who sold to London customers. It is difficult to be sure how much of this was made up of manufactured goods for resale, since the ledger figures usually do not specify such details, but the major part would have been of pig lead for the London works. For twelve months in 1809 to 1810, however, shipments to Walkers, Maltby & Co. included £13,200 of white lead and £10,000 of red lead. It is probable that sheet would have added at least a further £10,000 and shot, orange lead, etc. a few more thousands — perhaps suggesting annual sales through the London branch of about £40,000. An analysis of 100 pages of the 1810 to 1814 ledger gives a total of 114 customers of whom 51 (44.7 per cent) were in Northumberland and Durham, 40 (35.1 per cent) in Scotland, and 23 (20.2 per cent) south of the Tees.[35] Most of the customers, especially the local ones, bought in small amounts — red lead for glass works, white lead for painters, sheet for the building trade, etc., but there were a few large orders such as sheet to the value of £1,369 sold to the Prestonpans Vitriol Co. in 1814 for an extension to its sulphuric acid manufacturing capacity. A ledger for 1822 to 1824 gives a broadly similar pattern, although with an increase in customers in more distant markets. The proportion of Scottish customers is much the same at 35.2 per cent although their number has increased and more sales were made to the west coast and especially Glasgow in the absence of an effective Scottish competitor. The proportion of non-local English customers had risen considerably to 31.1 per cent and of these some were of significance. For instance, between May 1823 and July 1824 King & Co. of Southampton received £9,000 of lead goods, most of which was presumably destined for export. In addition there was an increase in direct overseas sales to agents in Denmark, Hamburg, Gibralter, New York, Quebec, Montreal, etc., but they were insignificant in total. While local customers may have accounted for a smaller proportion of the total, some of them, such as Abbot & Co. of Gateshead, who were purchasing almost £2,000 of lead goods p.a. in 1823 to 1824, were of importance. The local clients show the wide range of demand. They included coachmakers (such as Atkinson & Co. of Newcastle), potteries (Maling & Co. of Sunderland), glassmakers (Charles Attwood of Gateshead and the Cooksons of South Shields), plumbers and painters in large number, ironmongers, colourmen, dry-salters, druggists, gunmakers, etc.

As for the other works, there are statistics which show that the average capital employed at Elswick in the years 1815 to 1831 was £125,000, while the average annual profit was £7,144, giving a return on capital of 5.7 per cent, rather higher than at the other works. Elswick was, therefore, the largest single works (although its valuation was exceeded by the combined figure for Chester, Liverpool and Newcastle-under-Lyme) and in 1830 it was the largest individual producer of the original product, white lead, with an output of 1,231 tons.

In 1840 the partners began the purchase of the freehold of the site, paying £8,000 for the land adjoining the river and in 1858 the remainder of the original site was purchased for £17,800, while a further extension was made in 1864. It seems likely that the mid-nineteenth century was a profitable period for the partnership and that the purchase of the land was an effective way of using spare capital — in some senses a more rewarding investment than the purchase of railway shares and stock, which was a common practice for many partnerships at the time.

Also in 1840, the existence of a copy letter book for some months of that year gives some detailed evidence about the trade of the Elswick works. Significantly the fact that a single book lasts for only nine months as compared with the several years which are covered in the eighteenth-century letter books, gives some clue as to the extension of trade. As earlier, the pattern remains of a trade which was heavily concentrated in the eastern part of the country, with Scotland and the north-east of England prominent. The division of the country between the Walkers, Parker works appears to have become formalised. On 16 May an order from a Nottingham firm was sent to Chester with a covering note that since 'we do not visit Nottingham we conclude it to be intended for you'. Two days earlier the same distinction had been drawn with regard to an overseas market, although it seems less likely that this was a result of an agreement between the branches. Wm. Whittaker, consul at Santos, Brazil, had called at Elswick for 'consignments particularly of shot'. Not being involved with the Brazilian market, Elswick had recommended him to Chester with the comment in a letter to that branch 'we wish your Mr. Walker to see him himself if possible'.

The first letter in the book, dated 2 January, shows that the Elswick works had supplied white lead to Richard Grainger for the painting of the buildings which had been recently erected in the redevelopment of central Newcastle.[36] Grainger, however, was experiencing financial difficulties and Walkers, Parker wrote to him, 'We are favrd. with your letter of the 31 Ult. although attended with much inconvenience in doing so we will renew the acceptance of yours which falls due . . .' The bill of exchange, which Grainger was unable to meet, was for £500. A few days later another letter to Grainger shows that the Elswick works had benefited considerably from Grainger's orders for lead for Newcastle buildings, even if payment was delayed. In it two further bills, drawn at twelve months, for payment of deliveries of 'milled lead & pipe furnished you from August to Nov'r inclusive last year' for £1,400, are mentioned.[37]

At this time the trade was conducted by the usual series of sales directly to customers[38] and also to merchants in various towns. Orders were now received not by the managing partner, as in Fishwick's day, but by two travellers, Henry Patterson and E.B. Stamp, who were sent on tours to visit existing customers and obtain new ones. They were given a certain amount of flexibility as to prices at which they might take orders but they also received regular letters, while they were away from Elswick, which gave them fairly detailed guidelines. On 29 January Patterson, who was in Glasgow, was informed that 'the L'pool Houses are offering so very low say Sheets at £17 10s 0d & £17 15s 0d p. ton. You may offer at £18 2½% off but do not lose a good order @ £17 15s 0d this last price for a quantity for 20 or 30 Sheets we are not inclined to go under £18'. Occasionally, however, the freedom allowed the travellers in quoting prices rebounded. A letter to Patterson on 28 March noted that 'James & Co. have been complaining of a sale of yours', a sale which had obviously been made at a price below the agreed manufacturers' list.

By 1840 the Tyneside lead manufacturers had established a joint pricing agreement. On 27 March Walkers, Parker sent the following:

Circular. At a meeting held this day of Lead Manufacturers. Agreed that our prices be fixed as under with the understanding that no circulars are to be issued, or parties to be called upon. Invoices to be sent charged as under without any comments as to prices charged. Dry White Lead £28 Best White Paint £30 2nd £28 3rd £26 Orange

Lead £33 Red Lead £22 Litharge £20 Ground £22 Pat. Shot £22 with the usual allowances to the Foreign Houses.

This implies a fairly well established structure for eliminating price competition, although there is no evidence as to how long it had existed or for how long it would last. There was, no doubt, competition from small firms, especially in paint sales, and also from manufacturers in other regions, particularly at the boundaries of the Tyneside firms' main area, and this was to worsen as an effective railway network came into existence by 1850. It would seem likely that price lists, such as that given above, were considered at quarterly meetings of the Tyneside manufacturers, since on 25 June a fresh circular was sent, reducing the prices of white lead and paint by £2 per ton, and on 28 September, one of the last letters in the book gives a new price list circular. That these lists did not eliminate competition, even between Tyneside manufacturers, is clear from the complaint, mentioned earlier, from James & Co.[39] That this was not uncommon may be inferred from the fact that on the same day Walkers, Parker sent out the March price list, they wrote to a customer quoting some articles at prices between 10s and £2 per ton below the list. That this was not just an isolated instance is shown by annotations to the list of customers to which the September circular was sent, showing that some were given quotations up to 30s per ton below the list. While it is not possible to be certain, it seems likely that the agreement between the manufacturers was that they might continue existing discount terms to established customers but should use the common list to any new customers. That the price lists were used is clear from the fact that Walkers, Parker sent them to their travellers with instructions to sell at those prices. This could, however, present difficulties and, together with the March list of changes, Patterson was told not to inform customers who had recently purchased about the changes or 'they might claim the present years transactions to be made conformable thereto'. Indeed, one firm, Wm. Brown & Co., pointed out the fall in price of pig lead since their order for 80 casks of white lead had been placed and asked for a reduction. Walkers, Parker replied, 'we hardly expected that after the bargain was concluded an allowance from the price should be asked in consequence of an alteration in prices, giving the buyer *all* the chances of the Market'. Nevertheless, in view of the firms' long trading connection they agreed to reduce the price by £1 per ton for that portion of the order not yet delivered.

There were frequent difficulties with customers over the prices which they were being charged and it was usual for Walkers, Parker to turn the emphasis on to quality. To a complaint from Tigar, Champney & Co. they replied, for instance, on 2 April, 'we think the article offred you must be very inferior & not likely to pleas your Customers' and on 24 June to Smith & Young of Hull, 'if you can buy an *equal* article lower we must of course conform'. It seems likely that Walkers, Parker did not go out of their way to indulge in price competition but, as a respectable, large firm, responded to it in order to protect their established market. In doing so, however, they opened the way for the less scrupulous customer to force the price down.

This was not the only problem with customers, nor was Grainger unusual in causing problems over payment. In some cases there was no doubt about the probity of the customer and this may have led to accounts remaining outstanding for long periods. A letter to Cookson & Co. on 11 August, for instance, stated, 'We shall feel obliged by the payment of our Accts for Goods furnished your various establishments in the years 1837 1838 & 1839'. This was probably a large account and involved Walkers, Parker in providing considerable credit since the sales had obviously not been made against a bill of exchange which was negotiable. This was against the normal practice with large customers. On 10 August, for instance, 40 tons of white lead was sold to Blundell Spence & Co. at £23 per ton against a six months' acceptance at three months from the date of invoice. This was a long credit period but Walkers, Parker were only bearing the full cost for three months until they received the bill which could be discounted at their bankers. Even agreements such as this were sometimes fraught with difficulty, especially with small customers. For instance in January 1839 white lead to the value of £45 had been sold to John Patterson of Peebles. He had, however, kept the draft sent to him for acceptance, 'contrary to all mercantile custom', as Walkers, Parker subsequently complained when threatening legal action.

To some extent Walkers, Parker may have brought the problems on themselves by their apparent indolence towards the collection of small accounts. On 20 June 1840 they were still trying to obtain payment of goods to the value of £9 10s supplied to John Cunningham in 1834. There had been 'repeated applications' for payment but they had only just discovered that Cunningham had gone bankrupt two years ago and were now trying to get a dividend. Similarly on 10 July a letter to Henry Bond noted that no remittance had been received for white paint to the value of £4 10s 3d sent on 8 September 1836 and requested payment by return. Problems with small accounts may have been one reason why Walkers, Parker refused to deal with private customers. In the case of larger customers much greater care was taken and outstanding accounts were chased more vigorously, although not always with rapid results. Attempts were also made to establish the creditworthiness of new customers. Letters were written to friends and clients, who might know something about a potential customer and if this failed, as in the case of C. Kirk of Sleaford on 13 January, he was informed that 'As you are an entire Stranger to us, we require a satisfactory reference or previous payment before shipping your order, upon your satisfying us your order will be promptly executed.'[40]

While it is clear that payment for the vast majority of orders was received satisfactorily and that the letter book draws attention only to a minority which caused problems, the latter were costly (in time and effort, if not in eventual bad debts) and on top of this the satisfactory sales involved long and expensive credit. To survive these problems, as Walkers, Parker did, entailed a pricing policy which provided, above the normal costs of production, overheads and return on capital, an allowance for those extra costs involved in obtaining payment of accounts. In order to have a flexible credit policy, accounts outstanding for long periods and survive, it was inevitable that prices were higher than they might otherwise have been.

A final area on which the Elswick letter book of 1840 throws some light is that of overseas sales. Without being of great significance, these seem to have increased in importance as compared with earlier in the century. The chief overseas markets appear to have been North America and a few north European towns in Russia, Denmark, Holland and Germany. Some of these orders were considerable. For instance, on 18 March lead goods to the value of £2,000, including 20 tons of shot, were shipped to agents in Montreal, and on 13 March 59 tons of sheet and six tons of shot, valued at more than £1,200, to agents in Hamburg. That these were not just speculative ventures is shown by the fact that a letter of 19 May refers to a remittance from the Petersburg agents, Carr & Co., of about £2,000, covering sales from September 1839 to February 1840. The agents obviously kept Walkers, Parker informed of stocks since the letter commented, 'we have forwarded to your care . . . 55 sheets lead of the sizes which seem lowest in stock'. In view of the frequently stated modern comment that British exporters paid little attention to the requirements of the overseas market and expected foreign buyers to accept British styles and standards, it may be worth suggesting that Walkers, Parker made some effort to satisfy foreign customers. As an example, when an English firm, Watson, McNight & Co., ordered shot in 28 lb bags, Walkers, Parker replied, on 31 July:

> As you state that the order is intended for the So. American Market we think it right before executing your order to mention for your government that the shot usually sent there is put up in Bags of 32½ lb each equal to the arroba or quarter quintal: we have been given to understand that 28 lb Bags Meet with low sale & at lower prices.

Apart from these major areas of trade the letter book throws light on a few incidental details. There was a single complaint about the quality of the firm's white lead, which suggests that that product was usually of a very high standard, while there was one complaint about damaged sheets of lead, which produced an interesting reply. Walkers, Parker commented that the lead sheets were sometimes put into the holds of colliers and then coal piled on top of them! The slowness of delivery by coastal shipping also came in for some complaint, while the attitude towards the new postal service was ambivalent. An apology to one customer commented, 'we cannot imagine how it has happened not to reach you except that the post office people are not so careful as they were under the old system', while another, in Doncaster, was informed on Monday, 3 August, 'We shall be glad to send you what shot you may require by Thursday's steamer, & if you write by return it will be in time for her.'

After the 1840 volume we have little evidence on Elswick. In 1842 the works employed 96 men, 45 women and three boys under 18, which made it rather smaller than Chester but certainly among the largest lead works in the country. In the next decades the works continued to progress – in 1846 a new rolling mill, in 1851 four white lead stacks and a new slag-hearth and refinery, in 1855 a new steam engine, and so on. In this period the biggest question was perhaps the installation of smelting capacity. To a considerable extent the smelting of ore had always been controlled by the mines but from the 1840s imports of ore began to expand and there was some attraction to manufacturers to undertake their own smelting. In 1849 the minutes note that 'Langley Mills Smelting Works [south-west of Hexham in Northumberland] were offered to us, but declined on account of the high price, and the question of smelting was deferred.' In 1851 the topic again appeared with unsuccessful enquiries for land on which to build a smelt mill and in 1860 negotiations were actually commenced but never completed for purchase of land at Haltwhistle and Bardon Mill, on the Tyne west of

Hexham. Although a further three acres of land were purchased, adjacent to the Elswick site, in 1865, there is no evidence that smelting furnaces were erected. In 1884, when the works was put up for sale, it was of considerable extent, with a 930 ft frontage to the Tyne. There were 33 white lead stacks, two sets of red lead furnaces, a pipe mill, paint mill, shot tower, a 96 ins rolling mill, a 'Rozan' steam desilverising process, six hydraulic pipe presses and a 35 hp beam engine to provide power. In addition there were the partner's house, nine cottages and both men's and women's dining and bath rooms, since slowly industry was beginning to provide facilities for its employees.

Having dealt with the changes to the specific works owned by the Walkers' partnership we may now turn to look at the development of the firm as a whole in the period to the late-nineteenth century and how its structure was changed to cope with increasing size.

Although the firm took the form of a single partnership, each of the works which have been mentioned was run under a specific local title,[41] although 'the Walkers' partnership' and 'Walkers, Parker' (the final name for the partnership) have been used generically in this book to avoid confusion. With one exception the titles included the Walkers' name and also took the name of the local managing partner in order to reflect to customers the close attention paid to individual works. Inevitably this meant some change as partners died or moved but gradually, from the early-nineteenth century, as the partnership was monopolised by Walkers and Parkers and as the disadvantages to trade of changing the style of the firm became obvious, the local names were firmly established.

Initially Samuel Walker I, Fishwick and Ward probably had equal shares in the partnership but by 1786 at the latest the Walkers held a dominant share in the firm, which they were never subsequently to lose. Each of the four sons of Samuel Walker I had four shares, while Ward and Fishwick each held five and Maltby two. It is not necessary to go into the detail of the subsequent changes in partnership holdings,[42] except to mention those which had significant import. The changes invariably took place at a time when the previous partnership agreement had expired and, if a partner died during the period of an agreement, his shares were vested in the hands of his executors, themselves usually partners. The early partnership agreements were usually for ten or fourteen years, while from the middle of the nineteenth century they tended to be for shorter terms, of as little as three or four years, perhaps because such terms offered the greater flexibility which was necessary with the larger number of partners then involved.

The first significant development in partnership structure had come with the withdrawal of Fishwick when the partnership was renewed in 1800 and the death of Ward in the same year. These two had undertaken the early management at a time when the Walkers were not involved in day-to-day decisions but were concerned with the Rotherham ironworks. It is clear, however, that the problem of management succession had already been considered, since the 1786 partnership agreement had made provision for Samuel Walker Parker, a nephew of the Walker brothers,[43] to be introduced into the partnership. Although he did not receive a formal partnership share until 1802, he was given responsibility for the running of the Islington works when Ward left in 1792 to establish the works at Derby. At the same time his brother, William Parker, was appointed to assist Maltby in the management of Red Bull Wharf. This was a far-sighted way of solving the problem of management by appointing young blood, close members of the family who might be trusted (and at the same time providing for possibly needy nephews) while leaving the senior partners free from day-to-day management.[44]

William Parker was to make little impact on the partnership. His partnership share was always smaller than that of his elder brother and he remained at Red Bull Wharf, subordinate to Maltby. As such he appears to have been viewed with some suspicion by the Walkers during the partnership disturbances of 1824 to 1825 which concluded with the departure of Maltby. Although William Parker remained in the partnership he was subsequently to resign in 1832. By contrast Samuel Walker Parker played a very significant role in the partnership. In the late 1790s he was sent from Islington to Elswick to deputise for Fishwick and was formally made managing partner of that first and largest of the firm's works in 1802 − a position he was to retain until 1840. Long before then, however, he had followed the industrialists' pattern of divorcing living and working places. Initially Parker had lived at Low Elswick House, which had been built for Fishwick in the grounds of the Elswick works. Like the managing partner's house at Chester it was built with some style and in 1884 was described as 'A family residence with extensive pleasure grounds, greenhouses and good stabling.' At that time it had a hard tennis court, ice house, fernery, peach house and several greenhouses, and with its gardens, lawns and carriage drive was largely separated from the works. For some time this must have been satisfactory and so late as 1814:

Samuel Walker Parker esq. entertained a large party to supper etc. at his house at Low Elswick, near Newcastle. The Shot Tower was illuminated both inside and out with coloured lamps, round the galleries etc. on the outside and placed at intervals quite up the winding staircase in the inside; the tout ensemble had a fine effect. A considerable quantity of excellent fire-works was discharged on the field behind the works, and considerably heightened the pleasure of the scene. The Shot Tower was also illuminated, and had a flag flying on the following evening, and looked particularly well from Newcastle bridge.[45]

Illuminated at night the shot tower may have improved one's party atmosphere but its presence was probably rather more oppressive in the day. By 1818, when his daughter, Catherine, married her cousin, Joseph Need Walker,[46] another member of the lead partnership, S.W. Parker was living at Whitley House, Whitley Bay and he subsequently moved to the more rural delights of Scot's House, Boldon, which is described by Pevsner as 'a distinguished building'.[47] It is clear that the partnership's profitability, which had enabled Fishwick to rise in the world, had not disappointed his successor.

Little can be said about S.W. Parker, who does not appear to have taken a leading part in the Tyneside community. Like other industrialists, such as the Cooksons, he was a member of the Northern Counties Club formed in 1829, but there is no evidence of political or social activities, although in 1803, at the time of fears of an attempted invasion by Napoleon, he was appointed a 1st lieutenant in the Newcastle Loyal Associated Volunteer Infantry. Clearly he was an active and efficient businessman, since apart from running the Elswick works, he also had interests in the Tyneside shipping industry. He would appear to have been respected and in the 1840s was appointed by the Commissioners of Greenwich Hospital to fix the price for the sale of that organisation's refined lead. Although the Elswick firm was officially Walkers, Parker & Co. it was known locally as Parker's during this period and a Parker's Arms public house existed nearby.

In 1840 he was replaced as managing partner at Elswick by his son Samuel, although this seems to have been a less satisfactory appointment. The minutes of the partnership record that he was requested to retire in 1852 (although he was only aged 53) but he did not do so and it was not until 1860 that he was successfully persuaded to withdraw from the partnership.[48] The minutes are reticent about the reasons for this event but one may suppose that his management of Elswick was unsatisfactory. He was replaced at Elswick as managing partner by his nephew, Henry Parker, who had come to the works as a junior in 1847. His reign was relatively short, since he died in 1868 and this virtually terminated the Parker interest in the business. It had been intended that Henry Parker should be succeeded by his cousin, Frederick Henry, one of the sons of Samuel Parker, but he also had died in 1868. The only remaining Parker partner, Henry's father and Samuel's brother, was the Rev. Henry Parker, rector of Ilderton in Northumberland. The Parker share in the partnership accrued to him, but on his death in 1871, 'it was resolved to pay off the value of his 16 shares to his Executors, and thus terminate the direct interest of the Parkers in the concern'.[49]

The introduction of the Parkers into the partnership had temporarily solved the problem of management succession and on two or three other occasions during the nineteenth century the husbands of Walker sisters were brought into the lead partnership, presumably for the same reason. The major problem of the provision of overall direction for a rapidly growing firm was, however, solved by the gradually developing involvement of the Walkers in management. This probably commenced with Joseph Walker, the son of Samuel Walker I, whom we have seen to have been most involved with the lead partnership. In 1785 he had married Elizabeth, fourth and youngest daughter of the Nottingham hosier, Samuel Need,[50] and, although he continued a member of the Rotherham iron and steel partnership, his interests came to lie further south. It is he, together with Ward, who signed the annual accounts of the Derby works and in 1797, having resigned from the iron and steel partnership, he moved to Derbyshire. He lived at The Hall, Aston-on-Trent and became High Sheriff of the county.[51] With Ward's death in 1800 it was ideal that Joseph Walker was available to oversee the works and this offered the scope for the first direct involvement of a member of the Walker family in the management of the lead works, but Joseph died in 1801 at the age of 48.

There is no evidence on the management of the Derby and Chester works in the first decade of the nineteenth century but it seems likely that they were being run by the second generation of the Walker family (if we do not count the four-year involvement of Samuel Walker I as being a generation). During the 1810s several of the grandsons of Samuel Walker I received partnership shares and took up managing partnerships. Joseph Need Walker, for instance, the son of Joseph Walker, became a partner in 1814, with responsibility for the Derby works, where it would seem likely he had been involved for some time. Others of the new partners, such as Samuel Walker III, however, clearly took no active interest in the lead business. He was heavily involved in iron manufacturing, was briefly MP for Aldeburgh and resigned from the lead partnership in 1828.

Although both ownership of the partnership shares and its management were firmly in the control of the Walker family, the 1820s appears to have been an unsettled decade and one in which, but for the close family ties, the partnership might have broken up. There is no evidence as to why the partnership disagreements, which led to the resignation of Maltby and subsequently that of William Parker, arose and they may have been the result of personality clashes and the problems which the Walkers were experiencing with the Rotherham works.[52] It is at least possible, however, that the lack of financial success, shown in Table 4.1 which cumulates the figures given earlier for the individual works, may have been a factor.

Table 4.1: Walkers, Parker & Co., Capital Employed and Profitability, 1815-31

Year	Capital employed (£)	Profit/(Loss) (£)	Return on capital (%)
1815	484,421	(1,914)	−0.4
1816	479,058	19,487	4.1
1817	456,728	31,216	6.8
1818	478,504	50,824	10.6
1819	480,416	26,410	5.5
1820	473,143	1,478	0.3
1821	475,181	16,387	3.4
1822	471,656	13,272	2.8
1823	472,611	19,854	4.2
1824			
1825	405,305	75,128	18.5
1826	376,222	(28,564)	−7.6
1827	356,347	(8,890)	−2.5
1828	366,583	17,234	4.7
1829	351,943	(5,973)	−1.7
1830	329,398	(16,943)	−5.1
1831	309,166	(2,961)	−1.0
Total	6,766,682	206,045	
Average	422,918	12,878	3.0

Although we do not know how these figures were compiled, it is probably reasonable to assume that they are net of depreciation (in the form of charging repairs and renewals to revenue) and obtained after deducting an interest rate of 5 per cent on partners' capital. This would imply that they are net profit figures but if that assumption is untrue then they are even more disastrous than they would appear, since a return greater than 3 per cent could have been achieved by investment in government stock. Since it is noticeable that the worst returns occurred in the later 1820s, the profit failure could have had little to do with any internal disagreements in the pre-1825 partnership. It is likely that fluctuations in the price of pig lead had a considerable impact on profitability. The price, as shown in Appendix I, fell sharply from 1809 until 1817, which would probably have been a very unprofitable period given the high level of stocks and long production period involved in the white lead process. From 1817 until 1826 there was not a great deal of change except for peak prices in the cyclical peak years for the economy as a whole of 1818 and 1825, which saw the partnership make its highest profits of the period. Having averaged £22 to £23 per ton for most of this period the price fell to a low of £12 15s per ton in 1832, which undoubtedly influenced the partnership's profit figures in the late 1820s. It is clear, however, that profits were not uniquely correlated with the annual fluctuations in lead prices and the chief cause for the low long-run average level of profits in the period 1815 to 1831 can only be put down to a high level of competition in the supply of lead products. In the same way that the Walkers' partnership had extended its output we shall see in the next chapter that a number of other firms had also done so. Together with falling lead prices this made the period an unprofitable one, while the 1830s saw a balance between supply and demand rather more favourable to the producers and the 1840s was once again a period of expansion.

During the 1830s the firm was in the hands of a fairly narrow section of the Walker family. Of the four sons of Samuel Walker I, Thomas had died without male issue and no member of the family of Samuel II was still involved in the lead partnership. Between them the sons and one son-in-law of Joseph and Joshua, together with Samuel Walker Parker, son of a daughter of Samuel Walker I, controlled all the partnership shares. From the 1840s, however, there began a slight widening of the family holdings, perhaps helped by the improving financial position, which in itself led to the need for more managers, with the purchase of the Bagillt works and expansion at the others. Indeed the 1840 partnership agreement gives the first unequivocal evidence of the direct involvement of the Walkers in day-to-day management, listing the managing partners, in addition to Samuel Walker at Elswick, as E.S. Walker at Chester, J.N. Walker at Liverpool and Joshua II and P.A. Walker (the first of the third generation) at London.[53] It may be supposed that the family's direct involvement with lead had come about gradually, although withdrawal from the iron industry in

the early 1830s may have speeded it up. From the time of the involvement of Joseph Walker with the Derby works in the 1790s it seems
likely that some of the partners' involvement had deepened until they
became continuously involved. The evidence of a growing interest in
the affairs of the lead partnership, however, requires the qualification
that the Walkers appear to have taken increasingly long holidays as the
nineteenth century progressed[54] and that they undoubtedly depended
on salaried employees for the general running of the business during
these absences.

The improvement in the financial position which had led to the
purchase of Dee Bank in 1834 and the considerable expansion of the
Chester works under Sir E.S. Walker,[55] also led to an immediately
disastrous decision. Dependence on the supply of pig lead had always
been a matter for concern and sometimes a serious problem and this
may have been a factor behind the decision to invest in a smelting
works, with the purchase of Dee Bank. That decision made the partners
aware of their dependence on the supply of lead ore and apparently set
off a search for a secure supply. For some reason this was not found in
the most obvious area, north Wales, but in Shropshire, admittedly connected by water to the Dee. In the mid-1830s the partners took a share
in the lease of the Bog Mine and other lead mines nearby and invested a
great deal of capital in expansion, including deepening the shafts and
installing a large steam engine for pumping purposes. The prospects
were, no doubt, considerable but the historians of the Shropshire lead
field comment, 'Throughout most of the nineteenth century Bog
enjoyed a reputation for richness and profitability which was never
matched by its performance.'[56] Frank Walker, brother of P.A. and one
of the burgeoning third generation of Walkers, had in 1840 been placed
in charge of the Shropshire operation. He had just returned from some
unspecified activity in America and was sent back there in 1842 to
report on the running of American lead mines.[57] At this stage expense
appeared to be an unimportant object and in the following year he was
sent to Spain to report on lead production. These investigations realised
little recompense in the Bog Mine, 'the profits so eagerly hoped for
during these years of expenditure were not forthcoming, for disagreement among the shareholders led to the whole assets of the company
being put up for sale, as ordered by the Court of Chancery, in 1844'.[58]
On this occasion the reputation of the Bog Mine brought no takers for
the lease and, although some of the equipment was sold, Walkers,
Parker extricated themselves from the venture in 1845 with a total loss
of £78,613.[59] This should have been a lesson to the firm but, as we shall

see, both Walkers, Parker and other lead manufacturers were to continue to flirt with the control of their supplies of lead ore, the results usually being expensive and broken engagements rather than happily consummated marriages.[60]

The failure of the mining venture might have been expected to end Frank Walker's hopes of advancement but the Minutes note that in 1845 he was 'appointed to live at Birkenhead and visit Dee Bank weekly'. Hitherto Dee Bank had been run from Chester and it is difficult to see of what use a weekly visit could be, unless it reflected the usual level of works' supervision expected of partners. In any event the position was short-lived for in 1846 Frank Walker went to the Liverpool office, presumably to understudy J.N. Walker, and became a partner in 1849. Neither did the Shropshire venture have a serious impact on the partnership's finances, since in 1843 £1,000 per share was divided among the partners.[61] Although there are no profit figures for this period there can be no doubt that it was a very profitable one with rising demand enabling profit margins to be increased. Despite the high income which the partners were receiving it seems that some were dissatisfied, presumably because the profits achieved at some works were larger than at others. In 1847 it was agreed, therefore, that managing partners and their juniors should be paid partly in proportion to the profits made at their respective establishments. Payment by results did not last for long, however, since 'it was found to lead to the partners working for the advantage of their own branch, rather than for the general good of the whole concern',[62] and ceased in 1856. There is no evidence as to what undesirable practices caused this decision and since, as we have seen, the various works had fairly well defined sales areas, it is not easy to see why conflict occurred. It may be that the previous attitude of non-intervention in the sales area of another branch was breached or, perhaps, that those branches such as Elswick and Chester, which controlled much of the concern's purchases of pig lead, were, thereby, endeavouring to gain advantages.

The profitability of the concern is reflected in the fact that in 1849 the partners' deposits with the firm, on which they received interest at 5 per cent, stood at £170,000. In an attempt to discourage this surplus the interest rate was reduced to 4 per cent in 1852 and it was decided to pay off £200 per share p.a., double the previous rate. In addition the extension of the various works, which had been actively undertaken in the 1840s, was also continued with heavy capital expenditure in the following decade. The financial rewards were even spread to a social strata below that of the partners with the decision 'to give bonuses to

Clerks in good years', although, as the minutes admitted, 'this was not regularly carried out'.[63]

It may be that this laudable aim was unfulfilled as a result of a deterioration in profits. It is particularly unfortunate that there are no profit figures for the period from the early 1830s to the mid-1850s, a period for which it has been suggested that the partnership's profits were high, although all the other evidence which has been presented would confirm this assumption. From the mid-1850s, however, there are some profit figures, of rather dubious ancestry,[64] which are presented in Table 4.2.

Table 4.2: Walkers, Parker & Co., Output, Capital and Profitability, 1856-67

Year	Output of lead goods (tons)	Capital employed (£)[a]	Profit/(Loss) (£)	Return on capital (%)
1856	31,101	297,046	(11,085)	−3.7
1857	26,704	(317,320)	26,465	8.3
1858	30,178	(337,595)	20,027	5.9
1859	30,692	(357,869)	38,600	10.8
1860	30,491	378,144	33,803	8.9
1861	30,726	(390,681)	15,907	4.1
1862	30,755	(403,217)	31,759	7.9
1863	31,892	415,754	22,342	5.4
1864	33,130	(421,133)	36,083	8.6
1865	35,789	(426,513)	(16,614)	−3.9
1866	38,043	431,892	46,803	10.8
1867	39,802	(431,892)	39,988	9.3

Note: a. Since the figures given in the minutes for the years 1856, 1860, 1863 and 1866 imply an upward trend the arbitrary assumption has been made that this was strictly linear between each pair of given years and the intermediate figures (in brackets) so calculated to enable approximate figures for return on capital employed to be given.

With the exception of a fall from £25 to £21 10s per ton from 1855 to 1856, which may have had something to do with the loss in the latter year, the price of pig lead showed very little fluctuation over this period and should not have had much influence on profits. These do not appear to show any particular trend. They do not seem to correspond to trade cycle fluctuations, except that the highest annual profit occurs in the cyclical peak year of 1866, although it is immediately preceded by the largest loss. Neither is there evidence of correlation between output and the trade cycle, since, with the exception of a fall in the cyclical peak year of 1857, output is constant for the period to 1862 and then rises rapidly to 1867. While the run of statistics is too short to allow any certain conclusions to be drawn, it does seem that output and profits were weakly correlated. Overall the profit performance, although better than that for the period 1815 to 1831, is not very impressive and almost certainly worse than it had been in the 1840s. If this cannot be attributed to price changes in the raw material it seems likely that profit margins were under pressure from the competition of other firms at a time when costs, other than the price of the raw material, were rising.

Chiefly through the shareholdings of Sir E.S. and J.N. Walker, the second generation of the family retained power in the 1840s and 1850s, despite an increasing number of the third generation becoming partners, although with small holdings of shares. Even when those two retired from active business in the mid-1850s and were replaced as managing partners by the third generation, they still retained effective control through the size of their shareholdings. From the mid-1860s, however, the third generation had a large majority in the total shareholding and that, together with the higher profits in 1866 and 1867, may have been the cause of a resolution passed to increase the salaries of managing partners from £500 to £700 p.a. and of their juniors to £300. At this level managing partners were well but not extravagantly paid by what little evidence we have of contemporary standards (in the 1880s for instance the Cookson managing partners were paid £1,000 p.a.).[65] It was also decided to double the number of shares in the firm 'in order to enable the younger partners better to acquire shares', perhaps in response to earlier feelings about the long dominance of the second generation.

In the late 1860s there were some significant changes in the partnership structure. As a result of the deaths of two Parkers it was necessary to appoint a managing partner at Elswick and into the breach came Edward Joshua Walker (whose father had not been involved in the lead business). This was rapid promotion for a 31-year-old, the first member of the fourth generation to become a managing partner and as an emergency response to a difficult situation it was subsequently to be regretted. In 1869, E.C. Walker, the Chester managing partner also died and his brother A.O. was appointed as his replacement. The two major participants in the subsequent partnership disputes, E.J. and A.O. Walker had, therefore, been appointed to the direction of the partnership's two largest works in the same year, and from rivalry developed faction.

With the partnership approaching its centenary one might have expected signs of hardening of the arteries, a failure to keep up with modern techniques, since very few eighteenth-century firms lasted so long. There is, however, little evidence to suggest that this was so. Indeed the evidence of new developments is considerable. At Bagillt the new smelting works flue and chimney were erected and a steam crane installed at the wharf, while at Elswick a new pipe mill was erected, the approach to the works improved and a Rozan desilverising plant installed in 1874 in the same year that this was done by Cookson & Co., a much younger firm, while at Chester the capacity of the white lead plant was considerably extended. In addition the sending of E.J. Walker to the USA in 1872 'to enquire into the different processes of manufacturing lead in that country' implies a willingness to recognise the fact that the UK was not the only repository of new technology, which was positively unusual. As a result of this willingness to expand and continually modernise the works, output, as shown in Table 4.3, was considerably increased during the prosperous years of the later 1860s and early 1870s and was well maintained and even increased in the generally economically depressed years around 1880.

Table 4.3: Walkers, Parker & Co., Output of Lead Products[a], 1854-81 (tons)

1854	30,031[b]	1861	30,581	1868	40,781	1875	39,520
1855	28,768	1862	30,625	1869	38,732	1876	38,996
1856	29,281	1863	31,329	1870	38,214	1877	39,165
1857	29,148	1864	32,808	1871	44,253	1878	38,378
1858	28,025	1865	35,589	1872	40,880	1879	42,134
1859	31,219	1866	35,296	1873	40,726	1880	43,119
1860	30,768	1867	40,717	1874	41,516	1881	48,052[b]

Notes: a. Sales of pig lead are also included, fluctuating from a few hundred to over 4,000 tons p.a., but normally around 2,000 tons.   b. These are annualised from the actual figures available, 27,529 for the period Feb.-Dec. 1854 and 40,043 for Jan.-Oct. 1881 and are probably exaggerated since deliveries in the winter months were lower than in the rest of the year.

Perhaps more interesting than the totals are the figures from which they are derived, for deliveries of individual products, which are given in Appendix IV. These show deliveries of sheet rising from about 11,000 tons to about 16,000 to 17,000 tons over the period 1854 to 1881, pipe from about 5,000 to 9,000, shot from 2,500 to 3,000, white lead little changed, paint from 4,000 to 4,500 and red lead little changed. As compared with the position in the late-eighteenth century when white lead was by far the most important of the partnership's products, the products of the blue lead section of the industry had come to dominate the volume of production. As compared with the chemical products of white and red lead, the production of sheet and pipe was fairly simple and speedy and profit margins were therefore lower. The high volume of blue lead production does not, therefore, indicate its contribution to the firm's total profit. The considerable growth of sheet and pipe production was largely a result of the increasing demands of the building industry to meet the rapid growth of population with house, office and other buildings. In particular, in this period, considerable improvements in water supply and the extension of piped water to a growing proportion of new houses built, had a dramatic impact on the output of lead pipe. Sheet and pipe were relatively insensitive to the ordinary trade cycle (responding primarily to the building cycle with its different periodicity) and it was the buoyancy of demand for these products which enabled total output to ignore the impact of the cyclical trough of the late 1870s. By contrast demand for products such as shot, and white and red lead was affected by the trade cycle. Deliveries of the latter for instance fell from 2,500 tons in 1871 to well under 2,000 by 1876, as output of ironwork for capital goods, such as bridges and ships for which red lead provided the protective paintwork, fell off. The monthly figures also show that there was a considerable seasonality in deliveries. This was most obvious with shot for which there was a marked peak in deliveries in August and September and a trough in the early spring, coinciding with the demands of the game shooting season. The other lead products also showed some seasonality with a tendency towards high deliveries from May to October and low deliveries from November to March, reflecting the lower winter levels of activity of the building industry to which much of the total output went. As a result it was necessary to build up stocks during periods of slack demand in order to meet the higher demand of the rest of the year without having to increase capacity. The highest level of deliveries of sheet, pipe and shot was made from London (partly as a result of the fact that Lambeth produced no lead chemical products), while Elswick was responsible for the highest deliveries of white lead and Chester for paint (i.e., white lead ground in oil) and Chester and Elswick were more or less equal in deliveries of red lead until the later 1860s when Elswick became pre-eminent.

With this structure the firm seemed secure and in 1878 celebrated its centenary with each of the partners visiting the Paris Exhibition at the expense of the firm and A.O. Walker being asked to compile a record of the main events in the firm's history, which was published in 1879. Despite the general depression in the economy at the end of the

1870s the partnership was sufficiently well placed to resolve to raise the salaries of junior partners to £500 (from £300), although the clerical staff seem to have been less well treated. In 1879 the minutes noted that a new scheme for the payment of salaries had been introduced to obviate the continual requests for increases (presumably some kind of salary scale for clerks) and that 'any future application [for a salary increase] henceforth to be considered as a desire to leave the service of the Firm'. In the same context it was also agreed that 'No pensions to be granted under any circumstances.'[66]

At this point began the partnership dispute which was to lead, in 1889, to the sale of the partnership's assets to a limited company. Our knowledge of the dispute, which was chiefly between A.O. and E.J. Walker, is very much one-sided since it depends upon an up-dated version of the 1879 minutes, written by A.O. Walker,[67] although other evidence would suggest that A.O. was a trustworthy source and that E.J. was unreliable to say the least. The disagreement came to a head in November 1881 as a result of a series of disappointing annual profits at Chester, of which A.O. Walker was managing partner. It seems probable that the problem stemmed from the Dee Bank smelting works, which was unprofitable in the face of falling prices of imported ores. A.O. offered to resign but requested a valuation of the partnership's property, which was, according to his account, refused by the other partners because of the depressed state of trade. It is difficult to see the reasoning behind this, since in the circumstances a valuation would presumably have given a figure below the book value, which would have been disadvantageous to anyone leaving the partnership. In A.O. Walker's absence, E.J. proposed to a meeting of the partners that the former's resignation be accepted and that, if he continued to insist on a valuation, application should be made to the Court of Chancery to wind up the partnership.

This was eventually agreed and the Court of Chancery dissolved the partnership on 16 December 1881, the only peremptory action it was to take for some years. Although less notorious than that 'Monument of Chancery practice', *Jarndyce* v. *Jarndyce*,[68] *Walker* v. *Walker*[69] was not to be settled for twelve years and the legal profession was certainly a considerable beneficiary of the action. The initial action was between A.O. and Henry Walker as defendants and all the other partners as plaintiffs, but on 3 August 1882 the Court made an order that all the partners other than E.J. Walker should be struck out as plaintiffs and entered alongside A.O. and Henry as defendants.[70] The other partners had obviously recognised that they had taken, or been pushed into taking, the wrong side in the dispute. Their decision was probably encouraged by the discovery in May 1882 that E.J. Walker had overdrawn his account with the firm to the tune of £5,000, much of this having been withdrawn after his appointment by the Court as a Receiver for the firm.

Having made common cause the remaining partners decided to carry the firm on as a limited company as soon as the Chancery action was solved and, on the ground that the properties of the partnership were undervalued, to write the shares up to £4,000, against the £3,529 at which they stood in the proposed 1881 partnership, and offer E.J. Walker this value for his holding in order to get rid of him. The partners who had changed sides must now have been bitterly regretting their earlier decision to enter the law courts. Had the partnership agreement still been effective E.J. Walker might, quite justly, have been ejected from the partnership on these terms but now there was a different definition of justice and the Court of Chancery pursued it to the final destruction of the partnership. E.J. Walker, having opposed an offer of £3,800 per share to A.O. Walker in 1881, now refused to accept £4,000 himself, making it clear that personal greed lay behind much of his action, even if we discount A.O. Walker's claim that E.J. was demanding £6,000 per share and that if he did not receive it he would try to ruin the concern by litigation. E.J. was replaced as Receiver at Elswick by his junior, Ernest A. Walker, and removed from the managing partner's residence there, actions which no doubt strengthened his resolution not to be pushed out of the partnership. As a result the mills of Chancery ground slowly towards the valuation of the properties and their stocks before putting them up for sale. This was finally completed and an auction took place on 23 July 1884, after the sale had been extensively advertised and a catalogue of the premises produced. The only bid made was for the largely derelict Glasgow works but this was not accepted since it was below the reserve value which had been placed. This presumably came from an outsider but it is very difficult to see why the majority of the partners did not make a bid. Had they done so at the grossed up rate of £4,000 per share, which would presumably have been above the reserve, they would have achieved their earlier aim without E.J. Walker having any power to prevent it.

The sale having failed, the Court called a meeting of the parties involved, at which E.J. Walker made a bid of £160,000 for all the properties (this must have excluded the stock, which had been valued independently for the purposes of the 1884 auction at £235,000. It would therefore have put a total value of £395,000 on the business, some £50,000 below the figure implied by the other partners' offer to E.J. of £4,000 per share.) This offer having been opposed the Court ordered the properties to be put up for sale again and this was done on 20 May 1885 without obtaining a single bid. This is not surprising in the middle of a period known as 'the great depression' when prices, profits and interest rates were falling and foreign competition was becoming noticeable in many manufacturing sectors. It was unlikely that half a million pounds could be raised, except by the formation of a limited company and the selling of the shares to the public, and this was unattractive since the firm owed much of its value to the goodwill attached to the Walker family. In addition much of the practical knowledge of the lead business would have disappeared as soon as the Walker partners were bought out, although the importance of salaried management was increasing. In 1881, for instance, W.M. Hutchings with 'great experience in lead smelting at home and abroad [was] appointed practical Manager' at Dee Bank.[71]

In November 1887 E.J. Walker made a new offer to the Court, of £165,000 for all the works, a slight increase on his earlier terms. The defendants argued that this offer was unsatisfactory since it involved E.J. Walker taking over the business and running it on their capital until he had raised sufficient money to purchase their shares. By contrast they pointed to the fact that they were prepared to offer cash of £4,000 per share and were prepared to allow E.J. Walker his share of the increase in the value of the stocks since the last independent valuation. By this time E.J. Walker must have got himself into the hands of a company promoter since he amended his first offer to a figure of £493,000 (without knowing the change in the stock valuation it is impossible to say how large an increase this involved) and then further increased his offer to £500,000 together with £2,000 to cover the legal costs. This offer was considered by the partners and accepted and in particular they noted that a fall in the price of lead in late 1887 (reducing the value of the stock) made a fixed price offer attractive.

This argument was upset by the trade boom which began in 1888, the first marked sign of a let up in 'the great depression'. The market price of lead rose from £12 to £16 between the Novembers of 1887 and 1888, which enabled the partners 'to dispose of a good deal of stock at an unusual profit of £47,523' and also to increase overall profits. The white lead market improved, increased profits were made in desilverising, a contract for delivery over a two-year period of 6,000 tons of water pipe for Buenos Aires secured and 'general demand for manufactured goods was unusually brisk at this time'. Given the problems brought about by the legal conflict and the lack of security it involved, and the evidence of the dereliction of the Bagillt and Glasgow works, it might be thought that the partners had neglected the works and would not be able to cope with the upsurge in demand. This was not, however, the case. Although the business was nominally being run by the Court, in practice the partners (with the exception of E.J. Walker after the summer of 1882) had run it much as before. Repairs and renewals had not been neglected and the abandonment of Bagillt and Glasgow was only the dropping of peripheral activities, which had been taken on in a relatively prosperous period in the 1860s and had no significance for the major centres of the partnership. At Lambeth considerable sums were expended on the installation of hydraulic machinery, while at Dee Bank the Parkes desilverising process was adopted and a new blast furnace installed. In addition, on 7 October 1884, a lease of the Liverpool Lead Works in Paisley Street was taken, while the existing Liverpool office and warehouse were retained. There was also continued research into improved processes, in which A.O. Walker was much involved.[72] The most notable instance was his patent (no.11,120, 1884) of 'A process for Separating and Collecting Particles of Metal or Metallic Compounds Applicable for Condensing Fumes from Smelting Furnaces and for other Purposes.' This interest obviously stemmed from the problems of collecting the lead fume from the Dee Bank smelting furnaces, discussed earlier. Walker's patented proposal was for the use of a discharge of electricity in the flue leading from the furnaces which would cause the particles of lead to accumulate and both obviate losses of lead and the necessity for the existing long flue. This patent resulted from the experiments made by Professor Oliver Lodge[73] at Liverpool University on the effect of electric currents in attracting dust. By 1885 the process had been adopted at Dee Bank and the journal *Engineering* commented, that it was 'but one more of the many instances demonstrating the unexpected and often surprising manner in which "pure science" of one day may be very valuable "applied science" of the next'.[74] The process suffered from a number of problems, particularly lack of power, and it was not until shortly before the First World War that the electrical precipitation of fume was adopted on a large scale, especially in the USA.[75]

Far from having neglected the works the defendants appear to have run them normally, no doubt in the confidence that E.J. Walker was merely an aberration which would eventually disappear. A report, ordered by the Court, from the accountants, Chatteris, Nichols & Atkins, on the partnership's performance over the period 1875 to 1888 justifies the belief that the works had been well run financially. The accountants' report gave figures as in Table 4.4 showing a net profit over the period April 1875 to October 1888, after the deduction of partners' salaries, £73,000, and depreciation, £221,000 (which the accountants argued should have been reduced to £109,000), of £392,000. If the level of depreciation had been reduced then net profit would have been £578,000 or more than £40,000 p.a., instead of the actual figure of £29,000.

Despite the problems inherent in partnership structure it could, therefore, be said that the Walkers had made a fair success of running their lead business over more than a century, a period for which few such large partnerships had existed. They had overcome the problem of management, firstly by taking partners with a direct interest in running the business and then gradually by adapting to management themselves, judiciously ensuring that sufficient sons, nephews and, when necessary, sons-in-law were introduced to maintain that management at full strength. Growing prosperity and public school and ancient university educations failed to convince them that industry was socially unacceptable. True to their eighteenth-century beginnings in the hard work of the iron foundry they continued to recognise the source from which their prosperity stemmed and ensured that it did not dry up. Even though the partnership was well into the fourth generation by the 1880s there is no evidence of either the rags to rags in three generations syndrome, or that of failure of entrepreneurial drive by the time of the third generation family firm, stereotypes much loved by business historians examining the nineteenth century. There were, of course, sons of partners who showed no interest in the family firm but there were enough who did to make it reasonable to suggest that, but for the misfortune of the legal action, the firm would have been better run after 1889 had it remained in the Walkers' hands than it was by the limited company. Nor was it just a question that the young Walkers were a collection of enthusiastic amateurs. We have seen that they were prepared to learn from developments abroad, while the traditional English university education was beginning to be recognised as inadequate. As a junior A.O. Walker had been sent to Freiburg[76] to study metallurgy and at least one Walker son (intended for the partnership which he never got) was educated at the Royal School of Mines in the 1880s. In addition, works managers, subordinate to the managing partners, but with detailed knowledge of lead manufacturing, were appointed in the second half of the nineteenth century.

In the running of the partnership the Walkers had, of course, made

Table 4.4: Walkers, Parker & Co., Depreciation, Salaries and Profitability, 1875-88

Year to	Net profit			Partners' salaries	Expenditure written off to profit and loss account on								
					Land & buildings			Machinery & engines			Utensils		
	£	s	d	£	£	s	d	£	s	d	£	s	d
April 1875	67,677	12	1	4,510	13,079	16	6	9,645	16	6	8,286	3	0
1876	20,656	3	2	4,540	8,004	4	7	7,034	19	7	9,880	19	9
1877	48,030	5	1	4,910	3,688	1	7	7,697	11	8	5,156	4	5
1878	19,463	5	4	4,850	3,676	8	1	4,619	9	7	4,623	18	7
1879	12,731	15	4	4,900	3,633	10	0	5,082	2	10	3,434	15	9
1880	44,624	14	5	5,300	5,853	12	7	5,109	2	2	5,721	14	5
1881	1,090	17	7	5,600	4,365	4	3	4,638	14	3	4,104	13	7
1882	22,290	5	11	5,600	4,061	15	1	5,576	8	10	2,119	12	5
Oct. 1882	9,649	13	5	2,550	616	4	0	2,367	7	10	1,372	14	2
1883	13,225	0	10	5,100	1,771	9	0	5,304	2	8	4,425	18	9
1884	27,175	1	11	5,100	2,267	6	6	7,224	8	7	4,703	6	6
1885	23,805	6	6	5,100	1,915	15	3	6,449	5	4	5,778	19	10
1886	23,679	5	8	5,100	1,858	13	4	5,379	17	4	4,863	7	0
1887	13,248	3	10	5,100	2,290	7	6	5,021	3	3	4,965	12	11
1888	44,849	4	2	5,100	2,860	0	11	5,730	19	10	5,196	15	10
Total	392,196	15	3	73,360	59,942	9	2	86,881	10	3	74,634	16	11
Average	29,051	12	3										

a considerable amount of money. In the late-eighteenth century the Walker partners owed much of their prosperity to the Rotherham iron business but this was in decline from the end of the Napoleonic Wars and subsequent partners were largely dependent on inherited wealth and the current profits of the lead business. Already by the end of the eighteenth century lead and iron together had enabled the sons of Samuel Walker I to live in large houses and to begin to establish a position in society. Joseph and Thomas took country houses, respectively Aston Hall in Derbyshire and Berry Hill in Nottinghamshire and many of the next generation followed suit. Thomas II lived at Shotley Hall, Northumberland in the 1820s but by 1839 owned the estate of Ravenfield Park in Yorkshire, while his brother Joseph Need mortgaged his share in the partnership in 1814 to three of the other partners and his father-in-law for £69,876 which enabled him to buy the estate of Calderstone Hall, Childwall, on the outskirts of Liverpool. Subsequently, in 1853, he was to sell Eastwood, the house built near Rotherham by his father, and its estate of 300 acres for £30,000. The second generation appears to have reached the peak of landed and county success, with appointments as Deputy Lieutenant (for example, Henry Walker I of Blythe Hall, Nottinghamshire,[77] Sir E.S. Walker of Berry Hill, Nottinghamshire[78] and Joseph Need Walker of Calderstone Hall, Lancashire) for their respective county. Thereafter there were few further purchases of country estates, even with the depressed land prices of the period from the late 1870s and subsequent retirements tended to be to the south-east coast, with Hove an especial favourite. Perhaps this was a result of the lower opportunities for making a fortune in lead manufacturing, although it may also reflect a lesser entrepreneurial ability in the third generation.

Some support for the view that the second generation was the most financially successful comes from the evidence of probates. Before the late 1850s it is difficult to obtain accurate information on the value of estates at death but thereafter the gross value, rounded up to the nearest £5,000, is given. Of the major second generation partners Henry I left <£80,000 in 1860, Sir Edward Samuel <£100,000 in 1874, and, the largest estate of any Walker, Joseph Need left <£180,000 in 1865. In the next generation estates were considerably smaller, to an extent which other factors, such as the desire after 1894 to avoid the increased estate duty, could hardly justify. Frank left <£40,000 in 1872, E.C. <£25,000 in 1869, P.A. £33,801 in 1894, F.A. (son of Sir E.S.) £10,722 in 1910, Henry II (son of J.N.) £20,914 in 1911 and not until the death of A.O. in 1925 was a reasonably large estate again declared, at

£62,024.[79] The Walkers therefore made moderate fortunes out of the industry, at least as compared with two other family firms which we shall look at in the next chapter, in each of which one partner amassed a fortune of more than £½m.

Their fortunes were, however, sufficiently large to enable them to provide their children with whatever they thought best in nineteenth-century education. The evidence of high pretensions comes earliest with Joseph Need and Thomas II, who were sent to Repton in the 1800s, with the former going on to Eton. Thereafter the public school link became very strong; Joseph Need's sons were sent to Eton and his successors then built a family link from there to Brasenose College, Oxford; Sir E.S., E.C. and A.O. all went to Rugby and from there respectively to Cambridge, Durham and Edinburgh Universities. The list of public schools which benefited may easily be continued with Frank going to Charterhouse and the sons of Sir E.S. to Harrow but University was less common and if anything there was a turning away from this, as a frivolity, and a tendency to put sons into one of the works as a junior, 'to learn on the job'. This is just the opposite of the predictive model of declining interest in the family firm. It was, however, followed increasingly in the second half of the nineteenth century: in 1855 Horace, aged 17, to Liverpool; Frederick Adam, 18, to Chester and Frederick John, 16, to London in 1869; and Ernest Abney, 19, to Elswick in 1873.

The Walkers had done well out of lead but after 111 years the connection was nearly over. On 23 January 1889 the prospectus for a new limited company, Walkers, Parker & Co. Ltd was published and on 7 March, the balance of the purchase price having been paid, the old partners relinquished their position as Receivers and handed over the works to the limited company.

Notes

1. The lease is in M.L. Walker's collection.
2. John Nelson, *History, Topography and Antiquities of the Parish of St. Mary Islington* (1811), pp. 195-6. See also Clay, 'Blue Lead to White', and Stella Margetson, 'The Vanished Windmills of London', *Country Life*, 14 Nov. 1974, pp. 1442-4. For the technical details see R. Wailes, *The English Windmill* (1954).
3. Steam engines were, of course, erected in London in the late-eighteenth century, especially in the large breweries. The point is not that running costs *per se* controlled the geographical diffusion of the steam engine (although that was true to some extent), but that running costs controlled its geographical diffusion in particular industries. In brewing, for instance, the steam engine offered returns

which offset its high running costs in London, as it did in the lead industry in Newcastle but not in London in the 1790s. For the use of steam engines in the London porter breweries and an analysis of the comparative costs of steam and horse power there see P. Mathias, *The Brewing Industry in England 1700-1830* (Cambridge, 1959), pp. 82-93.

4. Nelson, *History of St. Mary Islington*, p. 196 and S. Lewis (jun.), *The History and Topography of the Parish of St. Mary Islington* (Islington, 1842), p. 307. The women were employed not because of superior resistance to lead poisoning but because they could be paid less than men for the tasks of filling and emptying the white lead stacks.

5. The figures are drawn from a document giving comparative statistics on the various partnership branches (the rate of return at Elswick, for instance, was 6 per cent). There is no evidence as to how the figures were derived and we do not know what was included in the capital figures or what was deducted before the profit figures were struck. It is unlikely that the figures are comparable with those quoted in Chapter 2 for Elswick in the 1780s and 1790s, since it seems likely that differing accounting conventions were employed in their compilation and that the capital figure was affected by the inflation of the period of the French wars. Subsequent figures for the capital employed at each of the works up to 1881 may be found in Appendix III.

6. The minutes state that Maltby 'appears to have owned this wharf as he brought no capital into the concern', John, *Walker Family*, p. 34. A document in M.L. Walker's collection, reciting the 1814 partnership agreement, however, states that Red Bull Wharf was the property of Joshua, Thomas and the heirs of Joseph (decd) Walker and was let by them to the partnership. It may, of course, be that the ownership had changed between 1785 and 1814. The freehold was sold in 1824 by Joshua Walker & Bros to the lead partnership for £8,634. The wharf was described as being, 'off 93 Upper Thames St.'.

7. Legal documents relating to the negotiations are in M.L. Walker's collection.

8. For further detail on the Islington firm and more information on the Walkers' partnership holdings see Rowe, History of Lead Manufacturing, Chapters IV and V.

9. For the period 1815 to 1831 bad debts for Red Bull Wharf and Lambeth were £54,797 out of a total of £94,250 for all the partnership works. This included the huge figure of £10,486 in 1826, possibly the result of a bank failure in the crisis of the previous year. Since Red Bull Wharf was only a trading centre it may have been carrying bad debts which were strictly attributable to other works. It may also be the case that Red Bull Wharf was responsible for the major part of the partnership's overseas trade, where bad debts might be expected to be high. It would not, however, appear to be the case that it was London which was the contributory factor since bad debts were no higher at Islington than for the provincial works. The overall level of bad debt figures was certainly very high when compared with subsequent periods when, presumably, methods of debt control had been more effectively worked out. Again for Walkers, Parker, bad debts averaged only £220 p.a. between 1889 and 1939 with a highest figure of £2,101 in 1915.

10. The agreement for the lease, dated 19 Nov. 1810, is in M.L. Walker's collection.

11. Assignment of lease, 2 Aug. 1825, in ibid.

12. The lease, which is in M.L. Walker's collection, was for 21 years and provided for rent increases after seven and 17 years. No other nineteenth-century lead works lease that I have seen provided for an increase in the rental during the term of the lease – even when the term was for a period as long as 99 years.

13. Documents in M.L. Walker's collection.

14. John, *Walker Family*, p. 41.

15. According to Hall, *Appeal to the Poor Miner*, p. 2, there was a Southwark works in 1818. I have been unable to find any other evidence which supports this reference. Walkers, Parker were certainly the occupiers of Orange Street from 1 Aug. 1824, according to an agreement in M.L. Walker's collection.

16. The Poor Rate Book (in Southwark Local Studies Library) shows the rateable value to have been only £180 in 1826.

17. Some documents relating to Orange Street leases are in M.L. Walker's collection.

18. Until well into the second half of the nineteenth century, when foreign imports of lead became important, one might argue that transport costs did not come into the decision, (this would not be quite true since both white and red lead gained weight as a result of the manufacturing process, about 20 per cent and 10 per cent respectively, and were therefore more expensive to transport than pig lead), and that since labour costs were lower on Tyneside it was cheaper to manufacture there and transport the finished product to London. Later in the century when both London and Newcastle were importing pig lead and had equal raw material costs, (in place of the earlier position where London took pig from Newcastle) one might have expected the situation to change and it to become cheaper to manufacture in London, but Walkers, Parker do not then show any signs of considering the erection of white and red lead plant in London.

19. The firm was sometimes referred to as Cox Bros and earlier it had been Cox, Poyser & Co. See Chapter 5 for further details on the firm.

20. John, *Walker Family*, pp. 34-5.

21. Pipe was originally made at Bersham, possibly by the successors to the ironmaster, John Wilkinson, who had patented a pipe-making process in 1790 (no. 1735). Chester erected its own machinery in 1812. The rolling mill was erected in 1803.

22. John, *Walker Family*, p. 35.

23. See Rhodes, London Lead Company in North Wales, pp. 7-8, 'The Flintshire shore of the Dee . . . became one of the first industrialised areas in Wales, and for a time in the eighteenth century the major lead smelting region in Britain . . .'. See also C.R. Williams, A Dissertation on the Industrial Changes and Developments in the County of Flint from 1815 to 1914 (University College of Wales, Aberystwyth, MA thesis, 1951).

24. Williams, Industrial Changes in the County of Flint, p. 74, 'By 1850 all the Flintshire smelting works, with one exception [Llanerchymor], had become the property of firms from Liverpool or Newcastle upon Tyne.' This development encouraged contemporaries to take an exaggerated view of the area's importance. S. Lewis in *A Topographical Dictionary of Wales*, vol. I (1st edn, 1834), unpaginated, entry under Bagillt, gave the area's output of lead as 100,000 tons at a time when UK output was less than 50,000 tons and he has been followed uncritically by modern authors. See, for instance, A.H. Dodd, *The Industrial Revolution in North Wales* (3rd edn, Cardiff, 1971), pp. 183-4, and Williams, Industrial Changes in the County of Flint, p. 89. By the fourth edition of his work (published in 1849), even Lewis had reduced local lead output to 25,000 tons, still an exaggeration.

25. John, *Walker Family*, p. 37. There is some confusion about this particular point, which is a reflection of the general confusion created by the existence of several smelting works in a small area, all of which at some time or other are described as Bagillt works. In the later-nineteenth century Walkers, Parker had two works, one at Bagillt and one at Dee Bank, although the latter was sometimes referred to as just Bagillt. The specific confusion over the 1834 purchase results

from the fact that Williams, Industrial Changes in the County of Flint, p. 39, lists three Bagillt works, the second of which is clearly the same as the one John says was taken over in 1834, while Williams (ibid., p. 87) states that it was the third works, owned by William Dutton, which Walkers, Parker took over in 1834. There is some detail on the works in T. Pennant, *History of the Parishes of Whiteford and Holywell* (1796) and there are plans of the works in Clwyd Record Office, D/DM 136/5(e), D/P/64 and D/DM 244/70.

26. John, *Walker Family*, p. 37.

27. *British Parliamentary Papers* (*BPP*), 1878 XLIV [c.2159], qq.13,247-13,412.

28. These and other valuations for the various works are given in Appendix III.

29. *BPP*, 1878 XLIV [c.2159], qq.13,252-3.

30. The principle was suggested by Bishop Watson in the eighteenth century and had been widely adopted. It involved the construction of brick or stone arched, horizontal chimneys or flues of considerable length ending in a tall, vertical chimney, which provided the draught to draw the furnace gases through the flues, on the sides of which the lead fume was deposited. The flues were sufficiently large to enable men to go through them and scrape off the lead deposit once or twice a year. It was estimated that the lead so saved soon paid for the cost of the flues. The board minutes of Walkers, Parker & Co. Ltd for 30 June 1902 note that 220 tons of lead were recovered from a recent cleaning of the Dee Bank flues, although it is not stated how long the accumulation period had been.

31. John, *Walker Family*, p. 51.

32. Ibid., pp. 42-3.

33. Williams, Industrial Changes in the County of Flint, p. 89.

34. Given that women were not allowed to be members of the Sick Club and that women were exclusively used for labour in filling and emptying white lead stacks, it seems likely that total employment at Elswick at this time was at least 100.

35. An analysis of a subsequent 100 pages from the same ledger confirms this picture: 124 customers, of whom 49 (39.5 per cent) were local, 43 (34.7 per cent) Scottish and 32 (25.8 per cent) south of the Tees. This shows a marked change from an analysis of a ledger for the late 1790s, which had shown 478 customers, of whom 251 (52.5 per cent) were local, 48 (10 per cent) Scottish and 179 (32.5 per cent) from the rest of England. Among the latter were customers over a much wider geographical area (Chester, Liverpool, Manchester and many more from the West Riding), than in the period 1810 to 1814, when much of this trade had been taken by the Chester branch. The expansion into Scotland would therefore seem to have been a response to the more limited English opportunities following the opening of Chester.

36. For Grainger and the building of central Newcastle see L. Wilkes and G. Dodds, *Tyneside Classical: the Newcastle of Grainger, Dobson and Clayton* (1964).

37. The affair with Grainger was a continuing saga. On 28 July Walkers, Parker refused his application to have a bill for £700 renewed, pointed out that no payment had been received for goods supplied more recently and requested £674 in settlement of deliveries made in April/May.

38. The exception seems to have been individuals who were not in a business in which they would require lead. There is a number of letters such as that to James Rhodes on 23 May 1840, which stated, 'we do not supply private persons with our articles' and recommended him to merchants from whom he could buy. It is likely that this was a policy of fairly recent date — since the 1820s ledger contains accounts involving small sales to individuals who are not described as being in business. Such a policy saved a large firm from small orders which were costly to service and, no doubt, had something to do with sales agreements with

merchants to whom Walkers, Parker sold for resale to private customers.

39. Nor was price competition the only element – efficiency was important. In response to an obvious grumble from Patterson, Walkers, Parker wrote to the traveller on 6 July, 'We note that James & Co.'s Circulars were in Glasgow at the time you rec'd ours, which shows their activity & the necessity for our being upon the alert we find them nearer home to lose no time advising where any change takes place.'

40. Out of the necessity for such information was born the commercial list, which appeared later in the century. One such, Estell & Co., *Newcastle District and Hull Commercial List* (1874), gave the Tyneside lead manufacturers the following ratings: Walkers, Parker A1 and 1 ('the highest ranks' and good for any amount of credit); Cookson & Co. 1, and Foster, Blackett and Wilson and Locke, Blackett & Co., both 1¼ ('high and good for any amount of credit'); while James & Co. were rated at 1½.

41. The following were the titles: (1) at Elswick – Walker (from 1782 Walkers), Fishwick, Ward & Co.; in 1800 Walkers, Ward, Maltby & Co.; from 1801 Walkers, Ward, Parker & Co. and this continued to be used, despite the death of Ward in 1800, until 1814 when it became Walkers, Parkers & Co.; but from 1822 until 1889 it was established as Walkers, Parker, Walker & Co. (2) at Islington – Walkers, Ward & Co.; by 1805 Walkers, Maltby & Co. and by 1814 Maltby, Parkers & Co. (3) at Red Bull Wharf, Lambeth and Southwark – Walkers, Maltby & Co.; by 1814 Walkers, Maltby, Parker & Co. and from 1824 to 1889 Walkers, Parker & Co. (4) at Derby – Joseph Walker & Co. (5) at Chester, Liverpool, Newcastle-under-Lyme, Bagillt and Glasgow – from 1799 Thomas Walker, Maltby & Co.; from 1814 Joshua Walker, Maltby & Co.; from 1825 Joshua Walker, Parker & Co. and by 1860 until 1889 Joseph Walker, Parker & Co.

42. Much of the information is available in John, *Walker Family*, while the changing partnership structure is discussed in Rowe, History of Lead Manufacturing, Chap. V.

43. For the Parker lineage see *Burke's Landed Gentry*, vol. III (18th edn, 1972), pp. 701-2.

44. The 1800 partnership agreement made it clear that the Walkers were not 'under any Obligation to attend to or act in the Management', which was to be the responsibility of Ward and Maltby who were to be paid respectively £150 (perhaps Ward by this time was close to being incapacitated) and £500 p.a. In the 1814 agreement Samuel Walker Parker and William Parker (each with a salary of £150 p.a.) were listed with Maltby as acting partners.

45. Sykes, *Local Records*, vol. II, part I, p. 80.

46. For the Walker lineage see *Burke's Landed Gentry*, vol. II (1969), pp. 624-7.

47. N. Pevsner, *The Buildings of England: Durham* (1953), p. 237. Ownership of the house clearly reflected a fairly high social position – the 1841 Census shows Parker to have had seven servants (two male and five female), plus a resident gardener and his family. There were in addition two women, Jane Robson (aged 20) and Susanne Thompson (45) for whom no occupations were given.

48. John, *Walker Family*, pp. 41 and 43.

49. Ibid., p. 47.

50. Joseph's elder brother, Joshua, had already married Susannah, Need's second daughter, and his younger brother, Thomas, had married Mary, Need's third daughter. Samuel Walker II and Need's eldest daughter were apparently immune. (There is however a footnote to the footnote. Need's eldest daughter, Hephzibah, had married Edward Abney and their only daughter, Elizabeth, subsequently married her cousin, Henry Walker, the son of Joshua Walker and

Susannah Need). This excellent example of the close ties between the non-conformist families who made their wealth out of the early stages of the industrial revolution, helps to explain why their firms were usually well placed to raise any additional capital or loans which they required, above their ability to plough back from profits. For Need see R.S. Fitton and A.P. Wadsworth, *The Strutts and the Arkwrights 1758-1830* (Manchester, 1958).

51. See M.L. Walker, 'William Wilberforce at Rotherham', *Transactions of the Hunter Archaeological Society*, viii, 2 (1960), pp. 50-66.

52. John, *Walker Family*, pp. 26-31.

53. Joshua, J.N., E.S. and P.A. Walker and Samuel Parker were to 'manage and conduct the trades or business of the said Copartnership' while Henry and Thomas Walker II and Samuel W. Parker were not 'under any obligation to attend to or act in the management or conduct of the Copartnership'.

54. 'Talking of holidays, I am surprised how much time off for them some of the Walkers took in pre-telephone days, whether for marine biology, mountaineering or shooting. I have the Game Book of the annual shooting and fishing holidays in Scotland, initiated by the three sons of Joseph Walker in 1819 and continued by Henry II up to 1898. Although the holidays lasted only about a month up to the mid 1830s, subsequently they rose to 1½ months and after 1859 they lasted 2¾ months or so for the rest of the time. J.N. Walker spent the full holiday period in Scotland up to 1864, as did his son Henry after 1850, his junior at Liverpool, except on rare occasions. Henry did not reduce his time away after he took charge at Liverpool (and Horace was probably away on the mountains at the same time). Other partners who joined them in Scotland were E.C. from Chester on ten occasions in the 1850s and 1860s and Samuel Parker five times.' Communication to the author from M.L. Walker. Frank Walker was an original member of the Alpine Club and his son, Horace, became its President in 1890-1. They both went away on climbing expeditions, Horace on one to the Caucasus in 1874.

55. He was knighted in 1841, presumably for services to the City of Chester, of which he was mayor in 1837-8 and again in 1848-9. In 1845 the partnership presented him with plate to the value of £150 'as a mark of approbation for the manner in which he has carried out the alterations and improvements at the Chester works'. He provides the earliest example of a graduate in the management of the lead companies, having been educated at Rugby and St John's College, Cambridge.

56. F. Brook and M. Allbutt, *The Shropshire Lead Mines* (Cheddleton, Staffs., 1973), pp. 44-7. The partners were Richard Cross and Joseph [Need] Walker, in a firm known as Walker, Cross & Co. J.N. Walker was presumably acting for Walkers, Parker & Co., while Cross probably had experience of lead mining.

57. John, *Walker Family*, p. 38.

58. Brook and Allbutt, *Shropshire Lead Mines*, p. 47.

59. John, *Walker Family*, p. 39.

60. In his report for 1927 the chairman of Walkers, Parker & Co. Ltd said, 'Naturally we are as Lead Consumers frequently approached with offers of Lead Mining Properties, but the history of our undertaking, which completes a century and a half of profitable industry in the year 1928, shows that serious losses have always resulted in any departure from our manufacturing business.' I am grateful to Miss O.S. Newman for pointing out to me that Sir E.S. Walker continued to invest in Shropshire mines after the failure of the Bog venture.

61. John, *Walker Family*, p. 39.

62. Ibid.

63. Ibid., p. 41.

64. They come from pencil annotations to a typescript copy of the 1879

minutes which was found in the Chester office. The output figures show a broadly similar trend over time to the delivery figures given in Table 4.3, but considerable annual variations, which do not as they should, equal zero over time. To the extent that the output figures in Table 4.2 are, therefore, inaccurate, the profit figures there must be suspect. On the assumption that accountancy techniques were fairly well advanced by this time, it has been assumed that these were net profit figures, after all charges, and that they should be related to the real capital of the partnership, excluding the deposits. This may be unrealistic since we do not really know what the deposits were. If they were used to finance stocks and credit sales, instead of relying on overdrafts and bills of exchange, they should be regarded as capital. Since the deposits could fluctuate considerably without compensating changes in capital, the assumption taken here may be considered justifiable. The figures for deposits and capital are given in Appendix III.

65. It is interesting to compare this with the figure of £16,700, the average salary of nine directors of ALM (excluding the chairman) in 1980. This was an increase of about 17 times for men who might be considered to be doing a similar job to the managing partners of the nineteenth century. By comparison, the average male shop floor wage in the 1860s might have been 25s a week while in 1980 it was about £110 (an increase of 88 times). Although these are simple and approximate comparisons and the differential between management and employees is still wide, it has obviously narrowed very considerably.

66. John, *Walker Family*, p. 48.

67. Ibid., pp. 49-54.

68. Charles Dickens, *Bleak House* (Chapman & Hall, autograph edn, n.d.), p. 532.

69. The action is described as Chancery Cases, *Walker* v. *Walker* (1881 W4839). There was also a subsidiary action *Walker* v. *Walker* (1891 W1241). These are the basic references from which it is possible to trace the records of the case in the Chancery papers in the Public Record Office.

70. Public Record Office (PRO), Court of Chancery, J15/1600, fo. 1533.

71. John, *Walker Family*, p. 50.

72. He was a noted amateur marine biologist and as early as 1859 published a paper, 'The Marine Fauna of the Dee Estuary'. His work in this field resulted in his being elected a Fellow of the Linnaean Society in 1874, though it was after the dissolution of the lead partnership that he published more than 30 scientific articles, chiefly on amphipods (sand hoppers), on which he was a recognised authority. I am grateful to M.L. Walker for these details.

73. Professor (later Sir) Oliver Lodge was a physicist and was appointed Professor of Experimental Physics at the newly formed University College of Liverpool in 1881, where he remained until he was appointed Principal of Birmingham University in 1900. He was best known for his work in electricity and as the inventor of the Lodge sparking plug. In his youth he had acted as an agent for Walkers, Parker & Co., Chester, in the sale of white lead in the Potteries. Presumably this is how he originally made a link with A.O. Walker, who was subsequently to provide the facilities (at Dee Bank) for the trial of Lodge's laboratory experiments on an industrial scale. See Sir O. Lodge, *Past Years, an Autobiography* (1931), especially pp. 24 and 174-5.

74. For the experiments see *Engineering*, 5 June 1885, pp. 627-8.

75. J.A. Smythe, *Lead* (1923), pp. 198-201.

76. While there he made an interesting discovery which he conveyed to the great metallurgist, Percy, who subsequently wrote, 'The attention of lead smelters and the public should be directed to the fact that a few years ago the Saxon Government practised the fraud of attaching the name of a great English lead-smelting firm to the shot manufactured by that Government. My friend, Mr.

Alfred Walker, a member of that firm, has communicated to me, for publication, the following statement on the subject. "In the year 1856 I was at Freiburg, and on visiting the shot manufactory I found that nearly all the bags of shot were sent out with the name of Walkers, Parker & Co., Newcastle-on-Tyne, printed on their bags. This shot manufactory was under Government management, as shown by the inscription over the door, 'Konigliche Sachsische Schrotgiesserei'." ' Percy, *Metallurgy of Lead*, note pp. 303-4.

77. Henry Walker I had moved on from Clifton House, Rotherham, which had been good enough for his father, even though a sale catalogue of the property, which had about 70 acres of land, could state in 1883 that 'although in such close proximity to the manufacturing town of Rotherham, and within the South Yorkshire Colliery District, the Estate being situate on the side of the town where there are no works, there is no annoyance from smoke'. Particulars of Clifton were given in a sale catalogue of the house, following Henry's death in 1860, 'The MANSION (one of the largest in the Neighbourhood), is Stone-built, in perfect repair, well arranged, and contains Dining Room, Drawing Room, Library, Gentleman's Morning Room, suitable Bed Rooms, Dressing Rooms, Dairy, Laundry, Housekeeper's and Butler's Rooms, with Servants' Offices, Stables, and Coach House, Vinery, Peach House, Pinery, Conservatory, Ice House, and all other requisites for a Gentleman's residence.' Sale catalogues in Rotherham Public Library. Even though Henry had moved to a more rural area the *Rotherham & Masbrough Advertiser* (28 Jan. 1860, p. 4) carried a black-edged column with an account of his funeral, with its lead coffin, and all the shops in the town closed.

78. Sir E.S. Walker acquired the Berry Hill estate after the death of the widow of his uncle, Thomas I.

79. Nor did the Parkers make large fortunes, even the Rev. Henry, dying in 1871 with the residuary Parker shareholding, left £45,000. Valuations of estates at death are not necessarily an accurate reflection of the money made during life. One good example may be Henry Walker II, whose estate of £20,914 in 1911 was considerably less than the value of his shares in the lead company alone in 1881, which was over £80,000. His son, Henry Francis Mostyn, born 29 Aug. 1880, turned out to be something of a waster and involved his father in paying off the heavy debts he regularly incurred. Another example is that of Edward Joshua Walker, who got what A.O. Walker at least would have regarded as his true deserts since he died in 1905 leaving an estate of £420 8s 3d.

5 THE FOUNDATION AND GROWTH OF OTHER LEAD MANUFACTURERS UP TO THE LATE-NINETEENTH CENTURY

'Butter is gold in the morning, silver at noon, and lead at night.'
Sixteenth-century proverb.

The development of other lead partnerships up to the late-nineteenth century (1) on Tyneside: Locke, Blackett & Co., Foster, Blackett & Wilson, and Cookson & Co. (2) in London: John Locke & Co. (subsequently Locke, Lancaster & Co.), and W.W. & R. Johnson & Sons (3) elsewhere: Cox & Co., Derby, Alexander, Fergusson & Co., Glasgow, and T.B. Campbell & Co., Glasgow.

Discussion of the development of the Walkers' partnership has taken the story over a wide geographical area and over almost a century beyond the dates at which others of the lead partnerships, which were to come together to form ALM, commenced. The turn of the eighteenth century was the time and London and Tyneside the chief places in which these developments were to occur.

Tyneside

We saw in Chapter 2 that the monopoly of white and red lead production on Tyneside, established by Walkers, Fishwick & Co., was soon broken by the formation of the partnership of James, Hind & Co. with works at the Ouseburn. While this firm provided significant competition it was never on the scale of that of Locke, Blackett & Co., set up at Gallowgate, close to the centre of Newcastle, at about the same time.

Locke, Blackett & Co.

John Locke[1] was a Londoner, whose family in the eighteenth century had a considerable connection with the service of the East India Company, and he was described as 'East India Company's Ships Husband'. It may well be that he recognised the demand made by that Company for lead products and decided that it would be a lucrative trade in which to get involved. The date for the founding of his firm is usually given as 1790 but there is no evidence of its existence before

1795[2] and from 1797 Locke appears in London directories as a lead merchant of 68 Mark Lane. At this stage the firm was acting as lead merchants but John Locke was obviously ambitious and decided that he would extend his interests to manufacturing lead. That he chose Tyneside to do this reinforces the view that the locational advantages of particular areas were changing and Locke clearly took the view of Tyneside that Walkers, Fishwick had done. This is worthy of note in view of his London base and the fact that the capital had become a major centre for lead manufacture by 1800. It is even more surprising since Locke was also in partnership with John Griffiths, who in 1792 had purchased the Bagillt smelting works of the London Lead Company.[3] While John Locke & Co. continued to purchase pig lead from Griffiths' Bagillt works it was obviously decided that north Wales was not a suitable area for manufacturing.

As early as September 1795 Locke had begun to show an interest in manufacturing lead on Tyneside. He asked Christopher Blackett (owner of Wylam Colliery and Newcastle agent for the sale of the produce of the Blackett/Beaumont lead mines in the northern Pennines), from whom he purchased pig lead, to make enquiries about Walkers, Fishwick's works at Elswick.[4] It was not, however, until October 1798 that Locke decided to act, since Blackett then wrote:

> I feel myself very highly honored by the offer you make me of taking a share in a white & red lead work with your house, and shall be very happy to accept it, provided I can manage the pecuniary part — at present I confess myself totally ignorant of the sum that may be required for such an undertaking. I will employ a Confidential Friend to enquire for a situation for the purpose, who can do it without any suspicion arising as to the intent of it.[5]

Thereafter developments came quickly with Blackett acting for Locke, who only made visits to Tyneside approximately once a year. By mid-November Blackett was negotiating for premises in Gallowgate, Newcastle, and the following month an agreement was signed to purchase about an acre of land there.

Shortage of capital may well have been a reason why Locke decided to take Blackett as partner, although it was obviously important to have someone to direct affairs on Tyneside and especially valuable to have Blackett with his links in the pig lead market. Before the end of 1798 Blackett could conclude a letter to Locke, 'Wishing Success to our undertaking' — the partnership was underway. Blackett immediately engaged for the purchase of the manure from the cavalry barracks on the Town Moor at '4d p. Horse each Week' and was inordinately pleased when he discovered that he had narrowly beaten Fishwick to the contract.[6] During the spring of 1799 work was going ahead with the building of white lead stacks at Gallowgate and Blackett was busy recruiting labour, chiefly from the Elswick works by offering higher wages than were paid by Walkers, Fishwick. In April, for instance, a red lead furnaceman was engaged at 20s per week plus house and firing and three other men with knowledge of various lead processes at 21s per week each plus house and firing. Blackett also succeeded in employing the woman who had superintended the setting of the white lead beds at Elswick. In all there was to be considerable dependence on knowledgeable workmen since the partners had no understanding of the processes.

The first two white lead stacks (of the six being built) were completed by mid-June 1799 and by the end of that month the first was filled with eight heats each containing three tons of lead. Then began the nervous wait to see whether the corrosion had been good. The first stack was emptied at the end of August after two months' corrosion and proved satisfactory. In the meantime the remaining stacks had been completed and work had commenced on the red lead buildings and furnaces, of which three were constructed. To Blackett's annoyance the furnaces took a long time to construct and there were considerable delays, although by June 1800 a Boulton and Watt engine had been installed to grind the red lead and a little had been made. Regular production of red lead was underway by October, by which time the firm had already commenced the production of lead shot. Again the initiative lay with Locke who had asked Blackett to obtain some samples of shot. Blackett had obtained some from a producer at Staindrop 'made in imitation of the Patent' (i.e. shot dropped from a tower). The interesting point is that Blackett argued that the fact that it was not very good 'may be accounted for in some measure by the shortness of the fall, which the person who sent it informs me was down an old *Coal Pit* 30 yards deep'.[7] Coming from a coal owner the underlining in the original was undoubtedly intended to convey a message and shortly afterwards an old coal pit on the Blackett estate at Wylam was prepared and after some initial problems the regular production of shot commenced in September 1800.

The works had been set up on the edge of the town, abutting the old town walls, in an area in which lead manufacture would cause no intrusion since there were already established there other offensive industrial activities such as tanneries and a slaughter house. Nevertheless in setting up without immediate access to water for transport Locke, Blackett's Gallowgate works was a rarity among lead works at this time and some of the problems attached (which must have included the negotiation of the steep bank to Newcastle quayside by horse-drawn waggons, containing loads of pig lead up and manufactured lead down) were not to be overcome even with the arrival of the railways, which improved the transport facilities of other urban works. The Gallowgate location was a peculiarly inappropriate one (which the immediate access to manure and subsequently to spent tan bark from a nearby tannery could do little to justify). Nevertheless the partners accepted the problems and the site was extended by further purchases of land in the 1830s and 1840s. In 1801, however, additional premises were taken in the Close, Newcastle, which provided the partnership with warehousing and its own shipping quay which would obviously facilitate the development of the business. Since both Locke and Blackett were busy men, with important interests elsewhere, the running of the business was from the first placed in the hands of a manager, George Burnett. He was clearly an able man, upon whom considerable responsibility was placed, not only to supervise the erection and subsequent management of the Gallowgate works but also to negotiate the purchase of the premises in the Close. In 1807 his efforts were rewarded by his being taken into the partnership.

As with so many businesses of the period it is difficult to interpret the statistics with confidence but the capital involved in the business appears in the accounts by 1803 at the latest at £24,000, a figure at which it remains until 1814 after which date the records cease. It seems reasonable to assume that this was the total capital put up by the partners (in the proportions of £19,200 by John Locke and Christopher, his brother, and £4,800 by Blackett) and that subsequent additions were financed from revenue and not capitalised. Table 5.1 presents such profit figures as are available with a rate of return calculated on the basis of a total capital employed of £24,000. As usual there is no evidence as to how profits were calculated but it is reasonable to assume that they were after management costs and repairs and renewals (given that the capital figure remained unchanged) and probably that they were after deduction of 5 per cent on the partners' capital, i.e. £1,200.

It seems clear that the partnership had soon been established on a profitable basis and that the extension of supply of lead products had been accommodated by rising demand rather than falling profit margins. Inevitably, however, the Elswick and Gallowgate works were attacking the same markets and Locke, Blackett certainly obtained a part of the orders of many previous customers of Walkers, Fishwick — their names appearing in both companies' books in the 1800s — especially in the local area. Of the 779 customers whose names appear in Locke, Blackett's sales ledger for the period 1800 to 1810, 221 (28.4 per cent) were in Northumberland and Durham, 156 (20.0 per cent) in Scotland and 400 (51.3 per cent) in the rest of England, with two overseas customers. As compared with the figures for Elswick given in Chapter 4 these show considerably smaller proportions of local and Scottish customers and a much larger one of those in the rest of England. This implies that although Walkers, Parker experienced the effect of competition from the new arrivals, they were fairly well entrenched in the local and Scottish areas and Locke, Blackett had to find some of their markets further afield. Whilst most of these customers were on the east coast of England and especially at Hull, there were more in Manchester, Liverpool, the west midlands and Bristol than had ever appeared in the Elswick ledgers. As with Elswick, however, London provided a large market and the turnover with John Locke & Co. was considerable, averaging about £23,000 p.a. in the period 1804 to 1814, although this included sales of pig lead.

Table 5.1: Locke, Blackett & Co., Profitability, 1804-14

Year to	Profit (Loss) (£ s d)			Return on capital (%)
30/4/1804	396	3	9	0.2
1805	4,240	5	1	17.7
1806	3,887	16	6	16.2
1807	3,252	19	2	13.6
30/6/1808 (14 months)	4,651	12	2	16.6
1809	4,803	2	0	20.0
1810	4,104	14	1	17.1
1811	(2,266	16	10)	−9.4
1812	2,518	14	4	10.5
1813	3,130	16	0	13.0
1814	4,821	1	3	20.1

The figures given in Table 5.2 for the output of manufactured lead by Locke, Blackett, give some scope for comparison with the figures for Elswick presented in Table 2.1, even though the latter are for a rather earlier period. Locke, Blackett's average sales of white lead were almost three-quarters of the average achieved by their Elswick rivals in the 1790s, and although sales of red lead were only about 50 per cent, they were selling as much shot per annum in the period 1804 to 1808 as Elswick had sold in 1799 to 1800, after the completion of the shot tower. The Wylam shot pit must, therefore, have offered a fairly efficient productive method.

Table 5.2: Locke, Blackett & Co., Sales of Lead Products, 1804-14

Period	White lead	Red lead	Yellow lead	Shot
		(£)		
2/7/1803-30/4/1804	8,000	5,100		4,700
30/4/1805	12,700	6,400	2,000	8,700
30/4/1806	16,700	4,900	2,000	7,100
30/4/1807	18,200	3,900	2,000	4,400
30/6/1808	21,500	3,900	2,100	5,200
30/6/1809	11,300	4,400	1,400	2,600
30/6/1810	18,000	7,200	1,600	4,800
30/6/1811	15,200	1,200	1,300	2,800
30/6/1812	15,100	4,900	1,000	2,800
30/6/1813	16,600	5,100	500	3,000
30/6/1814	19,600	5,600	300	2,800

The deaths of John Locke in 1816 and Christopher Blackett in 1829 probably had little impact on the firm, other than to bring members of the second generation into the partnership. Throughout these changes George Burnett, and then his son George, retained the position of managing partner and ran the works. Perhaps the most important action which they took, however, was their employment of James Leathart. Born at Alston,[8] on 25 November 1820, Leathart seems to have come from a modestly comfortable background. He is said to have been the son of a mining engineer[9] and lead prospector. Up to the age of 14 James was educated at Alston Grammar School and he then went to work 40 miles away in Newcastle where he became a junior at Locke, Blackett's. Letters from John Leathart to his son over the next few years would have pleased Samuel Smiles had he had access to them and give us some insight into the kind of life James led and how he reached his subsequent position. The letters are full of encouragements to 'personal exertion' and 'prudence & perseverence' and his father encouraged James to read in chemistry, mining and metallurgy in his spare time. Although often grumbling about his low salary and the long hours he had to work, James appears to have taken his father's advice and fitted himself for promotion.

Having been introduced to the Gallowgate works as an office junior, Leathart had rapidly moved into a position of junior management and was soon to become managing partner after partnership changes which were to take place in the later 1840s. In the meantime, there would however have been plenty for Leathart to do, since the works employed 40 to 50 women and a similar number of men in 1842.[10] Their activities are described in a detailed account of the works,[11] written in 1844, which describes the manufacture of red and white lead (with 'about a dozen' stacks), the dropping of shot into a pit and refining by the Pattinson process, all of which were undertaken at Gallowgate. The expansion of the firm's activities was, however, overcrowding the site and about 1846 an additional site was taken, on the Tyne at St Anthony's, with a 750 ft river frontage. It gave Leathart his first big opportunity since he was placed in charge of the erection of the works. It was developed on a considerable scale and for the first time the firm was to undertake the smelting of ore with the installation of ore hearths to deal with Spanish lead ores, which were beginning to come to the Tyne, as well as local ores. In addition a second Pattinson desilverising plant was installed — the Spanish ores being silver-rich — as were further red lead furnaces, while two new ventures, a lead rolling mill and pipe presses, took Locke, Blackett further away from their early basis in lead chemical manufacture. By the early 1850s there were nearly 100 employees at St Anthony's[12] and Locke, Blackett had capacity as great as that of Walkers, Parker at Elswick. Leathart had much to do with this expansion and he was subsequently to ensure that Locke, Blackett remained in the forefront of the Tyneside industry.

Over the ten years before the St Anthony's works was taken the firm appears to have been modestly profitable. A valuation[13] undertaken in September 1845 gives the total capital as a little over £52,000, of which £22,000 was in fixed capital and the remainder in stocks. The stated net profit divided between the partners over the previous ten years (after deductions which included the carrying of £9,000 to capital) was £37,772, which would imply a rate of return on the 1845 capital of about 7 per cent p.a. The problems attached to such a calculation make it a fragile tool on which to base any conclusions but it is a return which was well above that available on government stock and, since the partners were prepared to invest heavily in the creation of the St Anthony's works, clearly one which satisfied them.

It is possible that the 1845 valuation was prepared for a change in partnership shares in which Leathart was to become a partner but he had certainly attained this position by 1851. He and James Foster were to be joint managing partners, at annual salaries of £250 each, respectively responsible for manufacturing and commercial matters. They each had a one-eighth share in the partnership, whose total capital was to be £80,000, although in 1854 Foster and Leathart had each contributed only £4,278 5s 9d and each owed the difference from £10,000 to make up his full share. As often happened, managing and young partners were given partnership shares which were beyond their immediate financial means and allowed to pay for them out of future profits. At the end of 1856 Foster sold his share to the remaining partners and left the area. While there is no evidence as to what had brought about Foster's dissatisfaction, he will soon influence our story again.

From 1857 Leathart was in sole charge, a development he perhaps celebrated the following year by marrying Maria, eldest daughter of Thomas Hedley, the Tyneside soap manufacturer, and moving away from his previous residence at the Gallowgate works to one of the large Victorian terraced houses, on the edge of the Town Moor, 12 Framlington Place, which Newcastle provided for its comfortably-off business and professional men. They were to have 14 children and growing family and rising prosperity led to further moves which ended in 1869 at Bracken Dene in Low Fell, Gateshead. This was one of a group of large, detached houses, rather grander than Victorian villas, in their own considerable grounds, which provided another staging point in property ownership for Tyneside's successes. Here James Leathart developed 'The most important private collection of modern pictures in the neighbourhood of Newcastle',[14] and also had a playroom for his children built at the end of the garden in order that they should not disturb his contemplation of his pictures.

Under Leathart's management Locke, Blackett continued to develop and show considerable evidence of industrial dynamism. It was, for instance, the first Tyneside firm to replace the Pattinson process of desilverising lead with the Parkes process. This may have been a less than satisfactory move, given the fact that there were still problems with the latter process and it was followed by two other pieces of technological leadership which were decidedly unsuccessful. In December 1869 the firm took an exclusive licence for the Tyneside area of the Flach desilverising process, invented by a German who had taken a UK patent. The original licence was on a royalty basis of 2s per ton but the early success of the process led to the substitution of a fixed annual sum of £1,200 although this was subsequently reduced by stages to £500 as Locke, Blackett found the early promise of the process unfulfilled. Secondly, in July 1875 the firm took an exclusive licence for north-east England from the Sankey White Lead Co. Ltd to use the latter's new process for the manufacture of white lead. This was one of many aimed at replacing the old and tediously slow stack process.

Invented and patented by Edward Milner, it was, like many of its contemporaries, a quick process which commenced with litharge and a salt (sodium, potassium or aluminium chloride) and obtained lead carbonate by precipitation with carbon dioxide. Milner was a Warrington salt manufacturer and, like most of those who came to the problem of white lead manufacture from outside that industry, he failed to see the real problem. This was not to manufacture white lead more cheaply but how to produce a white lead which customers would accept as being equivalent to stack white lead. Unlike most of the new processes, Milner's at least reached volume production but, after some years of failing to produce a white lead with adequate covering power, both his own works and the plant at Locke, Blackett's were closed and reverted to stack production.

While there is, therefore, considerable evidence that Leathart was anxious to keep his firm in the forefront of manufacturing developments by adopting new technology there is doubt as to the soundness of his judgement.[15] While the other large lead manufacturing firms were equally aware of the necessity for technological advance and showed considerable interest in new processes, none of them was so willing as Locke, Blackett to adopt relatively unproven techniques and none had such obvious and large failures. The industry was a very conservative one (as were its customers), with well-established productive methods which were unlikely to be changed at all quickly by new techniques. The successful firm, therefore, was the one which concentrated on minimising the cost of existing techniques while being the second (or even third or later) to adopt a new process which was actually seen to be successful and profitable.

At the time at which Locke, Blackett had been adopting new technology, around the time of the boom period of the early 1870s, it also expanded physically with the purchase of John Warwick's lead smelting works at Wallsend. While this takeover may have seemed attractive at the time it was made, it was to be followed by a 20 year period of poor profits, rising competition and an increasing tendency for pig lead rather than lead ore to be imported as smelters were erected in Spain. In the long term Wallsend was another unsuccessful investment and the lease was disposed of in 1893.

The purchase of an additional works and increased investment in plant required more capital and in partnership renewals in 1864 and 1874 the existing partners brought in additional capital. By 1874 the total had reached £130,000, with the shares divided equally between J.A. Locke, Capt. C.E. Blackett and Leathart, who was again to be allowed to raise his share of the capital by drawing on his future share of profits. As managing partner, Leathart had his salary increased to £450 p.a. in 1864 and to £600 in 1874. The details of the new partnership arrangements do, however, suggest that the firm was financially unstable. The asset valuation included a balancing item of more than £31,000 for goodwill in 1864, which was unsatisfactory even in a period of growth and prosperity but which was potentially disastrous in the depressed years of the last quarter of the century when profits were in decline and unsatisfactory investment in plant had to be written off.

There is no detailed evidence for this period but, following the cyclical upswing of 1888 to 1889, which had increased demand for lead, the partners took the opportunity to liquidate their by now much-reduced assets by selling the company to the public in 1891. Blackett and Locke[16] withdrew completely but Leathart was to continue as managing director of the limited company, the prospectus for which gives some detail on the size and profitability of the old partnership. In 1890 the output of lead chemicals (white, red and orange lead and paints) had been 5,700 tons, which was perhaps one-half of that of Walkers, Parker and, therefore, suggests that Locke, Blackett's output of these products was much the same as that of the Elswick works. Blue lead output at 8,200 tons was less significant, comparing with an 1880 Walkers, Parker output (over three works in the case of blue lead) of about 30,000 tons, which would almost certainly have been higher in 1890. Finally silver output from Locke, Blackett in 1890 was ½m ounces compared with an Elswick figure of one million ounces in 1888. It seems fairly safe to say that the various Locke, Blackett works were by 1890 rather smaller in total output terms than the Elswick works of Walkers, Parker. For the three years to 1890 (almost certainly much more profitable ones than the previous ten years) the prospectus gave the profits of the old partnership, 'without deduction for interest on Capital or borrowed money, Managing Partner's salary, and depreciation (other than depreciation of loose plant and utensils)', as an average of £10,720 p.a. On the average of three good years the gross profits, therefore, only showed a rate of return of 10 per cent on the sale price of £105,000, which was itself below the asset valuation of the partnership. Net profits were unlikely to have reached 5 per cent in these years and it is, therefore, certain that the firm had previously been making considerable losses. This assumption is confirmed by the division of the purchase price of £105,000. The old partners received only £29,000 for their capital holdings, against the £105,000 at which they were nominally valued in 1874, while the bank overdraft took £23,000 and other loans £14,000, and creditors of the partnership (including the Northern Trust Ltd which received a fee of £3,000 plus £1,409 7s 3d expenses for selling the partnership's assets to the limited company) took a further £38,000.

After nearly 100 years the Locke, Blackett partnership had come to an end at almost the same time as that of their great Tyneside rivals, Walkers, Parker but it seems clear that the latter had been able to establish and then maintain through the depression years of the late 1870s and 1880s a much more financially viable organisation.

Foster, Blackett & Wilson (FB & W)

In 1862 James Foster, as we have seen joint managing partner in Locke, Blackett until 1856, returned to Tyneside, and with Robert Blackett[17] and John Wilson,[18] set up a rival lead manufacturing firm at Hebburn. There is little evidence on the firm but it was reasonably successful in its early years and a lease on additional land was taken in 1872 to accommodate expansion. With the exception of shot manufacture the firm was from the beginning intended to cover a wide range of lead manufactures; red and white lead, sheet and pipe were produced and desilverising undertaken. In 1895 the *Chemical Trade Journal* made a survey of the Tyneside lead industry and took the works of FB & W 'as a type [of the Tyneside firms], and in the following have given a description of the different processes as carried on by them at their works'.[19] While the selection of the firm may not have been made entirely on its manufacturing merits, it clearly was of some significance on Tyneside. The white lead works was described as the largest in the country and figures for the late 1880s give an average female employment (and women were usually used almost exclusively for filling and emptying the white lead stacks) of 140, while the stacks were said to hold nearly 3,000 tons of lead. Given a 50 per cent corrosion rate and assuming that each stack was emptied three times a year, this would suggest that the firm's annual output of white lead might have reached 5,000 tons (which compares with a figure of 5,700 tons of both red and white lead for Locke, Blackett in 1890 and around 8,000 tons of white lead which was produced by two of Walkers, Parker's works in 1880). Unlike Locke, Blackett there is nothing to suggest that FB & W was in the van of progress. There is no evidence of the adoption of new processes and in 1895 the firm was said to be probably the only one in the country which still used only the Pattinson process to desilverise pig lead. Given that the firm had a large market for its chemical sheet lead, and only the Pattinson process refined the pig lead to a sufficient purity to be used for this purpose, there was some justification for continuing with the process. The firm, however, had an enormous Pattinson capacity, three sets of pots, with twelve pots of twelve tons capacity in each set.

Shortly before this survey was undertaken the firm had been converted, on 1 March 1890, into a private limited company.[20] This was done purely to obtain the protection of limited liability, there being no attempt to bring in any outside money. The assets of the partnership were sold to the limited company for £118,325 and A.J. Foster (son of the founder) and the executors of James Foster held almost 6,000 of the 7,430 issued ordinary shares, nearly all of the remainder being held by Charles Frank Forster who had been made managing partner of the firm in 1877.

Cookson & Co.

In Chapter 2 we saw that the Cookson family had established an industrial presence on Tyneside in the early-eighteenth century. A century later their interests were widespread although they lay chiefly in glass manufacture. At this time Isaac Cookson III, great-grandson of the original Isaac who came to Tyneside, was beginning to withdraw from industry having purchased in 1832 the estate of Meldon Park in Northumberland. Shortly afterwards his partner, William Cuthbert, took the estate of Beaufront Castle, near Hexham. These might have been ideal conditions for the next generation to take over management but the glass industry was experiencing considerable change at this time and immediately after the abolition of the excise duty on glass Cookson & Cuthbert sold their works in 1846.[21] From the point of view of the two senior partners this was no doubt welcome, bringing a relief from responsibility together with cash assets to support their landed position. From the point of view of their sons, John and William Isaac Cookson and William Cuthbert II, however, it would have been less welcome. The proceeds of their shares in the partnership were unlikely to have provided a sufficient income on which to live. The sons, therefore, needed a way of making more money and had sufficient capital to enable them to finance new business undertakings.[22]

Of the three it was W.I. Cookson who was the most active. At the age of 20 he had been placed in Michael Faraday's laboratory for a period of about a year and scientific experiment, chiefly in chemistry and metallurgy but later also in electricity, remained an important interest. Already by 1844 he had begun business in chemical manufacture at Gateshead[23] and certainly from 1847 the firm, as W.I. Cookson & Co., was definitely making colours, especially Venetian red, and refining antimony. How and why the step to lead manufacture was made is unclear but it was commenced in 1851 when the three partners took a lease of land at Hayhole on the Tyne from the Duke of Northumberland. At this works the partners intended to manufacture white lead but it is clear that they intended lead manufacturing to be their major area of expansion, since they were shortly to purchase an additional works. This latter works, at Willington Quay (Howdon) on the Tyne, had been commenced in 1847 by Thomas Richardson and George Currie but was soon sold to William Hawthorn of the Tyneside firm of locomotive engineers. While, in the late 1840s, the Hawthorns may have had the money to diversify their interests, a slump in the demand for locomotives put them under financial pressure and led to the sale of the Howdon works.[24]

This was an opportunity for the Cooksons and Cuthbert to take an already established works within a mile of that at Hayhole. It was a large, integrated works, according to the sale particulars, with gas lighting throughout and rail sidings from the Newcastle and North Shields Railway to each of the manufacturing departments. The smelting department had two reverberatory and seven blast furnaces, while there were six calcining or improving furnaces and a desilverising house with three ranges of Pattinson pots capable of producing 120 tons of soft lead a week and 12 to 14 tons of enriched lead to be cupelled to obtain the silver. Finally, there were seven red lead furnaces and a steam engine to provide power. In addition there was land on which to expand.

On 31 July 1854 the leases of the Howdon works were assigned to the Cookson/Cuthbert partnership, which had within a few years become a potentially formidable competitor to the existing firms of Tyneside lead manufacturers. Nor was this the end of territorial expansion for on 1 March 1865 the partners surrendered the lease on the eastern part of the site, in exchange for a new one covering about an extra acre of land. Without any immediate use for the extra land the partnership was to sublet part of it but the remainder was used to rationalise the firm's geographical dispersion and the antimony and colour works was transferred from Gateshead to new buildings on the east Howdon site, a development completed in 1871. In addition, like many other firms, Cooksons provided accommodation for its more important workmen and 13 cottages were erected on Tyne Street, between the lead and antimony works.

In the meantime the Cooksons had been developing their manufacturing interests and from 1861 there are available profit figures to measure the success of this development, and these are given in Table 5.3. Unfortunately there are no separate figures for the Howdon lead works until 1870, but it is clear that the antimony and colour works was not particularly profitable[25] (although it was considered worthwhile to erect new works for these products at the end of the 1860s), while it was the lead works which made the real contributions to the Cooksons' prosperity. With interest on partners' capital assumed to be taken at 5 per cent p.a., the figures for capital employed show that the total investment in the lead business was small, not much more than one-tenth of that employed by Walkers, Parker in all their works and perhaps one-half of that employed by Locke, Blackett.

These profits were a useful addition to the inherited wealth and the income from the other industrial activities of W.I. Cookson (certainly the driving force behind the partnership, even though his brother, John, and William Cuthbert II held equal shares in the firm). From the late 1830s he had lived at 6 Eldon Square in the new Dobson/Grainger centre of Newcastle, but on his father's death in 1851, as his elder brother John moved to Meldon, he took over Benwell House, his brother's ex-residence. This was a pleasant, large house, with some acres of ground but it soon proved insufficiently grand for W.I., whose partners had both succeeded to their fathers' estates at Meldon and Beaufront. Retiring from active management in 1865, at the age of 53, he took the estates of Eslington Park, Yorkshire, followed by Denton Park, near Otley, and then finally moved to Worksop Manor, Nottinghamshire, a house which earlier in the century had belonged to the Duke of Norfolk. There he died on 1 November 1888, leaving a gross estate valued at £592,571.

It is probable that W.I. Cookson's partners had left him to be the dominant partner in the lead works from their commencement. John Cookson had taken over Meldon Park in 1851 and in 1873 owned 6,463 acres of land in Northumberland and it is certain that his time was devoted to landed matters. It is interesting to note that as the eldest son he was directed towards landed investment which was to prove less productive financially, even though W.I. Cookson may, in the 1850s, have felt that he was less fortunate than his brother. When John died, in 1892, he left £92,789, a considerable fortune, but one which pales into insignificance by comparison with that of his brother.[26] Similarly William Cuthbert II had taken over Beaufront Castle and its estates in 1852 and was presumably also much involved in landed affairs. W.I. Cookson's withdrawal from active business in 1865 would, therefore, have opened the way to new management. The Cooksons,

Table 5.3: W.I. Cookson & Co., Capital Employed and Profitability, 1861-73

Year	Lead Capital employed[a] (£)	Lead Interest on partners' capital (£)	Lead Profit/(Loss)[b] (£) Hayhole	Lead Profit/(Loss)[b] (£) Howdon	Lead Return on capital (%)	Antimony and Venetian red Capital employed[a] (£)	Antimony and Venetian red Interest on partners' capital (£)	Antimony and Venetian red Profit/(Loss)[b] (£)	Antimony and Venetian red Return on capital (%)
1861	39,280	1,964	2,184		5.6				
1862	41,160	2,058	2,297		5.6	30,940	1,547	941	3.0
1863	47,420	2,371	(3,930)		−8.3	31,940	1,597	1,971	6.2
1864	43,540	2,177	2,686		6.2	24,860	1,243	(555)	−2.2
1865	46,220	2,311	(341)		−0.7	24,300	1,215	889	3.7
1866	41,000	2,050	7,115		18.0	25,200	1,260	701	2.8
1867	48,120	2,406	3,549		7.4	25,900	1,295	(590)	−2.3
1868	54,660	2,733	4,413		8.1	25,300	1,265	4	0.0
1869	59,080	2,954	6,382		10.8	25,300	1,265	1,358	5.4
1870	72,320	3,616	7,820	(896)	9.6	22,020	1,101	4,182	19.0
1871	74,660	3,733	7,187	5,114	16.5	21,620	1,081	(1,387)	−6.4
1872	72,720	3,636	4,984	2,665	10.5	18,260	913	1,800	9.9
1873	72,740	3,637	630	5,580	8.5	18,280	914	720	3.9

Notes: a. These figures have been worked out on the assumption that the rate of interest on partners' capital was 5 per cent, the usual rate used in the period.   b. These figures are net of all deductions, including income tax.

however, adopted a slightly different technique to the Walkers, who tended to cling to active power while giving their sons partnership shares and training them to follow in their footsteps.

It was W.I. Cookson's sons who were brought in to take over the management when their father was ready to lay it down, since neither of John's sons took any interest in the business. His eldest son, William Bryan, having died at the age of 18 in 1859, W.I.'s second son, Norman Charles, who was also born in 1841, was educated at Harrow and brought into the firm in 1859. He was joined in 1863 by the third son, George John, who had been educated at Charterhouse but was also only 18 years of age when he joined the family firm. Although the brothers were only in their twenties they were largely responsible for the running of the business from the mid-1860s, but they were only salaried and did not receive partnerships. Although William Cuthbert II continued to draw a salary and obviously had some say in the management it seems likely that the developments which came from the late 1860s onwards, including the decision to transfer antimony and Venetian red manufacture from Gateshead to Howdon, were the result of the more dynamic management brought by the new blood.

It was not to be expected that N.C. and G.J. would for long tolerate doing all the work while the senior generation took most of the reward. On 19 October 1868 a memorandum was drawn up, which was aimed at giving the two brothers a more secure position, with a one-eighteenth share each in the partnership together with 5 per cent of the annual profits each in addition to their salaries of £300 p.a. each.[27] This was a satisfactory solution in the short term, since over the next four years the figure of 5 per cent on the net profit was worth an average of more than £400 p.a. each to N.C. and G.J., rather more than their salaries.

By 1872, however, the brothers were again feeling that their treatment was less than fair and N.C., in particular, was pressing for a new agreement. On 4 June a meeting of all five partners was held and N.C. and G.J. presented a memorandum of changes, of which the chief was that their partnership share should be doubled. While this was eventually agreed as from 1 January 1873, it is clear that relations between the partners were unstable.[28] The two juniors were anxious to obtain rewards commensurate with their responsibilities while the three seniors were reluctant to face changes. William Cuthbert II was anxious to protect the future position of his family in the partnership and it was his relationship with the junior partners which caused the major problem. On 30 November 1872 he wrote to 'My dear Willie' (W.I. Cookson) and commented 'I fear it is doomed that our families are not to act together for another generation as Norman has long determined to get rid of my supervision.' Temporarily the divisions were overcome by the increase in the junior partners' shares and the withdrawal of Cuthbert from all share in active management from 1 January 1873.

Early in 1876, however, the dispute broke out again with an angry but curiously plaintive letter from N.C. Cookson to his father. He complained about the level of recompense for the work he did and noted that 'in order to try and live economically I have given up carriage horses etc. as well as one of the three servants. We rarely have anyone to dine with us or stay with us and I have given up cigars and every other luxury I can.' Income was only part of the problem, however, and the letter goes on to show that Norman was angry that Cuthbert was trying to get his son, who had done little work in the lead business, a partnership. It would seem that N.C. and G.J. Cookson had decided that the time was right for a final showdown. They now wrote to Cuthbert requesting a salary increase and, on his refusal, threatened to resign. Clearly the business could not be run without them and negotiations now took place, chiefly between solicitors since Norman Cookson and William Cuthbert had passed beyond personal communications.[29]

Finally, on 6 April 1876, Cuthbert agreed to leave the partnership, selling his shares to his former partners at a valuation to be decided by two arbitrators, one chosen by each of the disputants. Those selected were James Foster (of FB & W) and W.W. Pattinson of the Felling Chemical Co. Their award valued Cuthbert's one-third share of the partnership at £32,600 and although there is no certain evidence as to how the additional share was divided among the Cooksons, a letter from W.I. to N.C. noted that John did not wish to increase his share and suggested that N.C. and G.J. took one-sixth each and gave W.I. a bill for the amount, on which they were to pay him 5 per cent interest p.a. Since, in addition, they were to receive increased salaries from the time of the dissolution of the partnership with Cuthbert, this was a very attractive settlement for the younger Cooksons and provided the basis from which they could go on to develop the business. Under their direction the firm was to rise to be the foremost lead manufacturing firm on the Tyne in terms of technology and at least the equal of Walkers, Parker or Locke, Blackett in terms of size.

It is the role played by Norman about which we know the most, since he was clearly of a dynamic and thrusting personality. Since he entered the firm at the age of 18 it is unlikely that he had any formal training in science and metallurgy but he clearly developed a considerable interest as a result of his practical involvement in the lead and antimony business and extended it to include electricity. He was responsible for a number of technical improvements in smelting of metals, lead manufacture and the design of secondary batteries and took out at least half a dozen patents for such improvements in the 1870s and 1880s. Perhaps more than any other partner in lead manufacture at the time he was technically competent in the business and capable of assessing the many new processes and developments which were being made. He had a considerable reputation both locally and nationally within the industry, wrote the lead section in the guide to north-east England published for the visit of the British Association to Newcastle in 1889 and, apart from A.O. Walker, was the only lead manufacturer to give evidence to the 1878 Royal Commission on Noxious Vapours.

One measure of the growth which was brought to the firm by the new management can be seen in the number of pages of invoice books used annually, rising from 1,289 in 1881 to 2,827 in 1893 and 5,272 in 1904. Similarly the number of staff on monthly salary rose from eight in 1875 to 16 in 1890, 28 in 1900 and 50 in 1908, while the monthly salary bill rose over the same period from £90 to £850. By the late 1890s the firm's employment was estimated at 400 to 500 with an annual wage bill of £30,000 to £40,000. Another measure is the introduction of new products and processes. In December 1876 a lead rolling mill was commenced at the Howdon works and was followed by the introduction of pipe manufacture in 1882. Meanwhile in 1874 the Pattinson process for desilverising lead had been replaced by the Rozan process. It adopted the same principle as the Pattinson process, that as melted lead cools the first crystals to form contain less silver than later ones and that when two-thirds of the original volume had crystallised it would contain only about one-half of the silver in the original. By the Rozan process, however, steam, or subsequently compressed air, was admitted by valves into the pots containing the molten lead and facilitated the separation of the liquid lead from the crystals whilst also oxidising trace metals such as antimony and arsenic. Norman Cookson claimed that, although the process required 'a large capital outlay' and 'a constant expense in repairs and renewals', there had been a considerable cost saving over the Pattinson process. It was no longer necessary to calcine the lead to remove impurities, while labour costs were one-fifth and fuel costs two-fifths of those required by the old process.[30] Mechanisation appears to have been a particular feature of developments at Cooksons from the 1870s. After a visit to the works the President of the Newcastle Chemical Society commented, 'The handling of large quantities of lead is done almost entirely by hydraulic machinery, which I have no doubt saves a large amount of money formerly spent on wages for hand labour.'[31]

The changes which came, however, were not only on the technical and production side; commercial matters were not neglected. A London office was opened in 1880, to improve marketing in the south of the country, and from the same date a series of branch offices was established in major towns, while overseas agents were appointed and some overseas offices, as in Paris, opened. A considerable sales network was thus established which enabled the firm to deepen and diversify its markets. As output expanded to meet this growing demand some rationalisation of product area became necessary. Although the dual production of lead and antimony was retained, in itself a highly unusual diversification, the manufacture of Venetian red was given up in 1890. It had never been a consistent profit earner and it would seem that in the cyclical upswing of the late 1880s the decision was taken to axe it and devote the resources to more promising sections of the firm's activities.

The final evidence for the expansion of the firm from the 1870s comes in the delivery figures presented in Table 5.4.[32]

Table 5.4: Cookson & Co., Deliveries of Lead Products, 1872-1910 (tons)

Year	Pig lead	Sheet	Pipe	Howdon Works total	Red lead	White lead[a]	Hayhole Works total[b]
1872	4,547	—	—	4,547	900	1,162	2,085
1875	2,971	—	—	2,971	1,470	1,087	2,587
1880	4,429	3,440	—	7,869	1,356	1,186	2,647
1885	?	3,490	555	?	?	?	?
1890	9,636	6,676	710	17,022	1,863	1,654	3,755
1895	11,936	6,863	927	19,726	1,867	3,857	5,891
1900	12,901	5,970	1,321	20,192	1,763	8,389	10,312
1905	16,886	4,154	1,354	22,394	2,016	9,974	12,126
1910	25,265	6,415		31,680	2,163	13,825	16,111

Notes: a. Includes both dry white lead and the dry white lead equivalent of the sales of paint (white lead ground in linseed oil).   b. This total includes small amounts of litharge and orange lead.

The early expansion came in the blue lead works at Howdon, with the decision to process, rather than merely sell, pig lead and the introduction of the rolling of sheets in 1876 and extrusion of pipe in 1882. In the 1870s the smelting capacity had increased, although from the early 1880s the flow of imported ore declined and was replaced by unrefined pig lead, which required an extension of desilverising capacity.[33] By the end of the 1880s it would seem that the output of pig lead from Howdon considerably exceeded the firm's manufacturing requirements and this may have turned thought again to the concept of expanding manufacturing rather than merely selling pig lead. Slowly from the late 1880s but with a sharp increase in 1893 and finally with a dramatic surge in 1899, the output of lead chemicals, particularly white lead, increased. The first stage undoubtedly involved the more intensive use of existing capacity in response to favourable trading conditions but new white lead stacks were built in the early 1890s and in 1898 Cooksons made the only successful nineteenth-century departure from the stack process when they installed plant to manufacture white lead by the chamber process.[34] In doing so they demonstrated the technical leadership which they had established in the industry in less than half a century and reinforced their position at a time when other firms were seriously affected by growing competition.

Although the chamber process was satisfactorily developed in Germany in the late-nineteenth century, the broad outline of the process had long been known. Indeed, so early as 1749 Sir James Creed had taken a patent (no. 651), which outlined the basic principles of the chamber process. However, a series of improvements over the next century had not brought it to satisfactory commercial production and Pulsifer wrote in 1888, that the attempts to work the process during the previous 50 years 'have universally resulted disastrously'.[35]

Broadly the chamber process could be defined as a controlled stack process. The chamber (at Cooksons) took the form of a brick building, approximately a 25 ft cube, containing square pillars of one foot section at five foot intervals. Brick and concrete arches connected the pillars, and these arches supported pieces of timber, over which were hung 'straps' of lead. The lead was cast in longer and thinner pieces than those used in the stack process, being approximately four to five feet long and five inches wide, with a weight of about four pounds. The aim here, as in all attempted improvements to the stack process, was to expose the maximum surface of the lead to the corrosive gases. Each chamber took 55 to 56 tons of lead at a time, much the same charge as would be placed in the stacks being built at the end of the nineteenth century. The distinctive feature of the chamber process was that temperature, vaporisation of the acid and the generation of carbon dioxide were under direct control.[36] A heating stove was used to produce and maintain the required temperature in the chamber, with six fires a day, every four hours. In addition acetic acid vapour, water vapour and carbon dioxide (the latter produced by burning coke in iron stoves) were introduced into the chamber through pipes.[37] As compared with the stack process the result of careful control was a much more even corrosion of the blue lead and no possibility of tan bark or hydrogen sulphide contaminating the white lead produced, while the corrosion level, at about 70 per cent, was higher. In addition the labour cost of filling and stripping a chamber was considerably less than for a stack[38] and, perhaps most crucially, the chamber process took considerably less time than the three-month corrosion period in the stack process. Initially Cooksons ran their chambers for 42 days but around 1909 to 1910 experiments were made in leaving them for 56 to 60 days and that was the usual period in the late 1930s.

Cooksons introduced the chamber process on a considerable scale, with more than 20 chambers, numbers 25 and 26 being added in 1905, and 27 and 28 in 1907. By 1900 output of chamber white lead was well over 4,000 tons p.a., at which level it exceeded stack output. Within a couple of years Cooksons had doubled their output of white lead and were to increase it still further over the next decade. With labour and other costs lower for the chamber process (in 1908 the works cost per ton was £2 18s 2d for chamber lead against £4 2s 7d for stack lead) it is not surprising that chamber output was more than 50 per cent greater than stack output in the years before the War. One might indeed have expected, given the cost differential, that stack production would have declined or ceased, but price was not the only factor.

Over the nineteenth century stack white lead had become a synonym for quality and customers in the paint trade had become so used to disappointment with the quality of alternatives that they were reluctant to purchase even satisfactory ones.[39] Locke, Blackett took advantage of this to advertise their white lead in a handbill, 'It is warranted absolutely genuine, free from all admixture of foreign or chamber white lead.' Even when chamber white lead was well established Cooksons still found difficulty in attracting customers away from the use of stack. An advertising booklet issued in 1923 stressed the advantages of chamber white lead, but, clearly recognising the conservatism of the market, concluded 'a most satisfactory compromise is sometimes effected by using a white lead of 50% Stack and 50% Chamber. Such a Lead seems to possess all the good points of each constituent, and few, if any of the bad ones.'

For the whole period in which Norman and George Cookson were developing the firm, we have figures for capital employed and profitability, as shown in Table 5.5.

Table 5.5: Cookson & Co., Capital Employed and Profitability, 1878-1905

Year	Interest on partners' capital (£)	Profit/(Loss) (£)	Year	Capital (£)	Profit/(Loss) (£)	Return on capital employed (%)
1878	5,590	(1,483)	18 months to 31/12/1889	90,000	37,400	27.7
1879	5,332	16,200	1890	166,000	26,530	16.0
1880	5,330	(325)	1891	172,000	37,991	22.1
1881	5,171	4,375	1892	183,000	36,405	19.9
1882	5,103	(3,548)	1893	164,000	25,609	15.7
1883	4,929	(3,580)	1894	161,000	18,210	11.3
1884	4,748	(1,014)	1895	164,000	27,075	16.6
1885	4,691	3,000	1896	182,000	17,363	9.5
18 months to			1897	174,000	9,592	5.5
30/6/1887	7,260	1,530	1898	175,000	22,834	13.1
30/6/1888[a]	4,921	11,700	1899	207,000	23,031	11.1
			1900	240,000	10,395	4.3
			1901	239,000	(45,001)	−18.8
			1902	182,000	2,369	1.3
			1903	177,000	11,886	6.7
			1904	182,000	(2,322)	−1.3
			1905	172,000	29,012	16.9

Note: a. The figures before and after 30 June 1888 are taken from different sources but the earlier series does continue until 1896 and shows a close correlation with the new series, which is preferred here because it contains figures for capital employed and the rate of return on capital. From 1889 the profit figures are struck before deducting interest on partners' capital but after deducting depreciation, bad debts, interest on loans, etc.

The most obvious feature of the table is the sharp distinction in profit-ability between the late 1870s and most of the 1880s with the period commencing with the boom of 1888 to 1890. In 1889 Norman Cookson commented that 'the last eight to ten years [have], to the trade in general, proved most unsatisfactory'[40] and it is clear from the table that he was speaking from personal experience. After the profitable years of the late 1860s and the boom of the early 1870s (see Table 5.3), Cooksons' profit figures give some support for the idea of a 'great depression' in the following 15 years. Between 1874 and 1877 the profit figures are incomplete, although there was a loss of more than £5,000 in 1874 and only a small profit in 1875. This would suggest that, with the exception of the unusually large and difficult to explain profit in 1879, there was an unprofitable uniformity about the period 1874 to 1887. It was certainly a period of falling prices which always made it difficult for the lead companies to make profits, since, by the time they were manufactured, their lead products were being sold at a time when their pig lead value was below that at which they had been purchased, often many months earlier. It will, however, be remembered (from Table 4.4) that Walkers, Parker had managed to remain profitable throughout this period and produce a respectable return on capital employed. Although Cooksons were an innovatory firm (especially in the 1870s) they do not appear to have been able to obtain the satis-factory increase in their market share in the depressed trading condi-tions, which would have justified their investment. By contrast, in the second period, commencing with the windfall gain brought by the rise in pig lead prices during 1888, considerable profits were made at a time when, as we shall see, many of the other lead companies were in deep financial difficulties. That this improvement continued after the 1888 to 1890 boom, in years which are often regarded as a continuation of the earlier depression, betokens a marked change in fortune. It is, how-ever, difficult to account convincingly for this change. The figures in Table 5.5 show a considerable increase in capital employed during the boom and the delivery figures in Table 5.4 show that it was possible to sell the increased output which was thus made available. To show what happened is not however to explain why it happened and the firm's dramatically changed competitiveness remains a mystery.

The profit performance of the later 1870s and 1880s may not, therefore, have given Norman Cookson the financial rewards which he expected from his new-found power with the withdrawal of William Cuthbert. Nevertheless it is certain that both he and his brother had a comfortable existence since they were involved in a number of other north-eastern industrial activities from which they obtained useful incomes. These included the Mickley, Cowpen and Wallsend and Hebburn colliery companies, while Norman was a considerable shareholder in and director of the Parsons Marine Steam Turbine Co. and director and subsequently chairman of both the Tyne General Ferry Co. and W.C. Gibson & Co. (later Adamsez of Scotswood). Norman Cookson, therefore, epitomised the family's long tradition of wide-ranging involvement in Tyneside industry. In addition he had a considerable involvement in scientific and social activities. He was president of the Northern Scientific Club in Newcastle, a member of the Linnaean Society and of the Royal Horticultural Society, of which he was a vice-president for some years. The latter involvement no doubt came from the fact that he had a national reputation for growing orchid hybrids.

It is unfortunate that we have no detailed evidence on Norman Cookson's income before 1889, the year in which the firm's profits revived from a long period of poor returns. It seems likely, however, that his other interests brought him the income to support those expenditures he was bewailing his inability to afford in the early 1870s. For a few years from 1889, however, his personal cash books have survived and give a detailed account of his income and expenditure. They show an already well-developed and expensive lifestyle, which the high profits of the time were more than able to support. Table 5.6 gives a breakdown of Norman's receipts from the firm of Cookson & Co.

Table 5.6: N.C. Cookson, Income, 1889-94[41]

Year	Salary (£)	Interest on capital (£)	Dividend (£)	Profit (£)	Total income (£)
1889	1,000	3,600	—	2,000	6,600
1890	1,200	5,200	8,700	8,600	23,700
1891	1,200	5,300	4,800	400	11,700
1892	1,200	5,300	22,000	13,500	42,000
1893	1,200	5,000	5,000	5,000	16,200
1894	1,200	5,100	3,000	2,000	11,300

In addition to his income from the lead company he received dividends from many investments — in 1889, for instance, the monthly dividend from the Mickley Coal Co. Ltd was either £300 or £450 and in 1890 it rose to £675 or £900 monthly, although in the subsequent depression income from this source fell to only about £2,000 p.a. in 1893 and 1894.

In the first couple of years of this period a significant proportion of

Norman's expenditure went on extensions and improvements to and furnishings for his house, Oakwood, Wylam, to which he had moved in 1879. In 1889 about £3,000 was so expended, including more than £500 on glasshouses for the growing of orchids and the following year nearly £4,000 was expended on extensions to Oakwood together with a further sum of £1,200 on plant and wiring to provide electric lighting. From this time onwards the building was in a state to be suitably furnished and more than £1,000 a year was expended on carpets, pictures and furniture, including in 1892 some £1,300 on carpets purchased during a visit by Norman and his wife to Egypt. Not only did they have Oakwood but they also leased a house in the country, for use during the summer months. When the accounts begin, in 1889, this was an unspecified house at Otterburn in Northumberland, but between 1891 and 1893 Castle Leod was leased from Lord Cromarty at a rent of £1,400 p.a. Finally, in 1894, the last year covered by the accounts, Bannoch Lodge was leased from Sir Robert Menzies for £2,000. The taking of houses such as these for a full year, even though they were only used by the Cooksons for July and August, involved considerable running expenses, including rates and wages for staff (around £75 per month in the summer and £30 in the winter). In 1892 the total of expenses incurred on the Castle Leod account was about £3,500. Other figures from the accounts show that Norman Cookson used his considerable income to support a very luxurious lifestyle. The expenditure of several hundred pounds on orchids was not uncommon, while at least that much would be spent annually on cigars and up to £1,000 p.a. on drink (with entries such as 6 July 1891 'champagne & claret £661' and 3 January 1893 'champagne etc. £600'). For the six years for which the account books are extant Norman's annual expenditure ranged from £11,287 to £16,592, but in spite of this very considerable level of outgoings, he recorded massive annual excesses of income over expenditure which show that he saved 57 per cent of the total income received in the six-year period.

As might be expected with income of this order, public school and Oxbridge became the norm for Cookson offspring, who followed this traditionally expected pattern of wealthy late-nineteenth century industrialists' families. It is, however, to be noted that the frequently drawn conclusion, that public school and university education drew potential industrialists away from involvement in industry, which came to be seen as rather unpalatable, and directed them towards the professions,[42] is no more true of the Cooksons than it was of the Walkers. Inevitably some members of the family found the academic life congenial, although others came back to become directors of Cooksons. Moreover their experience in the ancient universities was to induce them to begin, slowly in the period before the First World War but more rapidly thereafter, to bring non-Cookson family graduates into the firm's management and eventually onto the Board. If elite education did not dispossess the firm of its potential future leaders neither does Cooksons give any support to that other favourite canard, the decline of the third-generation firm. We shall see that Cooksons went from strength to strength under the direction of Norman's son, Clive, who was to become the first chairman of ALM, while one of Norman's grandsons, R.A. Cookson, who, unlike his father, did go into the family firm, despite the fact that his education followed that of his father at Harrow and Magdalen College, Oxford, also became chairman of ALM and, between 1962 and 1973, chairman of its holding company, Lead Industries Group.

That is, however, to take the story far beyond the profitability of Cooksons in the late-nineteenth and early-twentieth centuries. In 1905 the firm was turned into a limited company, although shares remained only in the hands of the Cookson family. Although Norman Cookson became managing director, effective control was to shift to his son Clive, even before Norman's death in 1909. The future lay with a new generation, a new organisational structure and with amalgamation rather than successful competition.

London

By the end of the eighteenth century London had developed as the most important lead manufacturing centre in the country, as a result of its easy access to fuel and raw materials by coastal shipping and the fact that it offered by far the largest market for manufactured lead products. Little is known about the early-established firms, although, from Fishwick's correspondence in the 1780s with his London partner, it would seem that Lancaster & Co., with works in Southwark, was the dominant firm, especially in the white lead trade. By the end of the century, however, the firm was in decline, two of the partners having died and competition from new firms having become noticeable. As on Tyneside, therefore, it was the newly created firms of the late-eighteenth century which came to dominance and such older established firms as there may have been which declined.

John Locke & Co.

We have already seen, in the discussion of the foundation of Locke, Blackett & Co., that John Locke had commenced as a lead merchant in London in 1795. In 1798 John Locke had six-tenths of the capital and John Griffiths the remaining four in a firm which was doing very well since the profit for the year to 30 June 1798 was £2,218 7s 4d and in the following two years rose to £4,400 and £5,600. The firm was both buying and selling pig lead and manufactured lead goods and had developed a considerable overseas trade, particularly with the East India Company, whose contracts appear often to have been shared with Walkers, Maltby & Co. In the early part of the nineteenth century there is no evidence that the firm had any manufacturing premises and it seems to have been satisfied with selling the lead products of Locke, Blackett, a firm with which it had an interlocking partnership. Although the Lockes continued to have a financial interest in the London firm, as with Locke, Blackett, they had little direct involvement and about 1818 William James Lancaster[43] was brought into the firm. In the following decades direction of the firm moved strongly into the hands of the Lancasters, reflected in a change of name in 1854 to Locke, Lancaster & Co., and in 1888 the Locke interest ceased with the death of J.A. Locke.

Evidence on this period is sparse but by the 1860s Locke, Lancaster were undoubtedly manufacturing as well as merchanting lead products, although it seems certain that their interests lay entirely in blue lead and that there was no production of lead chemicals. The only contemporary material comes in a sometimes acrimonious correspondence between the London and Newcastle firms, in the period 1869 to 1872, with letters passing between James Leathart and Samuel Lancaster when the negotiations were going smoothly and between Locke, Blackett & Co. and Locke, Lancaster in more stormy periods. In between these exchanges are letters from J.A. Locke, a director of both companies but whose contacts and sympathies seemed to lie with the Newcastle firm, to Samuel Lancaster in an attempt to act as mediator. The problem clearly stemmed from the long history of mutual links between the two firms, which had gradually become strained. In particular Locke, Lancaster, which had originally sold the Newcastle firm's pig lead, had undertaken its own smelting and desilverising operations and had sold its pig lead with the brand 'Locke, Blackett', which, of course, its customers had come to expect over the years. This antagonised the Newcastle firm which in its turn was criticised for supplying pig lead to other firms on more favourable terms than it charged Locke, Lancaster.

By the summer of 1872 matters had reached the point where J.A. Locke felt constrained to write to Locke, Lancaster:

> As the oldest member of either of our two Firms, springing originally from the same source, and connected as they have been for so many years both by family ties and mutual business interests, I naturally look upon the possibility of any severance of these ties with a considerable amount of pain.

Locke, Blackett had apparently threatened legal action over the continued use of its name on the pig lead made by the London firm. Through the good offices of J.A. Locke, the firms eventually retreated from the abyss of the law courts and in December 1872 signed an agreement which enabled the London firm to continue to use the disputed mark, while the Newcastle firm was to receive some financial recompense according to the extent of the London deliveries. It is, however, clear that a division had developed between the London and Newcastle firms since the death of John Locke in 1816 and it seems certain that nothing but a business relationship remained after the death in 1888 of the only remaining partner in both firms, J.A. Locke.

In 1871 Locke, Lancaster took the lease of premises in Bridge Road, Millwall, where they smelted lead ores, desilverised them, and produced manufactured lead. By 1886 the works had two blast furnaces and a third was subsequently added, the latter, according to H.C. Lancaster, the first mechanically-charged lead blast furnace in the country, adopting the 'bell and cone' principle which was common in iron smelting.[44] Output was considerable with a peak of 30,000 tons of bullion lead being treated a year and more than 100,000 ounces (about 3 tons) of silver being produced in one week by the firm's three cupellation furnaces. Such a works, unlike the white lead ones, was not labour intensive, and total employment, including management and office staff, was about 50 in the late-nineteenth century.

Among the large firms Locke, Lancaster was unusual in not having diversified from blue lead. Initially, in the early-nineteenth century with lead chemicals available from the Newcastle firm, the decision not to manufacture in London is understandable. Later, as the strength of the Newcastle link declined, it becomes more surprising, although it may be by the second half of the century that growing London competition made such a move unattractive. Perhaps also the relatively small numbers of the Lancaster family who went into the business may have limited expansion. From W.J. Lancaster and his brother, Samuel, in the 1850s and 1860s, control passed to John Locke Lancaster (one of twin sons of W.J., born in 1833) and his brother, Arthur Henry Lancaster born in 1841. They retained control until in 1894 the firm was amalgamated with another London company, which had strong interests in lead chemicals. Although there are no profit figures for the firm it is clear that the Lancasters did well out of it. W.J. left nearly £100,000 in 1866 and his brother £107,000 in 1883.

W.W. & R. Johnson & Sons

William Whittle Johnson was a painter and glazier in Ratcliff, east London, in the 1790s. Both of his activities brought him into contact with lead and he purchased both pig lead (for glazing bars) and white lead (for paint) from Walkers, Maltby & Co.[45] William Whittle I died in 1820 and the firm was carried on by his sons, William Whittle II (1793-1855) and Robert (1795-1863), who began to initiate changes which were to give it a large base in the London lead manufacturing industry. Plumbing was added to the partners' other activities, to which it was obviously closely connected, and led to a demand for increasing quantities of pig lead which the partners would manufacture into pipes at their own premises. From 24 June 1824 the partners leased numbers 4 and 5 Waterloo Place, Commercial Road, Limehouse, 'two substantially erected residences' and 'large manufacturing premises adjoining', where they were to begin to manufacture their own lead pipe. Probably at the time that the Commercial Road premises were taken the firm was commencing the major shift from being tradesmen into manufacturers supplying products for tradesmen and the original activities of painting and glazing were gradually being dropped.

About 1836 the firm took another step which placed it very firmly in the lead industry, with the taking of premises at 306 Burdett Road, to be known as Victoria Lead Works, with access to the Limehouse Cut which enabled lead to be brought into the works by barge from the Thames. Here stacks were erected and the manufacture of white lead commenced and subsequently, in 1860, the site was extended and further stacks built. As painters the Johnsons had obviously needed to purchase white lead as the base for the paints they needed and entry into white lead manufacture may be seen as an example of backward integration. However, it seems more likely that it was part of a deliberate move towards concentration on lead manufacture, which soon dominated the firm's activities. It became necessary to extend the Commercial

Road premises and further land was leased linking the site to Salmons Lane to the east, an apparently profitable expansion which was followed by the purchase of the freehold of the site in 1847. Rising profitability also enabled a rise in social position and Robert Johnson, for instance, was enabled to move away from east London to Hope House, which dominates the village of Little Burstead, Essex.

It is difficult to be precise about the size of the Johnson firm but the evidence suggests that it was probably smaller than the major Tyneside ones. An isolated figure giving a total wages bill of £91 for one week in August 1862 suggests an employment figure of around 100. There are no output figures but isolated references which suggest that there were eight white lead stacks and that the usual rate of filling them was one height a day, would suggest an output of about 1,200 tons of white lead per annum. It is probable that only white lead and pipe were produced in the early stages of the firm's lead activities but in 1862 a recently installed furnace was being used to smelt the returns from the stacks — although there is no evidence that the firm was doing primary smelting. Also in 1862 letters between the partners contain the first mention of rollings, which may suggest that a rolling mill had been installed at the Commercial Road premises for the production of sheet lead. By the late 1880s a further major development had been completed. This involved the conversion of the Commercial Road premises to the rolling of lead foil for the lining of tea-cases. As a result Johnsons developed a considerable stake in tea lead, the demand for which was rapidly developing as consumption of tea expanded as its price fell.

The firm could also lay claim to a brief period of notoriety as a result of one of Dickens' London perambulations. This had taken him into the east end where he had been told of the effects of lead poisoning on the women who worked at the lead mills. This was enough for Dickens to criticise the conditions in the lead works, which he implied was near Limehouse Church.[46] Since their Burdett Road works was close to St Ann's Church, Limehouse, the Johnsons felt aggrieved, wrote to Dickens stating that they took great pains to prevent their employees from contracting lead poisoning, and invited him to inspect the works. This he did and swiftly published an account of his visit which amounted to a virtual retraction of his earlier criticism.[47]

But I made it out to be indubitable that the owners of these lead mills honestly and sedulously try to reduce the dangers of the occupation to the lowest point. A washing place is provided for the women (I thought there might have been more towels) and a room in which they hang their clothes, and take their meals, and where they have a good fire-range and fire, and a female attendant to help them, and to watch that they do not neglect the cleansing of their hands before touching their food. An experienced medical attendant is provided for them, and any premonitory symptoms of lead poisoning are carefully treated. Their tea-pots and such things were set out on tables ready for their afternoon-meal, when I saw their room, and it had a homely look . . .

American inventiveness[48] would seem to indicate that before very long White Lead may be made entirely by machinery. The sooner the better. In the mean time, I parted from my two frank conductors [Robert II and Matthew Warton Johnson] over the mills, by telling them that they had nothing there to be concealed, and nothing to be blamed for.

The Johnsons had easily evaded the potential criticism and the firm was set in expansionary mood. The figures given in Table 5.7 show that from the mid-1860s there was a considerable increase in capital employed, while profitability[49] remained consistently very high until the figures cease in 1885, in contrast to the evidence we have seen for other firms of a much lower average profit with considerable fluctuations to include some losses in this period.

The evidence would suggest that the fortunes of the Johnson family rose considerably in the second half of the nineteenth century. From a capitalisation of only £11,000 in 1863, the firm was employing some £50,000 at the time of the withdrawal of William Whittle III in 1878 and the partners were withdrawing £3,000 to £4,000 p.a. in addition probably to a salary and interest on capital. It is not surprising that Robert Baines, born April 1862, son of Robert II, was educated at Oxford and that while the latter left £36,000 when he died in 1907, his son, dying at the Hope House in 1927 left £80,000. Since he returned from Oxford University to work for the firm, and subsequently become a director and then chairman from 1910 until his death, he provides further evidence that the educational system did not necessarily seduce the sons of successful industrialists away from industry. Other members of the family also made comfortable fortunes out of the business, Matthew Warton for instance leaving £65,000 on his death at Hove in 1914. The firm was not large (by comparison with Walkers, Parker and Cooksons) but it was successful – a success which it is difficult to attribute to any specific cause, in the lack of detailed evidence, although its involvement in the manufacture of tea-lead does differentiate it from

Table 5.7: W.W. & R. Johnson & Sons, Capital Employed and Profitability, 1863-85[50]

Year	Partners' capital (£) William Whittle III	Partners' capital (£) Robert II	Partners' capital (£) Matthew Warton	Total capital employed (£)	Profit (£)	Return on capital employed (%)	Partners' drawings (£) W.W.	Partners' drawings (£) R.	Partners' drawings (£) M.W.
1863	4,407	4,407	2,203	11,017	1,157	10.5	420	589	191
1864	4,434	4,266	2,274	10,974	1,647	15.0	500	737	509
1865	4,568	4,154	2,153	10,874	3,053	28.0	600	714	440
1866	5,137	4,588	2,448	12,174	4,089	33.5	632	764	486
1867	6,067	5,359	2,955	14,381	3,246	22.5	775	801	553
1868	6,543	5,773	3,182	15,498	5,725	37.0	778	865	527
1869	7,949	7,053	4,051	19,053	7,644	40.1	900	1,270	662
1870	9,955	8,645	5,265	23,866	8,081	33.8	1,200	2,380	791
1871	11,836	9,281	6,459	27,576	9,482	34.3	1,280	2,210	1,155
1872	14,187	10,574	7,653	32,414	12,798	39.4	1,570	2,501	1,267
1873	17,518	12,793	9,563	39,874	13,041	32.7	1,900	2,719	1,341
1874	20,637	14,856	11,461	46,954	9,301	19.8	2,400	2,950	1,372
1875	21,876	15,257	12,400	49,534	7,839	15.8	3,500	3,500	645
1876	26,480	14,531	13,716	49,728	14,455	29.0	6,050	4,150	1,750
1877	20,993	15,595	15,644	52,233	9,084	17.3	6,000	3,800	2,000
1878	18,430	15,027	16,061	49,518	7,100	14.3	6,700	4,150	2,100
1879	Retired	13,400	10,050	23,450	13,521	57.6		3,600	2,700
1880	from	17,526	13,145	30,671	5,315	17.3		4,100	3,075
1881	partnership	16,463	12,347	28,811	6,045	19.7		4,200	2,950
1882		15,718	11,988	27,706	6,282	22.6		3,600	2,900
1883		15,702	11,786	27,488	6,877	25.0		4,200	2,600
1884		15,431	12,134	27,565	8,698	31.5		5,100	3,825
1885		15,285	12,053	27,338	10,933	40.5		4,000	2,600
Average				28,680	7,627	26.6			

other firms. It does, however, seem likely that, in the years immediately after the last available figures for 1885, competition was becoming more noticeable and in 1892 Johnsons followed the trend in the lead industry and converted to limited company status, although for the moment it was to remain a family company.

In addition to Walkers, Parker, the Johnsons and Locke, Lancaster, there was a number of lead manufacturing companies active in London in the later-nineteenth century, although none of them was so large as the three mentioned and none was subsequently to be taken into ALM. They included sheet and pipe manufacturers, such as Grey & Marten, and white lead manufacturers such as T. & W. Farmiloe.

Other Areas

While by the second half of the nineteenth century the Tyne and London dominated the production of manufactured lead in the country there were works in other areas. In the discussion of Walkers, Parker we have seen that firm's significant presence in north Wales, Chester and Glasgow, while the dominant firm in Bristol, Sheldon, Bush, was discussed in passing in Chapter 3. There remains only one English firm, Cox Bros. & Co. of Derby, of any national significance to be discussed in order to complete this survey of the major nineteenth-century lead-manufacturing partnerships. In addition there were two Scottish firms which were subsequently to be drawn into ALM.

Cox & Co.

We have already seen, in Chapter 3, that the Cox family commenced a white lead works in Derby in 1806 and (in Chapter 4) that the firm took over Walkers, Parker's Derby works in 1839. There is no clear evidence as to when the firm commenced in the lead industry but, although tradition within the firm held that it was started in Nottingham in 1781,[51] it is likely that its lead manufacturing interests commenced in 1806. Subsequently in 1816 it was to purchase land in Nottingham which was to be used as a warehouse for distributing the firm's products in that town but it also became a manufacturing centre, with fixed capital of £12,000 invested there by the end of 1818, which included at least six white lead stacks. By this time the firm was operating on a considerable scale with the value of the partners' holdings totalling £93,000 after deducting the firm's debts. A subsequent statement at the end of 1825 shows that the capital had increased, with the working capital being stated as £83,400 after the deduction of £5,035 for interest on 'the whole of the Capital employ'd'. Assuming that interest was charged at 5 per cent this would imply total capital employed of £100,700. Figures for working capital in the period 1828 to 1830 give a range from £92,000 to £97,000, which suggests further expansion. As with Walkers, Parker in the same period, however, profits were low, £6,000 in 1826, £6,280 in 1827, £4,860 in 1828 and a loss of £1,970 in 1829.

Although these figures are less bad than those for Walkers, Parker's Derby works they are hardly satisfactory returns and it can only be assumed that profitability of Cox Bros & Co. (as the firm was known by the late 1830s) improved to justify the purchase of the Mill Hill works from Walkers, Parker in 1839. Other than that the purchase took place there is no detail as to the financial transaction, nor evidence as to the reasons for the transaction on either side. Indeed it is not until the 1880s that there is any further evidence on the firm.

By that time it appears at best to have been in a state of stagnation with total capital employed, at £92,000, less than it had been earlier in the century. This may have been a result of the fact that the interests of the senior partner, William Thomas Cox, had been in political and social affairs.[52] In any event the firm certainly did not prosper when it came into the hands of his children in 1882. It is almost unbelievable, but true, that after the initial balance sheet drawn up for the new partnership in 1882, the next one was not drawn up until 1894. There was of course nothing to compel a partnership to produce annual accounts or balance sheets but it is a fascinating reflection on the way in which the firm was run that for 12½ years all that appears to have been done was at six-monthly intervals to add interest at 5 per cent, on the partners' respective capitals at the beginning of the period, to those capitals. As a result the partners' capital rose from £59,000 to £86,000 but this was a totally nominal figure and was seen to be false when, for some reason, a set of accounts and balance sheet were drawn up in 1894. The result was the discovery that not only did the assets not cover the liability created by the credit of £43,000 of interest to the partners' accounts but there was a further loss of £3,000, making £46,000 in all which had to be written off the partners' capital. The firm had obviously been trading at a consistent loss and had been returning rather worse than −5 per cent on capital.

Not surprisingly, from the end of 1894 a regular system of audited accounts and balance sheets was introduced, and these suggest that changes in management were also introduced. While profits were small in 1895 and 1896 and there were losses in 1901 and 1908, which might be put down to falls in the price of pig lead, most years up to 1914 saw a return of about 10 per cent on capital employed.[53]

Alexander, Fergusson & Co., Glasgow

This firm was founded in 1854 by Alexander A. Fergusson and Henry Alexander.[54] Fergusson, who was related to the Tennants, had studied at Glasgow University before beginning work in Charles Tennant & Co.'s St Rollox chemical works. After ten years he left the firm and undertook further study of chemistry in Germany, before branching out into lead manufacture with a friend in premises in McAlpine Street, Glasgow. Initially the firm produced sheet and pipe and some paint but it was obviously not very successful and must have come close to closure, since the works was offered to Walkers, Parker in 1857 but refused.[55] This may have been a temporary problem brought on by shortage of credit, since the 1857 financial crisis hit the area severely with the closure of the Western Bank of Scotland with its 101 branches and the temporary suspension of payments by the City of Glasgow Bank. The storm was weathered, however, and the development of local engineering and shipbuilding industries provided increasing demand for lead pigments. Expansion to meet this demand faced problems of space at McAlpine Street and in 1874, when the manufacture of white lead was commenced and the paint trade was developed in a more extensive way, additional premises were taken at Ruchill, outside the city, on the Monkland Canal, a branch of the Forth and Clyde Canal. As with so many other of the works already discussed, this enabled pig lead to be brought and finished products to be moved out by water.

The firm appears to have taken advantage of the limited number of Scottish lead producers to build up a strong local reputation but it also established considerable overseas markets in the late-nineteenth century, especially in the Empire, with subsidiary companies set up in Canada, Australia and South Africa, and also in South America. Overseas markets were subsequently retained by regular visits from senior members of the firm in order to obtain sales. While 1909, when Australia, Egypt, India, New Zealand and South Africa were visited, was exceptional, at least one non-European country was visited a year. It would seem likely that the firm had expanded into foreign markets as being easier than challenging Newcastle manufacturers, with whom there was a price agreement, for a larger share of the Scottish market. Before an 1893 committee Fergusson was questioned and answered as follows:

Do Newcastle men compete with you – can they? – Oh, they do certainly.

On the Clyde? – Well, I should not say they can compete; they regulate our prices. We say our price is the same that they can put it here. If we began to fight I venture to say we could make it that they could not sell here; but we do not make enough for all the requirements, and far more comes from Newcastle of white lead than we supply from Glasgow.[56]

By the late-nineteenth century certainly, but probably earlier, the firm had entered lead smelting and was refining its pig lead by the Pattinson process. At this time the white lead works consisted of twelve stacks with an annual output of rather more than 1,000 tons. It was also in the van of progress since mechanical hoists were used 'by which the corroded lead is lifted from the stacks to the floor beneath', while the dishes of white lead pulp were taken to the drying stoves on a mechanical conveyor 'consisting of an endless chain provided with discs, on which the dishes are carried'.[57] The firm was also the first of the lead manufacturers to develop forward integration, manufacturing ready mixed paints (rather than selling white lead to be made up into paint).

T.B. Campbell & Co.

In 1830 T.B. Campbell set up works in Edinburgh as a pipe maker and metal merchant and in 1849 added a rolling mill for the manufacture of sheet lead. In 1869 a branch was opened in Glasgow at 29 Wellington Street, as a warehouse and saleroom for production from Edinburgh but further opportunity for expansion occurred in 1880, when the Clyde Lead Works, 126 Cornwall Street, Glasgow, was purchased. Campbells remained a blue lead company, producing sheet and pipe, was converted into a private limited company in 1901 and was taken over by Alexander, Fergusson & Co. Ltd and ALM in 1929.

By and large partnership structure had proved adequate for the lead companies for most of the nineteenth century. It had proved possible for those which were well run, such as Cooksons and Walkers, Parker, to raise sufficient capital for expansion internally, without any necessity to go to the market. In general also there had been no shortage of good management as a result of the continuation of partnership organisation and we shall certainly see that the performance of public limited companies was worse than that of the private companies which remained in the hands of the old partners.

Notes

1. The source of what little is known about John Locke is E.C. Fâche, *Lancaster & Locke. A Pedigree* (n.d.), pp. 11-12. I am grateful to Mrs D. Lancaster for drawing my attention to this work and sending me a photocopy.

2. In September 1795 Locke went into partnership with Mrs Joannah Freeman, whose husband Samuel, before his death in May 1795, had been one of the two major London lead merchants (the other was Thomas Preston) who purchased pig lead from the Blacketts of Tyneside. At the end of June 1796 Mrs Freeman left the partnership which was subsequently carried on as John Locke & Co. See NRO, Blackett (Wylam) Mss, Letter book 1794-6, Christopher Blackett to Freeman, Locke & Co., 28 Sept. 1795 and Christopher Blackett to John Locke, 9 July 1796.

3. Rhodes, London Lead Company in North Wales, p. 168.

4. NRO, Blackett (Wylam) Mss, Letter book 1794-6, Christopher Blackett to Freeman, Locke & Co., 28 Sept. and 3 Oct. 1795.

5. Ibid., Letter book 1796-8, Christopher Blackett to John Locke, 29 Oct. 1798.

6. Ibid., Letter book 1798-1800, same to the same, 26 Dec. 1798 and 1 Jan. 1799. Subsequent letters (in the autumn of 1799) show Blackett to have been less pleased when he discovered the supply of manure to be erratic since the barracks was sometimes largely empty of horses as troops were sent to other places. The manure was to be used for the stack white lead process since it was obviously either not possible or too expensive to obtain a licence from Fishwick to use his bark process – in which Blackett was clearly interested since he got Locke to check at the Patent Office for the date at which the patent protection ceased.

7. Ibid., same to the same, 23 Jan. 1799.

8. The chief source on Leathart is *Paintings and Drawings from the Leathart Collection 7th October – 18th November 1968* (Laing Art Gallery, Newcastle upon Tyne, 1968), introduction by Dr G.L. Leathart. I am grateful to Dr Leathart for the opportunity to consult and permission to quote from some letters to James Leathart, now in his possession.

9. This is probably too modern a title to have been applied in the early-nineteenth century. While the more important colliery viewers might deserve such a title, it seems likely that John Leathart was a working lead miner, with some ability, who did some consultancy work for mine owners. He may well have been the same man as the John Leithart of Alston Moor, described as a mine agent who published *Practical Observations . . . on Mineral Veins* in 1838.

10. *BPP*, 1843 [432] XV, Royal Commission on the Employment of Children, Appendix to the Second Report of the Commissioners, Trades and Manufactures, Part II, Report and Evidence of Sub-Commissioner J.R. Leifchild, p. 144. Since these figures were collected in 1842, in the trough of a severe cyclical downswing, they are likely to underestimate the average level of employment.

11. *Penny Magazine*, supplement Aug. 1844, pp. 337-42.

12. By 1859 employment is said to have reached 150 at St Anthony's when the works was described in W. White, *Northumberland and the Border* (1859), pp. 101-10.

13. Newcastle Central Library, Seymour Bell Collection, Portfolio 10 (Gallowgate).

14. *Athenaeum*, no. 2394, 13 Sept. 1873, p. 342.

15. It may not merely have been James Leathart's judgement, since after his death the firm had another unsuccessful venture into a new lead refining process. In March 1905 the firm agreed to take a licence from A.G. Betts to instal his

electrolytic process. This was certainly the earliest British adoption of the process and was undertaken at much the same time as it was adopted in America, at Trail, Canada and Grasselli, Indiana (I owe this reference to Dr J.K. Almond). Unlike the two American works which managed to survive the teething troubles, Locke, Blackett's licence was cancelled in 1908 in an acrimonious atmosphere. Undaunted the firm sent a representative to Spain in 1923 to investigate another electrolytic process. Perhaps fortunately his favourable report was shelved as Locke, Blackett was soon to be brought into the ALM merger.

16. J.A. Locke had died in 1888, leaving an estate of £21,000 gross. His death and the need to pay the value of his shares to his executors may have been a factor in the decision to sell the partnership to a limited company.

17. Although Robert Blackett became a partner in the firm it is possible that he was partly financed by Mrs Sarah Blackett, who in May 1863 sold her shareholding in Locke, Blackett. The two were presumably related and lived close to each other, he in Framlington Place and she in North Terrace, Newcastle. Blackett died in 1874 and his one-third share was purchased by James Foster for £48,500, which, compared with the original total capital of £36,000, shows the growth of the firm over little more than a decade. In 1877, when one-twelfth share was sold by Foster to C.F. Forster, the firm was valued at £120,000.

18. Wilson was a partner in the Tyneside chemical manufacturers, H.L. Pattinson & Co.

19. *The Lead Industries of the Tyne* (reprinted from the *Chemical Trade Journal*, 19 Jan. 1895), p. 4. At least two witnesses (J.S. MacArthur of Glasgow and Edward Dixon, managing director of Blundell, Spence & Co., colour manufacturers of Hull) to the lead committee of 1893, stated that Foster, Blackett made the best white lead. *BPP*, 1893-4 [c.7239-I] XVII.

20. PRO, BT31 4711/31057.

21. For the detail of these developments see C. Ross, The Development of the Glass Industry on the Rivers Tyne and Wear, 1700-1900 (University of Newcastle upon Tyne, PhD, 1982).

22. Interlocking marriages were a factor keeping them together. William Isaac Cookson had married Jane Anne, sister of William Cuthbert II, in 1839, while Cuthbert had married Mary, sister of John and William Isaac Cookson, in 1840.

23. There is no totally reliable evidence on the commencement of the business. This account is based on a typescript record of the firm, produced by an employee, probably in the 1920s. Dept of Palaeography, University of Durham, Cookson, Box III, 48 (hereinafter cited as 'Cookson', Box . . .).

24. For the Hawthorns see J.F. Clarke, *Power on Land and Sea . . . A History of R. & W. Hawthorn Leslie & Co. Ltd* (Hebburn, n.d.).

25. The following figures, for the years in which the profits (losses) on Venetian red and antimony were given separately, show that it was the former which was the more unprofitable:

Year	Venetian red (£)	Antimony (£)	Year	Venetian red (£)	Antimony (£)
1861	(286)	(21)	1870	(2,092)	6,275
1864	(989)	433	1871	177	(1,564)
1866	713	(12)	1872	(1,636)	3,435
1867	(1,004)	414	1873	(881)	1,601
1868	(82)	86	1874	904	(3,667)
1869	(1,957)	3,315	1875	1,105	2,312

These figures make the decision to continue with the production of Venetian red at Howdon, after transfer from Gateshead, rather strange. The only obvious explanation is that the interlinkages between antimony and Venetian red, the production of the latter being dependent on by-products of the smelting of the former which would be otherwise of little value, made it worth carrying them on in conjunction. The Venetian red profit figures may, therefore, have been struck after charges which would otherwise have been placed against the antimony account.

26. The comparison may not be entirely just since estates at death are to some extent dependent on consumption and generosity during life. Before 1894, however, there was very little reason for tax purposes to transfer assets to descendants before death. More importantly the value of landed estate was not included in probate before 1898. The annual rent of the Meldon estate in 1873 was £6,506, which (at 3 per cent, the usual figure for the return on agricultural land) implies a capital value of about £200,000. Since one would expect this value to have fallen during the agricultural depression of the 1870s and 1880s it seems that John Cookson's total assets at death were significantly less than those of his brother. Nevertheless the long run distinction has stemmed from the two brothers. John's descendants have remained landed proprietors and still live at Meldon Park. Their gross estates at death have been less than £100,000 (although in the twentieth century avoidance of death duties has made such figures of little meaning) and they have regarded the industrial branch of the family as having made the correct financial decision. The descendants of W.I. Cookson have remained in the lead business and the last, Roland Cookson, remained on the board of Lead Industries Group Ltd until 1980. Their estates have been valued at more than £100,000 and they have regarded the agricultural branch of the family as having made the correct financial decision. In the second half of the twentieth century with land prices rising, this latter view is, probably for the first time, correct. However, owners of land will recognise that millionaire status is in paper terms alone and that net cash flow is often negative.

27. Durham University, 'Cookson', Box III, 1.

28. Details of the dispute are contained in letters in ibid., Box III, 4-15.

29. The details are contained in letters in ibid., Box III, 18-26.

30. N. Cookson, 'On Rozan's Process for Desilverising Lead', *Trans. Newcastle Chemical Soc.*, iv (1877-80), pp. 171-6.

31. Ibid., p. 170.

32. Annual figures, for the deliveries of lead products for the period 1872-1910, and for deliveries to particular markets for the period 1887-1907, are given in Appendix V.

33. The supply of foreign ore fell, to be replaced by unrefined pig, as exporters in countries such as Spain developed their own smelting capacity. As a result of these declines the Howdon smelting capacity was reduced and the last ore treated in 1884.

34. Cooksons had been experimenting with chamber processes for some time. In 1876 they had taken an exclusive UK licence for the Brumleu process but soon abandoned it and in 1884 they paid F. Schmoll of Cologne for details of a chamber process which proved unsatisfactory. Paradoxically, for the successful introduction in 1898 there is no evidence as to where the idea came from or whether or not it was licensed from another manufacturer.

35. Pulsifer, *History of Lead*, pp. 279-80.

36. The process is outlined in Smythe, *Lead*, pp. 261-2 and Holley, *Lead and Zinc Pigments*, pp. 127-30.

37. The process varied slightly and of the post-1945 period Mr T.R. Barrow writes, 'Live steam was blown through water, contained in a constant-level vessel, situated below the Chamber, the hot water-vapour passing into the Chamber through earthenware pipes in the floor. This injection of hot water-vapour maintained a temperature of 70°C in the Chamber. The 4% strength acetic acid

entered the Chamber in the hot water-vapour. The carbon dioxide was obtained from scrubbed boiler flue-gas; these boilers were fired with very-low-sulphur content coke.' Communication to the author.

38. The Cookson figures give the wages cost per ton of white lead produced in 1909 as £2 9s 2d for the stack process and £1 15s 3¾d for the chamber process.

39. A laboratory experiment done by Cooksons in 1909 suggests that the covering power of their chamber white lead really was satisfactory. Five grammes of stack white lead, mixed in oil, from each of a number of leading makers covered between 176 and 208 sq. inches, while a sample of chamber white lead covered 256 sq. inches. Analysis showed that chamber white lead was chemically indistinguishable but had different physical properties from stack white lead.

40. In Wigham Richardson (ed.), *Visit of the British Association to Newcastle-upon-Tyne* (Newcastle, 1889), p. 118.

41. A comparison of the figures for Norman Cookson's share of the profits and dividends with the profit figures given in Table 5.5 does not show a strong correlation. Unfortunately there are no figures in the firm's books for division of profits to the directors, although Norman probably had an increasingly dominant holding. His father died in 1888 and his uncle, John, in 1892, having withdrawn from the partnership in 1887, their holdings presumably being transferred to Norman and his brother (who himself retired from active business in 1889, at the age of 44, to live at Trelissick, near Truro).

42. A recent example of this view is given by Paul Thompson, *The Edwardians. The Remaking of British Society* (1977), pp. 185-6. He comments, 'the expanding system of public schools offered a training for the industrialist's sons in the gentlemanly style of life. It was an education which concentrated on the production of rulers and professional men, rather than businessmen or scientists. The Empire provided administrative and military careers for these public schoolboys. Relatively open social mobility thus simply led to the diversion of the sons of the most successful entrepreneurs into a non-productive imperial ruling class. It was clearly difficult to develop science-based industries while any promising heir to industrial influence was put in the classics stream.' It should be clear that this is a caricature of the actual effect of the public schools, based on little evidence.

43. Fâche, *Locke and Lancaster*, pp. 12, 16. Mary (1777-1823), the eldest daughter of John Locke (sen.), had married William Norton Lancaster (1773-1838) a London merchant and insurance broker and the son of Robert Lancaster, a master mariner of Yarmouth. There is no evidence at all to connect the family with the London lead manufacturers, Lancaster & Co., already mentioned. Their eldest son, William James Lancaster, was born on 17 Sept. 1802 and died 18 April 1866.

44. The top of the blast furnace was an open cone, or hopper, into which a 'bell' was raised to fit into a seating at the base of the cone, thus sealing the furnace. A mechanical bucket system supplied ore, which was dropped into the top of the cone and lowering the 'bell' allowed the charging of the furnace to take place. It was claimed that this not only reduced the labour force on the furnace by one man but also reduced the amount of dust and fume with a consequent reduction in lead poisoning. The factory inspectorate of the Home Office was sufficiently impressed to obtain Locke, Lancaster's permission to circulate the details to other lead firms. The technique, more often known as 'bell and hopper', was first introduced with iron blast furnaces by George Parry in 1850, where its major advantage lay in facilitating the collection of blast furnace gases.

45. In one account, in 1807, he bought both town (i.e. London) and country white lead from Walkers, Maltby at 58s and 56s a cwt respectively, suggesting that

London-made white lead commanded a price premium over provincial produce, even though Fishwick claimed that Elswick produce was at least as good as that from Islington. Much of the information about the Johnsons comes from the collection of papers, letters and account books in the possession of R.J. Johnson, Threadgolds Farm, Great Braxted, Witham, Essex, to whom I am grateful for permission to consult and quote from the papers.

46. Dickens' original article is 'A Small Star in the East', *All the Year Round*, 19 Dec. 1868 and his account of the Johnson works is in ibid., 27 Feb. 1869, pp. 300-3. These were subsequently reprinted as Chaps. 30 and 34 of *The Uncommercial Traveller*. See also G.F. Young, 'Limehouse Luck or the Lead Mills Located', *The Dickensian*, vol. xxx, no. 231 (1934), pp. 173-6.

47. It is, however, noticeable that although the Johnsons claimed that the woman, mentioned in Dickens' earlier article, had returned to work, he made no retraction on this point. Subsequently in 1955 someone at ALM had the idea of using the Dickens connection for advertising purposes but their advertising consultants wisely replied that they were 'unanimous in saying that it would be a mistake to try and construct a prestige advertisement on material which grew out of an exposure by Dickens of lead poisoning'.

48. Dickens was obviously aware of the growing recognition that higher labour costs in the USA were leading to a substitution there of capital for labour, with subsequently rising labour productivity and falling costs which were enabling American producers to begin to compete with British goods in world markets.

49. The usual comment has to be made that it is not clear what 'profits' mean here, but the figures are sufficiently high that, even if they are not taken after deducting interest on partners' capital (which is unlikely) and charging depreciation, they show an unusually profitable concern. A partnership deed dated 30 March 1867 stated that plant and machinery was to be depreciated at 5 per cent p.a. and that partners were to be allowed 5 per cent on their capital. It does therefore seem probable that the profit figures were genuine. In addition the deed shows that the capital figure for 1867 included both fixed and working capital including the balance of book debts.

50. The figures are taken from a paper in R.J. Johnson's collection, which states that they were extracted from the firm's ledgers, which are no longer available. The difference between profit and partners' drawings in any given year is made up by the change in the level of capital employed between the beginning of that year and the next.

51. This is highly unlikely. *The Universal British Directory* (1791) gives no evidence of a lead works belonging to the firm in either Nottingham or Derby. It does, however, list Messrs Cox as wine and brandy merchants of Derby, and this fits with subsequent evidence. According to William Cox, *Pedigree of Cox of Derbyshire* (Derby, 1889), William Cox of Culland (born 31 July 1742, died 31 May 1827) was involved in the wine trade and subsequently built the shot tower.

52. Living at Spondon Hall he was Mayor of Derby, High Sheriff for Derbyshire and MP for Derby 1865-8.

53. This and all subsequent net profits are on an historical cost accounting basis.

54. W.M. Mackinlay, 'History of Glasgow Paint Manufacturers', *Oil and Colour Trades Journal*, 6 Feb. 1948, pp. 314-18 and *Oils, Colours and Drysalteries*, 15 June 1895, pp. 412-17.

55. John, *Walker Family*, p. 42.

56. *BPP*, 1893-4 [c.7239-I] XVII, qq. 3021-2.

57. *Oils, Colours and Drysalteries,* 15 June 1895, pp. 412-17.

6 THE DEVELOPMENT OF THE LEAD MANUFACTURING INDUSTRY IN THE NINETEENTH CENTURY

Dirty British coaster with a salt-caked smoke stack
Butting through the Channel in the mad March days,
With a cargo of Tyne coal,
Road-rail, pig-lead,
Firewood, iron-ware, and cheap tin trays.
                    John Masefield, 'Cargoes'

The development of the lead manufacturing industry in the nineteenth century: total output, imports of raw materials, exports of manufactured lead and competition from foreign imports; estimates of output of major products, white and red lead, lead sheet, pipe and shot; concentration of the industry both by region and in numbers of firms; profitability; cartelisation from the late-nineteenth century; technical change in the manufacture of the major lead products; the labour force, quality, employment levels and wages and conditions of work.

Having discussed the nineteenth-century development of the major lead partnerships we may now look at the overall development of the industry before 1914, and the considerable increase in output which took place. In lead, as with the other major non-ferrous metals required by industrialisation, tin and copper, Britain was largely self-sufficient until about 1850. Indeed in the eighteenth century she had been one of the world's major producers of these metals and considerable quantities were exported. There was, however, stagnation in exports of lead in the late-eighteenth century and decline in the second quarter of the nineteenth century. Despite some expansion, home output of lead ore was growing insufficiently rapidly to meet rising demand for the metal at home and to continue to supply export markets. From the late 1840s imports of lead began to be of significance and by the early 1870s imported lead exceeded the quantity mined at home and by the early 1900s it was ten times as great as the quantity of home produced lead.

The major statistics of these changes are presented in Table 6.1. Although there are no detailed and regular figures for home production before 1845 it is probable that output was increasing before that date to reach an output plateau of 60,000 to 70,000 tons p.a. between 1850

Table 6.1: UK Imports, Exports and Consumption of Lead, 1780-1914[1]

Period	(1) Average annual output of metallic lead from UK ores[a]	(2) Average annual imports of pig and sheet lead, and metallic lead from foreign ores[b]	(3) Average annual exports and re-exports of pig and sheet lead and lead shot[c]	(4) Average annual exports of red and white lead[d]	(5) Approx. annual home consumption of lead (cols. (1+2)−(3+4))[e]
1780-4		Negligible	13,800		
1785-9			16,100		
1790-4			14,100		
1795-9			12,500		
1800-4			11,200		
1820-4		3,300	13,500	900	
1825-9		900	12,000	1,200	
1830-4			9,800	1,100	
1835-9		2,400	12,500	1,400	
1840-4		2,200	17,500	1,900	
1845-9	54,400	5,600	14,300	2,400	43,700
1850-4	63,900	13,900	21,900	3,200	53,200
1855-9	67,500	13,600	20,100[f]	5,000	54,500
1860-4	66,600	26,600	28,100	5,500	58,300
1865-9	69,500	50,600	33,800	9,300	76,000
1870-4	63,200	72,000	42,100		82,800
1875-9	57,400	101,300	41,200		107,200
1880-4	47,900	110,600	41,400		106,800
1885-9	37,700	137,200	53,200		111,400
1890-4	30,900	183,900	67,600		136,900
1895-9	27,100	202,900	54,300		165,400

Table 6.1: (cont.)

Period	(1) Average annual output of metallic lead from UK ores[a]	(2) Average annual imports of pig and sheet lead, and metallic lead from foreign ores[b]	(3) Average annual exports and re-exports of pig and sheet lead and lead shot[c]	(4) Average annual exports of red and white lead[d]	(5) Approx. annual home consumption of lead (cols. (1+2)−(3+4))[e]
1900-4	20,400	243,400	48,900		204,600
1905-9	22,200	227,400	57,700		181,600
1910-14	19,200	227,100	58,800		177,200

Notes: a. There are no reliable figures before 1845 and it is highly probable that Hunt's early figures (before 1850) are underestimates.
b. The figures for the import of lead ore rise from negligible quantities before 1862 to peaks of about 30,000 tons p.a. in the mid-1880s and to 56,000 tons in 1896 after falling back to only 10,000 tons in 1893. After 1902 there is a further decline to between 10,000 and 20,000 tons up to 1914. The figures have been converted to metallic lead, on the assumption of 70 per cent lead to ore, and added to the figures for imports of pig and sheet lead.   c. Re-exports are given as negligible before 1824 and from 1854 to 1873 inclusive.   d. These figures are taken from the annual parliamentary returns for imports and exports of lead. The figures for exports of red and white lead cease in 1870.
e. These figures are no more than guesstimates. In particular they make no allowance for annual changes in stock although this is only likely slightly to impair their overall accuracy. More importantly, there are no export figures for shot, red lead and white lead after 1870. A deduction of 2,400 tons (the annual average for shot exports 1865-9) has been made to allow for shot exports after 1870. It is possible that this is an underestimate since shot exports were increasing before 1870 and might have been expected to do so to some extent thereafter, although one would expect them to meet increasing competition towards the end of the century. From the late 1860s imports of white lead began to increase and it has been assumed that the net balance of exports over imports remained roughly constant at the level of the 1860s of about 4,000 tons p.a. (which the figures in Table 6.2 show to be only marginally higher than the estimates for the years 1903-13). There are no figures for imports and exports of red lead after 1870 and the assumption has been made that net exports remained at the level of the average of 1865-9, which is probably an underestimate. Finally, the figures for white and red lead exports have been reduced by an average of 15 per cent (e.g. type metal, pewter, etc.) on which it is not possible to obtain any information. It seems likely that the final figures exaggerate the level of lead consumption in the UK. Both Hunt's *Mineral Statistics* and the annual parliamentary returns which succeeded them contain figures for lead available for home consumption. There is no evidence as to how they were arrived at but they show broad similarities to the figures given here.   f. Exports of shot are no longer included after 1856. From 1857-70, however, the figures are available from parliamentary returns and the export of shot for those years, although excluded from column (3) has been deducted before arriving at the figures in column (5).

and 1875. In the final quarter of the century the general decline in wholesale prices was intensified in the case of lead by the rapidly increasing supply of foreign ores and pig lead which caused a fall in price from £22 per ton in 1875 to under £10 in 1893-4. Taken in conjunction with the facts that many British mines were becoming exhausted and that the cost of discovering new veins was high, it is not surprising that mines closed with great regularity in the late-nineteenth century and that home production supplied an inconsiderable proportion of total consumption. Imports came chiefly from Spain, although there was some diversification towards the end of the century, especially with supplies of bullion lead from Australia and Mexico. These provided the raw material for large desilverising operations and the considerable outputs of silver which many of the lead manufacturers produced. Gradually, however, foreign producers extended their own desilverising operations;[2] by 1914 most imported lead had already been desilverised and after the War desilverising was to be of relatively limited importance.

Towards the end of the nineteenth century there is fragmentary evidence of a trade association consisting of the five major British desilverisers, Cooksons; Locke, Blackett; Locke, Lancaster; Walkers, Parker and H.J. Enthoven & Son of London, the latter being involved only in smelting and refining. The association was probably developed in response to the declining supply of bullion lead from abroad and as an attempt to provide a more powerful purchasing organisation which would reduce costs. As a result of the declining availability of foreign bullion lead in the years before 1914 it is probable that desilverising became less attractive to the British manufacturers, since profitability depended crucially on throughput. A notebook of desilverising costs, probably relating to the Elswick works, gives monthly figures over the period 1900 to 1914 showing that the cost of desilverising varied from 11s 2d to 46s 9d per ton of lead, largely dependent on the throughput, which ranged from 300 to 1,600 tons a month.

With the growing importance of imported raw materials, increasing complexity of business transactions and the appearance of a well-established group of firms engaged in the manufacture of lead products, there developed a metals market. The London Metal Exchange as a terminal market dates back to the mid-nineteenth century. Informal meetings of those dealing in metals became formalised in 1877 with the establishment of the London Metal Exchange Co. Ltd, of which organisation Robert and Matthew Warton Johnson were founder members, while Norman Cookson was admitted a member in 1878.

Before 1914 iron was traded on the market but since 1918 the Exchange has only dealt with the major non-ferrous metals, copper, lead, tin and zinc. There were two price fixings daily, with transactions either for cash (maturing the following day) or for three months' settlement. The Exchange provided the lead companies with the opportunity to balance sales of lead products and purchases of primary and secondary lead. On a rising market demand for lead products would increase more rapidly than the companies' normal supplies of pig lead and they would buy on the Exchange against their rising sales and vice versa on a falling market — an operation known as 'hedging'.

Alongside the growth in imports of raw materials came growth in the sales of British manufactured lead products in foreign markets, as large-scale production enabled a reduction in costs. In the late-eighteenth century export sales were small and individual firms often seemed to regard the occasional shipment abroad as a speculation. From early in the nineteenth century, however, overseas sales were beginning to play a significant role in the concerns of the white lead manufacturers, the section of the industry in which large-scale production was most prominent. Pulsifer states[3] that the first factory for the manufacture of white lead in the USA was set up in Philadelphia about 1804, but

> The new enterprise excited the jealousy and ire of the English manufacturers who had supplied the markets of the country, and there is a tradition that the destruction of the factory by fire, soon after operations were begun, was the work of a young Englishman, who applied for work a few days before the disaster, and who left for home in a vessel which sailed for London early in the morning of the day following the fire. The factory was not rebuilt until 1808 or 1809. During its erection an agent of the English white-lead factories frequently warned Wetherill [the owner] of the danger of the under-taking, and confidentially informed him of his orders to crush the new enterprise. Nothing daunted, Wetherill completed the factory, and operations were begun; but, true to his instructions, the agent of the English companies immediately put the price of his commodities to such a point that absolute ruin to Wetherill seemed inevitable.

In 1810 imports to the USA are said to have been 1,150 tons[4] while the output of Wetherill's factory was only 369 tons. However, the Anglo-American War of 1812-14 freed the American company from British competition and enabled Wetherill to establish his firm. He was soon joined by other American producers and any British dominance of the market was broken.

From 1815 there are parliamentary returns of the annual exports of lead and these are presented in full in Appendix VI. They show that exports of lead chemicals were small in the years after 1815, rising slowly from about 1,000 tons p.a. to about 2,000 tons in 1840 and 3,000 in 1850, with white lead making up about two-thirds of the total in the early years but only about one-half by 1850 as exports of red lead caught up. Figures of this order accounted for only a small proportion of the total output of the industry. In 1830, for instance, Walkers, Parker's output was five times the level of British exports of white lead. The majority of exports in this period were included under the classification 'pig lead, rolled lead and shot' and it is probable that pig lead, rather than manufactured lead, made up the vast proportion. Pig lead went chiefly to Europe, with Holland, France, Italy and Russia being notable markets, while China and the East Indies were also of importance. Total British lead exports ranged from 15,000-20,000 tons p.a. in the period 1815 to 1820 but fell to 10,000-12,000 tons in the 1830s as home demand took up more of available supplies. With the rise in imports of foreign lead from the 1840s, total export volume was able to increase once again, reaching 20,000-30,000 tons p.a. in the 1850s and 60,000 tons by 1870. It was in this period in the mid-nineteenth century that the exports of manufactured lead reached significant proportions. Exports of red lead, usually less than 500 tons p.a. before 1840 were never less than 1,000 tons after 1850 and averaged 3,000 tons in the 1860s. Shot exports began to exceed 1,000 tons in the mid-1840s and reached 3,000 tons by 1870, while those of white lead rose from just over 1,000 tons in 1840 to 6,000 tons by 1870. Sheet and pipe exports were not separated from those for pig lead until 1862, when they were 4,000 tons for sheet and 1,500 tons for pipe, rising respectively to 6,000 and 2,500 by 1870 and growth continued through the 1870s and into the mid-1880s, by which time their joint contribution was around 15,000 tons p.a.

The trend in exports was towards concentration on markets away from the industrialised countries of Europe and North America, where competition was increasingly fierce, and where tariff barriers were rising from the late 1870s. Russia, China, India and Australia became important and many of the companies were appointing agents further and further afield. In 1893, for instance, Walkers, Parker appointed an agent in Japan and in 1906 they appointed H. Quane & Co. of Christchurch, New Zealand, as agents for the sale of white lead on a commission of 25s a ton.

Within a few decades of their serious penetration of export markets from the 1840s, British manufacturers were finding themselves in competition in their home market with German and, by the early-twentieth century, American lead manufacturers. Unfortunately there are few statistics on the importation of lead products before 1914 but the literary evidence is clear that foreign competition in the home market developed from about 1870. In the 1830s and 1840s imports of white lead were of the order of 20-30 tons annually, chiefly coming from Italy and Germany, but in 1867 there was a sudden increase from less than 100 tons annually to more than 1,000. Of the 1850s and 1860s a subsequent manager of Cooksons noted, 'as there was no foreign competition and a steadily increasing demand both for home and abroad, buyers were seeking sellers all the time, and managers of sales departments were unknown'.[5] By the turn of the century competition had become serious. A.J. Foster in 1905 noted that white lead imports had reached 11,000 tons p.a., as against a British output of 40,000-50,000 tons.[6] He commented that the chief sources were Germany, Holland and Belgium, which were 'dumping' their surplus production on the British market. At the time of Joseph Chamberlain's tariff reform campaign it is not surprising that Foster favoured a tariff on imported lead products. The annual report of Locke, Blackett & Co. Ltd dealt with the same theme in 1906:

In spite of the activity of trade in Germany, where business is at present much more prosperous than here, large quantities of manufactured lead continue to be sent, duty free, from that country into England, depressing prices here; while German manufacturers are protected by a substantial tariff from competition from outside.

Table 6.2: UK Imports and Exports of White Lead, 1903-13 (metric tons)[7]

Year	Imports	Exports	Excess of exports over imports	Year	Imports	Exports	Excess of exports over imports
1903	14,700	12,500	−2,200	1909	12,900	15,400	2,500
1904	12,600	12,800	200	1910	11,700	16,400	4,700
1905	12,400	14,300	1,900	1911	13,600	16,300	2,700
1906	11,500	16,100	4,600	1912	14,200	18,700	4,500
1907	12,000	16,200	4,200	1913	13,059	14,968	1,909
1908	12,900	15,400	2,500				

Although most of the blame for imports was laid at Germany's door, imports of white lead from the USA reached 4,000 tons in 1914. Table 6.2 shows the extent of the problem in the years before 1914.

The stabilisation of the position in the years immediately before the War was largely the result of the setting up of an international White Lead Convention. This followed the pattern of international market sharing agreements for many commodities, which had been established from the late-nineteenth century. Both Continental and American white lead manufacturers were included in the international agreement arranged in 1911, in which production and trade quotas were agreed with fines and compensations for surpluses/deficits and the price of white lead was increased. Although there were difficulties with individual firms both in the UK and in Belgium, the international convention seems to have worked smoothly up to 1914. It was based on market shares on an average of the years 1907 to 1910. The average annual sale of white lead in the UK market was 63,020 tons, of which home producers supplied 49,995 (79.3 per cent), the German producers (with whom were included those of other continental countries) 9,075 tons (14.4 per cent) and the American 3,950 (6.3 per cent). In its first year, 1912, the convention worked quite well to these figures. As a result of price rises total deliveries fell to 57,338 tons, the British producers were 853 tons in excess of their quota and the Americans 14 tons, while the Germans had a deficit of 867 tons. With the advent of war in 1914 the international convention broke up and the withdrawal of continental supplies opened the way for profitable opportunities for British manufacturers in the home market. Overseas competition was, however, to return as a problem in the 1920s.

While it is possible to produce reliable figures for UK production and imports of metallic lead and fairly detailed figures for exports, the same cannot be said for UK output of lead products. The total output of the various products of a small industry, such as lead manufacture, was not something which interested Victorian statisticians. Even had it done so it would have been difficult to have obtained an accurate answer because of the considerable number of producers, many of them small and widely spread over the country, and the considerable range of products.

The starting point for any assessment of output of lead products must be the estimated figures for annual consumption of pig lead. From 1845 these are given in column 5 of Table 6.1. There are no accurate figures for UK output of pig lead before that date but a figure of about 40,000 tons p.a. at the beginning of the nineteenth century might not be considered unreasonable. Allowing for exports, this would imply a home manufacturing consumption of a little under 30,000 tons p.a. By 1850 this had risen to a figure of about 50,000 tons, to about 100,000 by the 1870s and 200,000 by the early 1900s.[8] The chief objection to the annual consumption figures given in Table 6.1 is that they make no allowance for secondary consumption, the reuse of old lead. There is no serious evidence on this subject but, although scrap metal had been remelted for centuries, it seems unlikely that it would have offered a significant source of supply in the early-nineteenth century. The supply of secondary metal would have depended chiefly on the demolition of buildings in which lead roofing and lead pipe had been used to a considerable extent. It is improbable that this occurred on any scale, although as the century progressed and old city centre buildings were demolished, in order to make way for railways and commercial premises, it would have become more common. As that happened, however, the price of pig lead was falling, as a result of the extension of supply of imports, and this would have made the recovery of scrap lead less attractive.[9] Since there are no figures the problem is entirely academic but while allowing that secondary supply may have expanded in the later-nineteenth century it seems not unreasonable to suggest that the figures given in Table 6.1 give a fairly accurate guide to annual production.

To turn to an attempt to assess the output of specific lead products is to offer greater hostages to fortune. The output of white lead is the best documented area and, since from the early-nineteenth century its manufacture was beginning to be dominated by large firms, it offers fewest problems. We know that the four works of Walkers, Parker produced about 3,500 tons of white lead in 1830. Since London was the accepted centre of the trade at that time and Walkers, Parker had only the small Southwark white lead works there, it is clear that the firm could not have dominated national output. On the Tyne, Locke, Blackett and James & Co. were significant producers while there were firms in many towns such as Bristol, where Riddle & Co. had five stacks. Total output is unlikely to have been less than 10,000 tons p.a., while 15,000 would be an upper bound. At that level white lead would have accounted for about 20-25 per cent of total lead consumption. By the mid-1850s Walkers, Parker's output had doubled to 7,000-8,000 tons p.a., a growth which probably reflected an increasing share in the trade. There had been a number of closures of small works in the previous decades, including the only firm in York, probably as the development of the railways brought more effective competition from the larger works. In addition many of the companies which were to become of importance, Cooksons, Alexander, Fergusson & Co. and FB & W, had not been formed or were, as yet, only small. Walkers, Parker may, therefore, have accounted for one-third of total output, which was probably in the range 20,000-25,000 tons, at least one-third of total consumption of lead. Writing in 1888 Pulsifer[10] gave a figure of 50,000 tons as an estimate of UK output in the mid-1880s, given by a principal British manufacturer. Implying that white lead took not far short of 50 per cent of total lead consumption, this is obviously too high a figure. The output of Walkers, Parker in the late 1870s was still only about 8,000 tons, having fallen back from the 9,000 reached in the late 1860s. It is improbable that any of the older-established firms had a better performance and the new ones did not become large producers at all quickly – Cooksons' output being less than 1,500 tons p.a. in the 1880s. While some increase had taken place since the 1850s it is probable that it was checked in the depression years of the late 1870s and mid-1880s and that total output was of the order of 30,000-35,000 tons in that period. The figure of around 50,000 tons annual output is again quoted by several sources for the early 1900s and for this period it seems not unreasonable. The output of Walkers, Parker was still about 8,000 tons p.a.[11] but other manufacturers had expanded apace. In particular Cooksons' output had reached 4,500 tons just before the introduction of the chamber process in 1898, 7,500 tons in 1899 and 10,000 by 1905, while that of the Mersey White Lead Co. Ltd (a firm which had only commenced production in 1889) reached 5,500 tons by 1897.

Although there were some fluctuations, white lead maintained a share of total lead consumption of about 25 per cent throughout the century and was, therefore, a very significant part of the industry. It is, however, the case, as with other basic manufactures, that German and American outputs were growing more rapidly than British. In the early-nineteenth century Britain dominated world production with at least 50 per cent of the total, but by 1886 German output was of the order of 40,000-50,000 tons (of which only 25,000 was for its domestic market), while output in the USA reached 100,000 tons in the early 1900s.[12]

Figures for red lead are much more difficult to establish. Although it was usually produced by the large white lead manufacturers it was also made by many small producers since the capital required was relatively small and there were few economies of scale. It is, therefore, impossible to produce an accurate guess from the output of known producers. It is certain that output expanded as iron become more widely used as a construction material, since red lead was chiefly used as the major constituent of anti-corrosion paints. It was certainly less significant in total output than white lead. Walkers, Parker, for instance, produced an average of about 2,000 tons p.a. from the 1850s to the 1880s (one-quarter of the level of their output of white lead), while Cooksons produced 1,500-2,000 tons p.a. in the 1890s and 1900s. A reasonable guess for the 1900s might be a total output of 12,000-15,000 tons.

Since sales of litharge were considerably smaller than those of red lead it seems certain that the production of lead chemicals (which also included such commodities as lead acetate) could only have averaged about one-third of total lead consumption. Blue lead, that is to say metallic lead, products must have made up the remainder. Of these sheet lead was undoubtedly the most significant, used chiefly by the building trade for weather protection. It is not possible to produce any early-nineteenth-century figures, but Walkers, Parker's output was averaging 11,000 tons p.a. in the 1850s and 17,000 tons in the late 1870s, rather more than one-third of its total output of lead products in each period. Most of the large lead manufacturers were involved in lead rolling by this time, but in addition there was probably a declining number of small producers who cast rather than rolled sheet lead. Total output in the 1850s probably exceeded 25,000 tons (more than 40 per cent of total lead consumption) and perhaps reached 50,000 by the late 1870s, with output closely linked to fluctuations in housebuilding.[13]

Inseparably linked to the rising consumption of lead sheet as a result of building developments was the demand for lead pipe. It is probable that this was modest in the early-nineteenth century, when running water supplies and sewage systems were uncommon but from the time of the public health inquiries of the 1840s provision of these facilities began to increase and there was a dramatic upsurge in the last quarter of the century.[14] To an even greater extent than the production of sheet, the manufacture of lead pipe was open to small capitals and many plumbers bought pipe presses and manufactured to their own requirements. Their total output is impossible to assess. Of the large firms, however, Walkers, Parker were producing some 4,000 tons of pipe p.a. from the 1850s, a figure which had, in fact, fallen slightly by the late 1870s. Unlike sheet where the firm might have accounted for as much as 40 per cent of total production in the 1850s, and not much less in the 1870s, its proportion of pipe production must have fallen considerably, possibly from 40 per cent of an output which might have been 10,000 tons in the 1850s to less than 20 per cent of an output which might well have been expected to have doubled by the late 1870s. With the rapidly increasing demand from the building industry, output had soared by the early-twentieth century. The Elswick works alone was producing 3,500 tons p.a., as much as the whole firm in the late 1870s, and output for Walkers, Parker as a whole probably approached 10,000 tons. Some of the large firms had, however, grown slowly and Cooksons, having first laid down a pipe press in 1882, were only producing about 1,500 tons p.a. in the early 1900s, rather more than Locke, Blackett's output at the same time. There must, therefore, have been a considerable expansion in the output of the specialist producers, outside the integrated firms, in order to account for the total figures of sheet and pipe output given in Table 6.3, which are unfortunately not available separately. After 1929 the figures are broken down between sheet and pipe, with the latter accounting for about 55 per cent of the total. If this proportion held for the 1900-1914 period, the output of pipe would have been about 80,000 tons in the 1903 peak, declining to 45,000 tons by 1914.[15] It would, therefore, have shown a more significant rise over the nineteenth century than any other lead product, entirely the result of changing standards in water supply and sewerage.

Table 6.3: UK Production of Lead Sheet and Pipe, 1901-18 (tons)[16]

1901	144,263	1906	128,159	1911	106,888	1916	64,075
1902	144,999	1907	118,610	1912	98,696	1917	48,735
1903	150,409	1908	120,976	1913	89,504	1918	32,697
1904	140,986	1909	113,515	1914	84,195		
1905	133,665	1910	113,660	1915	75,011		

It is clear that production of sheet and pipe had come to play a dominant role within the total output of lead products by the turn of the nineteenth century, especially in periods of building boom. In the years 1901 to 1904, for instance, they accounted for about 70 per cent of the home consumption of pig lead, and as much as 50 per cent in 1911 to 1914 during the slump in the housing cycle.

The remaining blue lead products accounted for a relatively small proportion of total consumption. Lead shot was obviously of some significance but there are few figures from which we might make a reliable estimate of total output. The process of manufacture in the shot tower was highly labour intensive and output was limited by the rate at which molten metal could be poured – probably creating a maximum capacity of 40 tons a week or 2,000 tons a year. That would, however, require very long working days and continuous working throughout the year, while in practice normal output might have been expected to have been of the order of 25 tons a week. Taking the number of shot towers known to exist and allowing for other shot-makers using the older process, it is likely that maximum capacity was around 20,000 tons p.a. and normal working capacity perhaps 12,500 tons. It is, however, obvious from the evidence in Chapter 3 of the unsuccessful careers of several of the towers, that the capacity exceeded demand and it is doubtful if output reached 5,000 tons in the first half of the nineteenth century. By the 1850s the basis for statistical estimate becomes a little less weak. Walkers, Parker (with three shot towers, the dominant firm in the trade) were producing about 2,750 tons p.a. rising to 3,500-4,000 between 1865 and 1875 before falling back to only a little over 3,000 tons in the late 1870s. With a decline in the number of working towers since the early-nineteenth century, it is difficult to believe that total output was much more than twice that of Walkers, Parker. This would give a range of 6,000-8,000 tons over the period, confirming that it was unlikely that these figures were reached earlier, especially since exports of shot were rising from 1,000 tons p.a. in the mid-1840s to 2,000 tons in the late 1850s and 3,000 tons in the late 1860s. Growth in output in the 1850s and 1860s would, therefore, seem to have been the result of growing penetration of overseas markets. In the late-nineteenth century, however, rising home demand became a factor of sufficient importance to encourage the setting up of new shot manufacturing firms, such as the Newcastle Chilled Shot Co. Ltd, and in the early-twentieth century to induce a cartridge manufacturer such as Eley to build a tower to supply its own needs. There are no figures for output but it may by that time have reached 10,000 tons p.a.

Of the many other uses for lead, each of them individually making a small demand, there is only one, the rolling of tea lead, for which an output figure can be given. F.C. Hill, of W.W. & R. Johnson & Co., who was responsible for the formation of a pooling agreement in the tea-lead trade, stated that the total output was at one time 15,000 tons p.a. and he was probably referring to a date within the last two decades of the nineteenth century. It is not clear whether tea lead would have been included in the figures for sheet output, quoted earlier, but by the early-twentieth century tea-lead output was already in decline as production began to expand in the Indian sub-continent.

It would be attractive to end this section with a summary of the individual production estimates to check against the figures for total lead available for home consumption given in Table 6.1. The estimates which have been made are, however, so fragile that the results are not likely to be conclusive. For what they are worth they suggest that the output estimates given in that table are too high for the 1850s (since it is unlikely that there was much secondary lead then available to cover the difference) and just about correct for the late 1900s, when allowing for an increase in miscellaneous uses of lead and a larger but still small supply of secondary material.[17]

One of the obvious things that comes out of the discussion of the output of specific products is that the nineteenth century saw the domination of the lead industry, as compared with its eighteenth-century structure, by large firms. Walkers, Parker in particular had a major share, although this began to decline in the second half of the century with the establishment of firms such as Cooksons and the expansion of others such as Locke, Blackett. In the period 1855-9, however, Walkers, Parker delivered an average annual total of 28,323 tons of manufactured lead products, no less than 46 per cent of the tonnage of pig lead available for home consumption, although by the years 1875-9 its share had fallen to 33 per cent. It is, therefore, prob-able that the level of concentration in the industry was at its highest in the mid-nineteenth century (as a result of the huge size of Walkers, Parker) with a gradual decline until 1914 when the largest firm (still Walkers, Parker) had perhaps 20 per cent of the market but was approached by a number of medium-sized firms such as Cooksons with around 10 per cent each. The result of this change was a move from a position in which Walkers, Parker almost certainly exercised price leadership and had considerable quasi-monopoly power to one of severe competition with, as we shall see, resultant moves towards cartelisation.

It should be noticed that the dominant position which Walkers, Parker had achieved in the industry preceded and was independent of the development of a national railway network, which is so often taken as the factor which enabled the expansion of large firms and broke down the regional monopolies of small local firms. There can be no doubt that many small, inland lead manufacturers, especially those close to supplies of lead ore, were more viable before than after the development of the railways but it was water transport which enabled Walkers, Parker and, indeed, other firms to expand. It is noticeable that nearly all the lead works erected in the period 1780-1840 had imme-diate access to either a navigable river or canal and those which did not (such as Locke, Blackett's Gallowgate works) were only a short distance from such facilities. Water transport considerably reduced the cost of moving heavy goods and enabled the penetration of distant markets. Walkers, Parker had achieved mid-nineteenth century dominance on the basis of three water-based works with a good geographical dispersion which enabled the firm to supply a large part of the home market.

The result of this development was the demise of many small producers in coastal towns and those which could be supplied by river and canal from the ports. This competition intensified in the later-nineteenth century with the development of a number of medium-sized producers. Their integrated structure, from smelting to a variety of manufactured products, perhaps gave them an advantage over the small firms which were often dependent on a single product, although small-scale pipe producers flourished on the boom in demand for lead pipe which they were able to exploit locally.

As well as concentration in the size of firms, the nineteenth century also saw growing regional concentration in the lead industry with the Tyne and the Thames as the two major centres, reflecting opposing principles of location. The Tyneside industry had developed on the basis of local supplies of raw materials and the attractions of manufacturing to increase the value of the final product to be shipped. The industry on the Thames had developed on the basis of the nearness to a large final market. That both sectors had been able to flourish during the nineteenth century was partly a result of the fact that raw material costs accounted for a large proportion of final price (there was, therefore, little conflict between the two principles) but it was also due to the fact that within wide limits locational advantages were less significant than other factors, especially entrepreneurial ability, in ultimate success.

Table 6.4: Shipments of Lead Products from the River Tyne, 1862-1913 (tons)[18]

	1862	1873	1881	1890	1900	1913
White lead and paint	7,500	13,000	18,900	18,167	23,879	23,848
Red lead	4,500	7,000				
Litharge	800	1,000	864	755	769	598
Sheet	4,500	8,000	8,775	13,281	11,034	8,922
Pipe	1,500	2,500				
Shot	750	1,000	799	901	979	1,011

Since a large proportion of the London output was used in the immediate area there are no useful statistics available to show the region's output but for Tyneside there are figures for the shipment of lead goods from the river which account for a considerable proportion of that area's output. These are given for selected years in Table 6.4 and show considerable growth up to the boom of 1873 with stagnation for more than a decade thereafter, except in the output of sheet and pipe which rose under the impact of improvements in public health. From the very end of the century there was an increase in the level of shipments of chemicals, largely under the influence of Cooksons' adoption of the chamber process. In the years 1901 to 1910 shipments of lead products from the Tyne averaged more than 40,000 tons p.a. and accounted for more than 20 per cent of the total consumption of pig lead in the country. To this, of course, ought to be added the unquantifiable figures for local consumption on Tyneside and those deliveries made by rail, the latter undoubtedly an increasing proportion of local output in the late-nineteenth century.

Although the Tyneside industry had been originally built up on the basis of local ore, the area was well based to receive the growing imports of lead ore and pig lead from Europe. It was also well placed to supply European countries with manufactured lead and it is noticeable that even Germany was a significant customer, taking about 500 tons of lead chemicals annually in the 1880s and more than 1,000 tons in the 1900s. Russia took about 1,000 tons annually in the 1880s and 1890s, declining thereafter to a few hundred, but recovering by 1914. Jointly Norway and Sweden took about 700-800 tons annually throughout the period, rising dramatically to more than 3,000 in 1915. Other European countries consistently took a few hundred tons of lead chemicals annually and together Belgium and Holland took almost 2,000 tons in 1911. For much of the period, however, it was Canada which provided the largest single market, averaging over 1,000 tons and taking more than 2,000 in peak years such as 1886-7 and 1906-7.

In sheet and pipe competition appears to have been more severe, although some hundreds of tons were sent from the Tyne to Canada annually. Germany took 1,000 tons in 1882 but little more than 100 by the 1900s, while Russian imports fell from 2,000-3,000 tons to only a few hundred. Reasonable stability in total shipments was achieved by increasing sales to countries such as Norway and Sweden, although demand there was beginning to fall off in the years immediately before the war.

We may next consider the profitability of the lead industry, since with the exceptions of the 1830s and 1840s, there are profit figures for at least one of the firms in each decade of the nineteenth century, although it is difficult to draw more than tentative conclusions from them.[19] There appear to have been long periods of greater or lesser prosperity, which had nothing to do with the business cycle, although cyclical peaks and troughs did sometimes leave a mark on the trend. It appears that the Napoleonic War years were profitable ones but that they were followed in the period from 1815 to 1830 by a long period of low profits, if the Walkers, Parker figures were typical. Nevertheless even this period of serious competition and poor profitability was interrupted by good profits in the cyclical peak years of 1818 and 1825. With no figures for the next two decades it is impossible to say much with any confidence. In Chapter 4, however, it was argued that the late 1830s and 1840s was a very profitable period for Walkers, Parker on the basis of the evidence of that firm's expansion and ability to finance the large loss made on the mining venture. It is possible that such profitability was generally true of the industry, since demand was rising but there appears to have been no major increase in the number of firms in the industry – competition may, therefore, have become less severe.

Prosperity appears to have gone on at least into the early 1850s, with continued expansion in existing firms and the development of new ones such as Cooksons. It may be that this expansion brought about a temporary decline in profitability in the second half of the 1850s and the first half of the 1860s (a period in which neither Walkers, Parker nor Cooksons produced good profit figures). Demand appears again to have caught up with supply and profit figures are high for the late 1860s and into the early 1870s. Thereafter the broad pattern should be low profitability in the late 1870s and 1880s with a recovery from the late 1880s, if the general arguments about the 'Great Depression' are to be believed. This was certainly the experience of some firms, as is evidenced by the comments of Norman Cookson, quoted earlier, and the profit figures for his firm. As an efficient and progressive company one would expect Cooksons to have weathered the 'Great Depression' better than most but their experience was considerably worse than that of Walkers, Parker, a century old concern, which remained profitable throughout the period. Similarly Johnsons appear to have been only marginally affected and made a higher return on capital employed in the depression year of 1885 than at the peak of the boom of the early 1870s.

It is possible, had we possessed comparative profit figures for a number of firms in the earlier part of the century, that we would have found a similar disparity in performance. It may well be that individual factors with regard to each firm were of greater significance for profitability than the more general economic circumstances facing them all. Since there was considerable similarity between most of the large firms in terms of product range it is hardly likely that what they produced was of importance in determining profitability. At the margin the proportion of total output made up of particular products may have made some difference to profitability, and a balance of output between chemicals and blue lead products almost certainly offset swings in profitability (high profits for some companies in the cyclical trough year of 1879 must have been the result of the fact that the building cycle peaked in London in that year). It is probable that inter-firm differences were largely a result of differing managerial abilities but it is difficult to see why Cooksons, which shows evidence of good management, should have been unprofitable in the late 1870s and 1880s. Without detailed figures for production costs, overheads and prices realised it is impossible to offer conclusions. It must, however, be said that relative profitability was (and is) a highly complex matter, dependent on the relative mix of such factors as location, entrepreneurial ability (or lack thereof), up-to-date technology, etc.

We might also ask what happened to the profits which were made. The evidence given to date has shown that the ploughing back of profits, as in most industries, was the way in which the lead manufacturing companies grew in the nineteenth century. Virtually nothing has been said about banks for the simple reason that the surviving records are reticent on the subject. While there must have been considerable dependence on banks for the discounting of bills there is no evidence that either short- or long-term loans played any significant part in the development of either Walkers, Parker or Cooksons (the firms for which there are the most detailed records). Significantly both of these firms developed from already prosperous backgrounds in other industrial fields and were, therefore, more likely to be internally self-financing from partners' capital. For the other firms the evidence is insufficiently good to be conclusive but the lack of any reference to significant borrowing is surprising.

If plough-back was one way of using profits, the other major use was consumption. Mark Girouard has put the position both succinctly and delightfully:[20]

> Once a Victorian merchant, manufacturer or professional man had made a sufficient fortune he was faced with the dilemma of whether or not to set out to establish his family in the landed gentry. Not all

who had the means took the plunge; it was, after all, a debatable advancement to cease being a great man in Manchester, Halifax or Newcastle and to take up a doubtful position as one of the new rich knocking at the door of county society ... The growth of knowledge and technology was undermining all accepted beliefs and values. But if man were descended from apes and Genesis was a fairy story, at least the stately homes of England were still stately, something solid and traditional to fall back on. Moreover to the English merchant or industrialist, working twelve hours a day, disciplining and denying himself, fighting for survival in the commercial jungle, there was increasingly present the vision of a quiet harbour at the end — an estate in the country, a glistening new country house with thick carpets and plate glass windows, the grateful villagers at the doors of their picturesque cottages, touching their caps to their new landlord, J.P., High Sheriff perhaps, with his sons at Eton and Christ Church and his clean blooming daughters teaching in the Sunday School.

We have seen that all the lead manufacturers responded to the temptation to greater or lesser extent, since there was a gradient of possible responses to the siren call of land, not all of them requiring a complete break with industry. It is noticeable that the two families which made the strongest claim to the establishment of landed status (the Cooksons and Walkers) had done so early in the nineteenth century more on the basis of other industries than on lead. They both appear in the first edition of *Burke's Landed Gentry* of 1846 and no other lead manufacturer subsequently made that haven (of one manufacturer Girouard wrote, 'he never made the haven to which all Victorian new rich aspired, *Burke's Landed Gentry*'[21]). Nevertheless lead profits sustained the landed bases of these families. Sir E.S. Walker, who had appeared in *Burke's*, as Walker of Berry Hill, Nottinghamshire, did not retire from the management of the Chester works until 1852 and John Cookson of Meldon had a one-third share in the Cookson firm until 1887. Other members of the families went on to establish comfortable landed estates (to a considerable extent the industrialists did not adopt primogeniture and did not, therefore, build up powerful heritages but spread their money to their sons, who then established numerous small estates and wealth holdings in various parts of the country). W.I. Cookson went to live at Worksop Manor and Henry Walker II moved to Perdiswell Hall in Worcestershire for instance, where they undertook the role outlined by Girouard, while remaining partners in their respective firms.

While such movements as this, 'anxious to get as far away as possible from the scene where their money had been made'[22] (as Girouard wrote of the 'cottentot grandees'), certainly occurred among the lead manufacturers, the more common move was to a local house in the country from which one could oversee one's business. This was a response not only to rising wealth but also to the increasing social unacceptability of living (and, therefore, entertaining) cheek-by-jowl with one's works. It offered the pleasures of a country retreat with power over the local village and perhaps tenantry, without a loss of control over the business from which one's fortune came. It was also possible to experience the pleasures of a landed estate well before retirement and then in later years slip gradually into the full-time life of the landed gentleman without any of the traumas involved in the sudden switch. Of this pattern the examples are numerous from the James family at Wylam, the Fosters at Stocksfield and Norman Cookson at Wylam, on Tyneside, to the Johnsons at Little Burstead in Essex and the Lancasters at Send in Surrey. Lead manufacturing brought conspicuous consumption, in varying degrees, to the leaders of all the major firms in the nineteenth century and it is noticeable that none of them went out of business in that period.

In the very late-nineteenth century, however, profitability declined as competition once again became more severe (now made more serious by the rise of foreign competition). This led to the first tentative attempts by the large firms to curb competition, although the merger wave which hit much of the economy in the 1890s was largely absent from the lead industry. Competition was to be lessened by means of the convention of manufacturers to control output and prices.

The first step in this direction was claimed by F.C. Hill: 'In *1878* I made a Price arrangement in *Tea Lead* which grew into a Pooling Quota Association January 1st 1885. This was, I think the first Pooling Association in the Metal Trade, and lasted for 43 years . . .'.[23] In itself the tea lead agreement was not of great significance but Hill worked for W.W. & R. Johnson and it was, therefore, likely that the success of control in one branch of the lead industry would lead to attempts to reduce competition in others. In 1883 the London white lead manufacturers came to a price agreement and in 1892 that was formalised into a nation-wide White Lead Conference. This set up minimum selling prices according to the areas of the country the producer and purchaser were in, the type of purchaser and quantity he was taking, with strict control of discounts, drawbacks and credit. No attempt was made to control the output of white lead with the exception that members bound themselves not to increase their existing capacity without permission. Most of the major English manufacturers joined the Conference,[24] although significant exceptions were Cooksons, who were just beginning to expand their output, and the Mersey White Lead Co. Ltd (Mersey) of Warrington, a newly-formed company which was also in the process of expansion. The influence of the firm of W.W. & R. Johnson was considerable, since it had taken the initiative in February 1892 in inviting other firms to a conference and Robert Johnson II and F.C. Hill were subsequently appointed chairman and secretary of the Conference.

Although there were numerous minor amendments to the Rules of the Conference it appears to have worked satisfactorily for a year or so but by the beginning of 1894 there were the rumblings of potential trouble. The Conference had set prices at a level which encouraged purchasers to switch their trade to outside firms. In January Walkers, Parker threatened to resign unless they were given a free hand to deal with outside competition or were subsidised for the losses they were experiencing as a result of competition from Mersey. The Conference approved price reductions and staggered on for another year without resignations but Walkers, Parker withdrew in 1895 and it is probable that the defection of its largest single member brought the Conference to a close. Under renewed competition prices fell further and many firms were making losses around the turn of the century. Low profitability, especially for paint dealers, led to an increase in adulteration in the late 1890s with the addition of such materials as chalk and gypsum, which of course brought white lead into disrepute and reflected on the manufacturers. Although in serious competition with one another they could still see the benefits of combination on a matter which affected the good name of their generic product and the White Lead Corroders' section of the London Chamber of Commerce took action to deal with the problem. The white lead manufacturers provided financial guarantees to support the action and in 1897 a travelling inspector was appointed at an estimated annual cost of £750.

Competition among the white lead manufacturers caused further deteriorations in profits (even Mersey made a loss in 1901), and attempts were made to resurrect the Conference. Nothing seems to have come of an attempt in 1901, less because of the unwillingness of the manufacturers to come to an agreement (although Cooksons were in the middle of expansion of their chamber output), than because of the recognition that imports would increase if any attempt were made to increase prices. In November 1902 the most far-reaching proposals yet suggested reached printed 'Heads of Agreement' form and included not only minimum selling prices but also a quota pool by which each member would receive a fixed output quota and pay fines for any excess tonnage and be paid from the pool for any deficiency in output. Negotiations again broke down but during 1904 the first attempts were made to involve foreign manufacturers in a cartel. In March the Walkers, Parker minutes note that there had been a meeting at which the German manufacturers had agreed to limit exports to England if 'the English corroders also limit their output'. For some months negotiations went on but they eventually foundered on the inability to obtain agreement from the Belgian producers. Sporadic negotiations took place for some time but the turning point came in 1911, with board meetings of several companies devoting more time to discussing the details of a potential cartel than any other subject. Negotiations had been established with the major foreign producers and alongside these preparations for an international convention to fix quotas and control prices for the white lead trade between the major trading countries, went negotiations for a UK convention. Prices and consequently profitability were so low that no firm wished to sink the delicate craft whose launch was underway, but several had special positions which they wished to negotiate. Cooksons wanted to establish special treatment to safeguard their large capital expenditure on chamber plant, while Rowe Bros of Liverpool were holding out for a larger quota for their virtually untried 'quick' process and Mersey, which had stood outside all previous negotiations, had still to be dealt with.[25]

Long before the final agreement was signed a series of price increases introduced by those firms approving the scheme, showed the others the benefits which might be reaped. In August 1911 the price of white lead was raised by 30s a ton, by a further £1 in September and 10s in October. These increases must have had a considerable effect on the laggards, virtually all the UK manufacturers signed the pooling agreement at a meeting on 22 November and the international convention was settled early in 1912. The result of the agreement was a reduction in total output as a result of the imposition of quotas – Walkers, Parker estimated that they would lose about 860 tons (about 10 per cent of their previous output) as a result of accepting a quota of 15 per cent of the agreed total home output of 50,000 tons. It is possible that some of the smaller producers were squeezed even further, although since there are no details of their previous outputs this must remain uncertain.[26] The benefits of the price increases must more than have offset declining output for most firms, since the cartel functioned continuously from

1911 (although the international agreements went into abeyance during the First World War). Although, like all pooling agreements, helping to keep inefficient firms in production, the cartel did not seriously penalise the more efficient. Cooksons, for instance, with their relatively low-cost chamber production, were able to exceed their quota and pay the cost of fines and expenses (more than £5,000 in 1912) out of the ensuing profit. At that Company's Annual General Meeting on 23 April 1912 the directors felt able to express the hope 'as a result of recent arrangements in the trade that better figures for White Lead would be obtainable in the future'. They were quite justified — profits from lead chemical production rose from about £9,000 in 1911 to £27,000 in 1913.

Considerable space has been devoted to the development of the White Lead Convention because of its subsequent importance in dominating the market during the inter-war period and moving the industry from a highly competitive and unprofitable basis to a soundly profitable one. Given that the major firms involved in the negotiations had an integrated product structure it was inevitable that they should see advantages in cartelising the trade in other lead products. As a result, by the commencement of the First World War the free competition of the nineteenth century was virtually a dead letter throughout the lead industry.

Negotiations for a red lead or lead oxide convention appear to have commenced in 1910 at the same time as the major white lead negotiations were beginning and frequently with the same participants. Representatives of Lindgens & Sons and Bergmann & Simon, the chief firms in the German red lead combine, came to England to agree terms for quotas in the English and Continental markets but the negotiations broke down because of the objections of Walkers, Parker, which was currently exploiting a new method of production. Without Walkers, Parker, the most significant firm in the trade (they eventually received a quota of 37.5 per cent of UK output when a cartel was established), there was no future for the negotiations. Had Walkers, Parker been a single-product company they might have maintained an independent position but the safeguarding of their position with other lead products and pressure brought to bear by other manufacturers eventually forced them to negotiate. Further discussions in 1912 led to agreement in principle on a convention, although it was not until September 1914 that the final negotiations commenced and January 1915 when an agreement, excluding Continental producers, came into force. As with the White Lead Convention it was based on output quotas and controlled prices, with fines for deliveries in excess of quota and payments to those firms which were below quota.

For some time before 1911, when the United Kingdom Lead Manufacturers Association (UKLMA) was formed, there had been regional associations of blue lead manufacturers, issuing price lists which were supposed to be binding on their members. There is little evidence on these associations but they appear to have suffered from the competition of non-members and an inability to obtain agreement between the regional bodies, each of which wished to have special arrangements in its own area, while individual firms had grown up on a basis of selling where they could throughout the country. There was much complaint about 'illicit allowances made by the travellers of certain firms' and frequent disagreement between the regional associations – the Midland Association, for instance, passing a resolution in September 1906 that 'special prices be put into force to keep out Newcastle competition'.[27] At the end of 1910 a general meeting of the regional associations was held at which unity made some headway. The regions had already been pulled together in a 'Defence Committee' to deal with foreign competition and the larger firms were involved in the white lead negotiations. The problem for blue lead was chiefly that there were many small firms with low production costs, especially in the manufacture of pipe, which could see no advantage in joining a trade association which would limit sales.

The principle which was accepted by the meetings of regional associations, obviously heavily influenced by the large firms which were also involved in white lead, was that a pooling agreement, based on quotas for each firm, was essential. After many months of wearisome negotiations a provisional pool was established from 1 August 1911, with quotas based on average outputs over the previous ten years, and this worked sufficiently satisfactorily to be renewed for 1912. Unlike red and white lead, where the volume of international trade was such that a UK convention would not work without the simultaneous arrangement of an international pooling agreement, blue lead was to some extent isolated by transport costs from foreign competition. Nevertheless as soon as it was heard that the Belgian producers had formed an association, the UKLMA resolved, on 1 November 1912, 'to take advantage of this fact to at once secure and execute orders in Belgium, in retaliation as a first step towards negotiations which would doubtless follow in regard to Foreign competition . . .'. It was agreed that the Belgian sales should be made at below cost (a probable loss of 5s to 10s per ton was envisaged) and that firms should be compensated from the Defence Fund for such sales. The eventual cost was about £1,000 in 1913, which would suggest that a considerable tonnage was despatched, but it was fruitless since an international agreement did not materialise, as a result of the breakdown of the Belgians' home convention. Late in 1913 A.J. Foster, managing director of FB & W and chairman of all three major UK lead conventions, had been to the Continent for discussions with the Belgian and German producers. As a result of these talks the Continental producers had reduced their demands, in exchange for a complete cessation of their exports of sheet and pipe to the UK, from a compensation payment of £18,200 p.a. to one of £11,000. The British producers felt that this was still excessive and doubted the Europeans' ability to control exports and the UKLMA therefore decided 'that margins all over the United Kingdom be immediately reduced about £1 per ton ... in order to make it unremunerative for foreign lead to be imported'. It was expected that this would enable better terms to be made but the advent of war meant that concern about foreign competition could be set aside for the duration. In any event the problem had not been a serious one because of the largely local nature of the sheet and pipe market. Imports in the nineteen months, November 1910 to May 1912 were only 5,855 tons, less than 5 per cent of home output as compared with about 25 per cent for white lead.

A more serious problem was the internal bickerings between sheet and pipe manufacturers in the UK. In the preparations for the 1911 agreement the London accountants, Wenham Bros, had been appointed to establish the various firms' sales over the previous ten years. A.E. Wenham noted that all of the firms 'had very special circumstances, which in their opinion entitled them to a very much larger share of the business in the future than was shewn by the returns', and he suggested a fine for excess sales over quota 'so high that it would not pay anybody to work in excess'. This idea was rejected and from an unsatisfactory beginning the UKLMA was to be dogged by the problem of establishing acceptable quotas, especially for the small firms. There were also problems with small dynamic firms, with a commitment to *laissez-faire* doctrines, which remained outside the Association. In particular W.T. Glover & Co. Ltd of Manchester and Cohen of Glasgow caused problems by undercutting prices. Since their impact was local, members in such areas bore the brunt of the competition and were tempted to break with the Association and lower prices in order to overcome their problem. The Association did, however, take steps to control outside competition — for instance on 19 September 1911 it agreed to pay an annual sum of £600 to the leading British manufacturer of rolling mills and pipe presses not to sell machinery to non-members.

Problems were made worse by falling demand for sheet and pipe as the building recession deepened. In 1912 sales by UKLMA members fell below 100,000 tons and in 1913 to 90,000, as against 127,000 in 1910. Increasing competition for falling orders did bring some outsiders to heel and in September 1913 W.T. Glover wrote to the UKLMA offering terms on which his firm was prepared to join the Association. It was not, however, only negotiations with traditional outsiders such as the Glovers which caused problems but also the rise of completely new firms in blue lead manufacture, clearly either ignorant of the existing overcapacity or anxious to chance their arms. In May 1914, for instance, James & Co. Ltd, the Tyneside lead chemicals firm, advertised for a sheet mill and pipe press[28] and was eventually given a quota of 200 tons (which suggests that only a pipe press was installed).

Despite the problems the UKLMA was successful, as measured by its impact on prices. The agreed margin over pig lead prices at which sales were to be made had risen from £3 10s per ton to £4 10s in May 1914 and, with the withdrawal of foreign competition, to £5 10s in October 1914. The rising margin was a valuable sweetener to (as well as being a partial cause of) the falling level of demand.

Having discussed the major factors affecting the level of output in the lead industry, we may now look at its technology, which did not experience dramatic change during the nineteenth century. Since the influx of imported ores involved more of the manufacturers in smelting activities, we need to be reminded of the position in lead smelting. The major development, the introduction of the reverberatory furnace at the end of the seventeenth century, had been widely adopted in north Wales and Derbyshire in the eighteenth but only began to be introduced in the northern Pennines and on Tyneside on any scale in the nineteenth century. It would appear that, apart from historical influences and conservatism, the chief factors determining the choice of ore hearth or reverberatory were the types of ore and available fuel in a particular region.[29] In the later-nineteenth century water-jacketed furnaces began to be introduced but, as Norman Cookson wrote in 1889,[30] not to any great extent, 'the supplies of imported and poor ore having diminished', and rich local ores were better treated in the ore hearth.

More significant developments came in the treatment and refining of metallic lead. Most lead ores contained silver but it was only possible in the early-nineteenth century economically to recover it if the silver content was at least eight ounces per ton of ore. This clearly meant that a considerable amount of silver was being wasted and there was some experiment to obtain an alternative to the existing cupellation process, which would enable less rich ores to be desilverised. The solution was finally obtained by a Tyneside chemist, Hugh Lee Pattinson. Quite by accident he noticed, when cooling some lead in a crucible, that the surface lead crystallised first and on analysing these crystals he discovered that they contained a lower percentage of silver than the remainder.[31] This discovery provided the basis for the process which Pattinson was to patent in 1833. It was a very simple enrichment process consisting of a series of large iron melting pots. Pig lead containing silver was placed in the middle pot and as the lead cooled the crystals on the surface were skimmed and moved to another pot. The process steadily developed through the refilling of the central pot, enriched lead progressing through the pots to one side and reduced lead to the other. The result was refined lead at one end of the line of pots and silver-rich lead, containing perhaps 300 ounces per ton, at the other. The latter was then taken away for the silver to be recovered by the old method of cupellation, in which the lead was oxidised to litharge, leaving the precious metal.[32] The advantages of the new process were not only that it could treat ores containing so little as two to three ounces of silver per ton, but also that it considerably reduced the fuel costs of treating richer ores. Pattinson's process was rapidly innovated and spread throughout the world.

In 1850, however, a new process was patented, by Alexander Parkes of Birmingham, in which zinc was added to molten lead and formed an alloy with the silver in the form of a surface crust on the molten lead. This was removed and underwent further treatment. It is often said that the Parkes replaced the Pattinson process but while that is true it did not occur on a large scale until the end of the century. There were difficulties in the working of the Parkes process, chiefly in separating the zinc and silver, early established plants were not profitable and it was not until the 1880s that it began to spread. In 1889 N.C. Cookson, writing about Tyneside, stated 'Only one large firm [Locke, Blackett] work the zinc process on an extensive scale, the world-famous Pattinson process of separating lead and silver by crystallising being still that used most generally ...'.[33] By 1910, however, it was said of Pattinson, 'This is ceasing to be used ...', and of Parkes, 'This is coming into general use ...', while in 1923 Smythe could write that the Parkes process 'has practically replaced all the older methods'.[34]

As a result of the introduction of the Pattinson process, the attraction of refining pig lead became considerable and most of the lead manufacturers added silver refining to their other activities. The scale of output was considerable. In 1863, when the British Association visited Newcastle, Walkers, Parker exhibited a cake of silver made at the Elswick works and valued at £8,000. In 1873 it was estimated that 1.5m ounces of silver, with a value of about £400,000, were annually extracted from lead on the Tyne,[35] while in 1888 Walkers, Parker alone produced one million ounces of silver at Elswick, most of it coming from imported bullion lead.

If desilverising saw successful developments in technology, then the area which saw the most experimentation but least success was the production of white lead. Between 1786 and 1866 there were no fewer than 83 patents and a further 23 between 1877 and 1883, taken out for new or improved white lead processes and the pace of invention increased in the last decades of the nineteenth century with each volume of the *Journal of the Society of Chemical Industry*, from its first publication in 1882, reporting on several new developments. We have seen that the problem lay in the existing stack process, with its high cost, slow rate of production and lack of control. As demand expanded in the nineteenth century and competition between manufacturers increased, the potential prize for the person who could produce white lead more cheaply and quickly became considerable. The activity took two forms, attempts to improve the efficiency of the existing stack process and the invention of entirely new (or at least modifications of unsuccessful old) processes. It is noticeable that it was existing manufacturers who were responsible for the former, while they were entirely missing from the serried ranks of the patentees of new processes, the latter, frequently chemists, coming entirely from outside the white lead industry.

Among the changes introduced by the white lead manufacturers were alterations to the method of setting the stacks – the amount of tan bark (and some manure still used to get the fermentation going), how wet it was, how much was reused and how much new; the size, shape and thickness of the pieces of lead to be corroded were altered and control over the movement of air in the stacks was attempted by introducing vent pipes and varying their position. All the changes were the result of blind empiricism, for science seems to have had little to offer in the way of theoretical instruction about the chemistry of lead. Towards the end of the nineteenth century, however, the picture began to change as firms such as Cooksons employed trained chemists to analyse their products,[36] while in the early 1900s Locke, Lancaster & Johnson employed K.W. Goadby on a consultancy basis. He was lecturer in bacteriology at the National Dental Hospital at the time and was to become an international expert on lead poisoning.[37] His main work, using glass jars as experimental stacks in his laboratory, was concerned with bacteriological activity in the bark and no doubt came as a revelation to the firm's management. He stressed that it was bacteriological activity which was responsible for successful corrosion and that this depended on correct conditions. Where there was insufficient air and the stacks became too hot the bacteria died and corrosion was poor but where there was too much air the process was cooled and the bacteria did not multiply — effective and controlled ventilation, particularly to the centres of the stacks, was essential. Moisture was also important to the growth of bacteria and the spraying of the bark with water was introduced. For fermentation to be satisfactory bacteria needed to be spread evenly and Goadby was able to provide scientific backing for the empirical practice of using old and new bark in the stack. The old bark contained high levels of bacteria which the new bark provided with feedstock, while the bacteria active in new and old bark were different. In the new bark lower temperature bacteria provided the initial temperature rise while in the old bark the higher temperature bacteria were active causing the breakdown of cellulose to form carbon dioxide. It was, therefore, essential that old and new bark were thoroughly mixed to obtain an even fermentation. For the first time there was now a sound understanding of what was happening in the stack which it was possible to use in order to make empirical improvements.

The alternative solution was to give the stack process up as inefficient and inherently uncontrollable and search for an alternative — an approach which was more attractive to outsiders than those in the industry. The ideas ranged from the chemically absurd to the theoretically sound but largely impractical.[38] Many of the new processes were revised versions of Sir James Creed's chamber proposal of 1749. Other proposals came from some very famous names, including John Wilkinson, the eighteenth-century ironmaster and Hugh Lee Pattinson. The latter entered the lists of experimenters with patents in 1840 (8,627) and 1841 (9,102) but meeting with little success in the production of basic lead carbonate he turned to finding a satisfactory alternative and in 1849 patented a process for the manufacture of lead oxychloride. This development deserves to be better known as one of the first reasonably successful alternative pigments to white lead, since in the early 1870s Pattinson's Washington Chemical Co. was manufacturing about 800 tons p.a.[39] Doubtless each new process which reached the production stage caused some concern to the existing manufacturers. In 1840 Walkers, Parker wrote to an agent, Henry Patterson, 'Do you hear anything of the Birmingham or Patent W. Lead or are you at all interfered with by that Article being offered for Sale?' and a few weeks later, 'we are putting up sheet Lead 20/- or 30/- p. Ton but have no intention of touching W. Lead or paint, out of regard to our *patent* opponents'. Such problems, although occurring regularly, were individually short-lived.

There were other processes, notably precipitation ones such as the Sankey process adopted by Locke, Blackett in 1875, which reached works production but did not last very long. There were also, in the last two decades of the nineteenth century, under the stimulus of the controversy over lead poisoning, renewed attempts to produce an alternative to lead carbonate which would be less poisonous and the most common was lead sulphate. There had been earlier developments of this product but none met any success[40] until the 1880s when J.B. Hannay developed a process, which was taken up by several companies and widely advertised as the solution to the problem. In the wake of a major denunciation of lead poisoning in the *Daily Chronicle*, the White Lead Co. Ltd of Possilpark, Glasgow, organised a visit of some thirty journalists from London to see its works. The result was a series of laudatory newspaper articles, of which that in *The Times*, one of the more conservative, ended, 'It may, therefore, reasonably be anticipated that the Hannay process will in time supersede the ordinary system of manufacture, the workers in which long since enlisted the sympathies of Dickens in his "Uncommercial Traveller".'[41] The White Lead Co. had achieved maximum publicity for its product at a time when the rival product was under severe criticism and, as one result, the Chief Inspector of Factories wanted to know about the process. He was informed by his Glasgow inspector that the visit had been 'to advertise the Company which has not up till now been successful' and, in rather wearied overtones, that 'The manufacture of Sulphate of Lead as a substitute for the Carbonate (the ordinary White Lead of Commerce) is by no means a new thing. It was tried thirty years ago and more than one fortune has been lost in the attempt to make it a success.'[42] In practice the journalists had been deceived and the only one of them who was an expert, a reporter for the *Oil & Colourman's Journal*,[43] wrote a critical article pointing out that the colour of the sulphate was poor and that there were several other companies making similar white lead substitutes which were not particularly successful.

In the late-nineteenth century all the objective accounts agree that in

England there was no serious alternative to stack white lead. In 1889 N.C. Cookson, whose firm had experimented with several of the new processes, commented that stack white lead

> still more than holds its own in competition with various chamber, precipitating and other systems. On all these alternative methods — many of which reappear under a new guise every few years — large sums of money have been spent; but I think the general conclusion arrived at among manufacturers is that white lead can only be made cheaply and rapidly at the expense of quality, and that the old Dutch process is the only one by which white lead can be produced of good colour, body and density combined and of the necessary chemical constitution to saponify properly when ground in oil.[44]

Changes were soon to come, however, and in 1909 Holley could refer to 'quick processes that have been found by experience to be economically successful'.[45] We have already seen the introduction of a chamber process by Cooksons in 1898 and this remained the only really successful new development in this country.[46] There was, however, the setting up of the Brimsdown Lead Co. Ltd to manufacture white lead by a process developed by Prof. Bischof, which produced basic lead carbonate, beginning with litharge. Production costs were, however, high and we shall see that Brimsdown was never a profitable company. Most of the new and certainly the successful developments were made in the United States. Of these the least successful was the Bailey chamber process, but the Carter and Rowley processes were subsequently operated on a considerable scale.[47] They reduced the time for corrosion to a few days by granulating the lead and therefore offering a large area on which the acid and carbonic acid gas, which were directly controlled, could operate. They were cheaper than the stack process but not entirely successful in producing a lead carbonate which would compete in quality with stack white lead. At the beginning of the First World War then, stack white lead still dominated the industry and estimates for the early 1920s give a world output of 275,000 tons of white lead, of which 180,000 was produced by stack, 45,000 by chamber and the remainder by a variety of other processes.

In the manufacture of the other major lead chemical, red lead, there was little change until the very end of the nineteenth century. The two-stage process in a reverberatory furnace, first oxidising lead to massicot and then (after cooling and grinding) roasting it until red lead was obtained, remained the normal practice.[48] It was a time-consuming process which depended on heavy manual labour (especially 'rabbling', the stirring of the lead in the furnace with an iron bar — a process akin to puddling in the iron industry and aimed at continually opening new areas of lead to the action of air). Perhaps the only difference by the late-nineteenth century was that furnaces were larger, taking a charge of about 30 cwt, than they had been in the late-eighteenth century when 16-18 cwt was the norm. By contrast an account published in 1912 stated

> Red lead can be, and is, now made on an extensive scale in such a way that all operations, from commencement with pig-lead to the final packing, are carried out by mechanical means so entirely closed in that the worker does not come into contact with the material . . . The pig lead is melted, stirred, and mixed in a covered-in melting pot. The massicot which is formed is drawn off by an exhaust into a hopper, from the bottom of which it is fed mechanically on to the floor of the furnace. Mechanical rabbles stir it from the centre to the outside of the furnace floor, from where it is conveyed, under negative pressure, to the hopper of a grinding mill. From here it is again similarly led into another furnace. The exhaust pipe from this furnace collects the finished product, carrying it mechanically to a hopper which automatically feeds the red lead into casks. Negative pressure throughout prevents escape of dust.[49]

It is difficult to see why it had not been possible successfully to introduce a mechanical furnace before this date and indeed John Percy, writing at the end of the 1860s, mentioned a mechanical red lead furnace, with a central shaft fitted with blades but this had obviously not proved practicable. Even the Americans, usually by this time in advance of Britain in the introduction of labour-saving devices, had failed to solve the problem and in 1909 Holley described the usual two-stage hand furnace as the equipment for manufacturing red lead in the USA.[50] Unusually, therefore, it was Britain which led the way in the new technology, although the story of its development and eventual innovation is a very tangled one.[51]

It begins with the setting up at Runcorn in December 1897 of one of the many companies to produce white lead by a new process, Matthews' Lancashire, Cheshire and North Wales District White Lead Co. Ltd, which deserved to fail if only because of the length of its title and promptly did so in September 1901.[52] In the meantime G.V. Barton, who was employed by that company, had developed his own process for manufacturing lead oxide, which was to become known in lead manufacturing throughout the English-speaking world as the Barton pot. This was patented in 1898 (nos. 17,178 and 21,830) and in 1901 the Runcorn White Lead Co. Ltd, with Barton as manager, was set up to carry on the Matthews' works, but it went bankrupt in October 1903.

In the Barton pot melted lead was stirred mechanically in the presence of air and steam, which led to the formation of lead oxide which was skimmed mechanically from the surface of the metal. Initially the process converted only about two-thirds of the metal to oxide and was, therefore, uncommercial but in fresh patents in 1902 and 1908 Barton added improvements which led to almost complete oxidation. Barton now tried to obtain licensees for his process but with little success until 1911 when the rights were purchased by Rowe Bros Ltd of Liverpool, a well-established builders' and plumbers' merchants. By this time the idea had been around for some time and it was perhaps inevitable that it would be improved by someone else.[53] W. Eckford, at this time Walkers, Parker's Bagillt and Chester works manager, who was very conveniently placed to hear what was going on at Runcorn and Liverpool, developed what was to be known as the Eckford pot. This was a modification of the Barton pot, by which a pig of lead was fed into a well at the bottom and to one side of the pot. The level of molten lead in the pot was by this means maintained continuously, while in the Barton pot liquid lead was fed in by a pipe at the top of the pot. The Eckford pot also contained a mechanical stirrer of slightly different design from Barton's. In 1911 Eckford received royalty payments from his employers which imply that about 2,500 tons of red lead was produced by his process. Clearly Eckford had solved the problem of producing a saleable product, and this spurred on other UK manufacturers.

Cooksons had taken a share in the Barton patents and, in conjunction with Rowe Bros, set up the Three Castles Lead Co. Ltd to exploit the patents. Experimental plant was set up overseas as well as at home and by May 1913 Cooksons felt strong enough to offer the USA rights to the major American company, National Lead Co. The offer was backed with the threat of Cooksons' intention of erecting a full production plant in the USA but 'we should probably be willing to consider an arrangement for licensing or even for outright sale . . .'.[54] National Lead must have felt that they were under a squeeze from the British companies since they were also offered the Eckford patent by Walkers, Parker. In an attempt to get free they obtained legal advice but this was unhelpful and National Lead had to purchase the American rights to both sets of patents in order to avoid potential competition.

Back in the UK, while Walkers, Parker had their process in production from 1910, it was not until 1912 that Cooksons installed Barton pots at Hayhole. Relations between these two main producers were touchy and there were regular threats of legal action for patent infringement, but it will be remembered that this was the time at which the lead manufacturers were drawing together in pooling associations and the possibility of a rift over red lead threatened the fragile security which had been achieved in white and blue lead manufacture. Mutual differences were therefore shelved, Walkers, Parker continued to use the Eckford process unchallenged and Cooksons, after complex disagreements in 1915 with their partners, Rowe Bros, agreed to continue to pay them a licence fee and everyone ended up moderately happy.

While the development of mechanical red lead production was an area of technology in which British manufacturers were in the van, there were no other areas which saw major changes. The rolling mill was well established in principle and merely experienced change in detail, while in pipe manufacture the extrusion press developed from manual to gas-powered, hydraulic operation and also, no doubt, changed in the detail of design. In general, however, power supply remained of traditional forms — human muscle and steam engine. Beam engines remained in operation into the twentieth century to provide power for blast furnaces, grinding of pigments and turning of rolling mills. In the second half of the nineteenth century, however, gas engines began to be used where small amounts of power were required for a specific purpose — pipe presses were often powered by gas engines (although subsequently converted to electric drive in the years before the First World War). Larger gas engines, four of 70 hp each, were installed as the chief power source at Brimsdown Lead Co. Ltd in 1902-3 and several works used gas engines to drive dynamos to provide electricity for lighting purposes from the 1890s. Gas lighting seems generally to have been replaced by electricity between 1895 and 1905 although the Elswick works was still lit by gas in 1911 and the Gallowgate works of Locke, Blackett as late as 1922. While electricity was widely used for lighting it was only gradually introduced for power purposes from about 1900.

Finally, in this survey of the industry over the nineteenth century, we may look briefly at the labour force, on which there is not a great deal of evidence. Total employment in lead manufacturing, even in the early-twentieth century, was only a few thousand. The Census of Production in 1907 gave employment in white lead manufacture as

1,737 persons and since this was the most labour-intensive branch of the industry it is unlikely that total employment exceeded 5,000.[55] Individual firms numbered their employment in hundreds, although Walkers, Parker, with several plants in different parts of the country, might have employed as many as 1,000 at most and certainly had a works employment of 750 in 1898.[56] From the beginning of the nineteenth century, when a works such as Elswick with perhaps 100 employees was unusual, the common plant size for the major firms had reached 100-150 by the middle of the century. At the time of the 1842 report of the Commission on the Employment of Children, Elswick had 96 male and 45 female employees, while employment at the Chester works of Walkers, Parker fluctuated between 150 and 179 between March 1839 and July 1840. By the end of the century an employment level of around 250-300 was common – FB & W employed 167 males and 118 females at their Hebburn works in 1897.[57]

It is clear that most of the labour was unskilled work, with a limited amount of semi-skilled work (more the result of practice and dexterity than innate ability) as in the rabbling of red lead furnaces and a very small amount of skilled work – the chief smelting and desilverising jobs, for instance, and coopering (a considerable output of casks and barrels for packing red and white lead was largely produced internally). Of the unskilled work most was heavy manual labour, moving pigs of lead to melting pots, charging furnaces and loading heavy finished products on to waggons and barges. Work for women was especially of this kind, being chiefly concerned with filling and emptying white lead stacks, each of which might contain 100 tons of materials. Women were employed in the work because they could do it equally as well as men, while they could be paid less.[58] In an area such as Tyneside, with limited job opportunities for women, the white lead factories provided a significant place for female employment. Even there, however, they appear to have attracted only the lowest quality of labour, especially as the dangers of lead poisoning became more widely recognised. For many working-class girls, to go into service with its much longer hours, drudgery and low wages was infinitely preferable to working in the 'white lead'. At the lowest end of society, however, the white lead works were often the only alternative harbour to destitution and the workhouse for a single woman or one with a husband unemployed or on low earnings and a large, young family. Frequently those who ended up on poor relief as a result of lead poisoning were described as 'deserted women' – a clear reason for their need to take any available employment. Beyond the need to earn, however, there was often a fatalism about their attitudes to the work which points to a kind of 'prison mentality' engendered by the circumstances in which they lived. Awareness of the dangers of lead poisoning brought no attempt to escape from its influence. Isabella Rushton, aged 23, who was blind from the effects of lead poisoning said

> I have never worked at any other place but Walker Parker & Cos [Elswick] I had two sisters who worked there both died from the effects of the white lead. They never worked anywhere else. My oldest sister (Ann Rushton) died first — she was about 27 when she died, that is about 6 years ago. The other one Margaret Rushton died about 4 years ago she was turned 24.[59]

It seems likely that the lead works attracted some of the poorest quality of labour, although they were also, undoubtedly, the resort of some of the casualties of society. While Sir William Chaytor was putting the case too strongly when he informed his superiors in London, 'I have enquired into more than one case of lead poisoning always with the same result. Leadworkers as a rule are of the lowest order, drink all they make & do not get proper meals of solid food besides leading otherwise debauched lives',[60] there was some truth in his claims which are supported by many shrewder commentators. Charles Booth commented that:

> The labour employed is mostly casual; the men are taken on and paid off day by day . . . The white lead works are looked upon as the last resource of the starving — of the helpless and lowest class of the unskilled — but there are those who, having once 'tasted the lead colic', prefer to apply to the Guardians rather than repeat the experience.[61]

and T.M. Legge, HM Medical Inspector of Factories, had much the same to say in 1898.[62]

Almost inevitably employers commented critically on their labour. Just as two examples we may take the comments of William Sloane of Walkers, Parker at Chester and George Burnett of Locke, Blackett to the 1842 Commission — 'we are obliged to put up with the refuse of the people — Those who offer their services are usually loose characters' and 'They belong in general to the very lowest class.'[63] The problem was more marked in some firms than others and Legge drew attention to the fact that some firms employed a fairly permanent labour force (usually of higher quality) while others depended on casual labour. At FB & W, for instance, Legge noted an average employment of 167 males and 118 females in 1897 but in August 1898 he found that 292 males and 443 females had been employed at some time during the previous eight months.[64] By contrast a firm such as Cooksons had much lower turnover and much less severe incidence of lead poisoning as a result. It was not, however, merely a distinction between the practices of different firms, since there was also a distinction within firms between different jobs. In the more skilled occupations long service and family involvement over several generations, with a particular firm, were much more common than was casual labour.

As a small employer requiring limited specific skills, the lead industry largely followed wage levels set in the surrounding area, although average wage levels tended to be low both because of the high level of female employment and the large proportion of unskilled workers. Women employed in setting and emptying the white lead stacks were the only large group of workers paid by the piece, most other employees being paid by the shift or by the early-twentieth century by the hour. Women appear to have started work at 6.00 a.m. and worked usually until they had done a prescribed amount of work — early afternoon (1.30 p.m. or 2 p.m.) being frequently given as the time at which they ceased work. For this stint they were paid in 1842 at Locke, Blackett's 6s 3d per week according to the managing partner and about 10s per week at the Ouseburn works of James & Co. Precise earnings figures may be given for the women employed at Walkers, Parker's at Chester in 1839-40, for which period a payments book is extant. They were paid weekly for a week of six days, although there was a tendency for some to work less than a complete week which pulls down the average earnings. For three casually selected pay days from the 73 covered by the volume the average wage was: 15 March 1839 for 26 women 7s 9d; 4 October 1839 for 23 women 7s 4d; 31 July 1840 for 24 women 7s 3d. It is therefore probably not unreasonable to state that a woman working a full six-day week of perhaps seven hours per day, could take home eight shillings a week. Higher earnings could be made by working longer hours when demand was high, but such periods, like the converse of slack demand and less than normal working, do not seem to have been common. By the 1880s earnings had increased by about 50 per cent. Isabella Rushton told the Gateshead Poor Law Guardians, 'I made 12s upstanding wage and [with] what we could make overtime I would make 16s or 17s a week or thereabouts.' These figures, which relate to Walkers, Parker at Elswick, are confirmed by the very precise figure for the daily average earnings of all women employed at FB & W during 1892 of 2s 4$^{11}/_{16}$d.[65]

For women, who were chiefly employed only in a single activity, it is not too difficult to quote an average wage which bore some relationship to actual take-home earnings but when we turn to male employees the position becomes much more complex because of the wider range of jobs. The 1842 Commission gave a range of 16s to 22s per week for men's wages for James & Co. on the Tyne. Although these would appear to be rather high, they receive some support from detailed figures for the Chester works of Walkers, Parker which give the following average weekly wages: 15 March 1839 for 117 men 15s 7d; 4 October 1839 for 131 men 15s 5d; 31 July 1840 for 141 men 17s 7d. Since these figures cover all male employees, including a few boys and those working occasional days, they imply an average figure of close to £1 a week for full-time adult males. This figure is confirmed by the evidence of a wages book for Locke, Blackett & Co., St Anthony's works, an all-male works without the unskilled white lead workers, for 1851-2. There the average wage was £1 a week, with a range from 15s to 25s. Many employees received a regular wage on the basis of a six-shift week with only very rare overtime. The smelters received 4s 2½d per shift (although the slag hearth smelter received 4s 6d) and did only five shifts a week, while separators (working Pattinson desilverising pots) received 3s 9d and usually did six shifts a week. Labourers received 2s 4d to 2s 8d per day, while red lead workers, receiving rather more than labourers but less than skilled men (between 2s 10d and 3s 8d) seem to have done six or seven shifts regularly. The single foreman, Thomas Craig, received 27s a week throughout the period without variation. The length of the shift may have depended on the requirements of the work, as in the tending of a smelting or a red lead furnace, but the general evidence suggests that the normal day for male workers in the lead industry was one of twelve hours with 1½ hours off for meals, identical to that imposed by legislation in the textile manufacturing industries.

When Booth wrote, in the 1890s, giving figures for London, where one would expect to find high wages,[66] he stated that leading hands in blue lead works received up to 35s a week, ordinary men 24s to 28s but second-class labourers less than this.[67] There are no figures available from the firms' records for the late-nineteenth and early-twentieth century from which one might check such figures. After the First World War, however, an earnings survey at Cooksons stated that labourers received 5d an hour pre-war and skilled workers 4s 6d to 6s per day. The evidence is clearly scanty although it does suggest that the lead manufacturers paid what was the going rate in their area, since they had few requirements for particular skills. It also seems likely that earnings in the industry grew more slowly than the national average in the late-nineteenth century. Perhaps this could partly be put down to the fact that 'There is no trade organisation in this industry.'[68] Being a small industry, without a homogeneous group of skilled workers, lead manufacture had not developed its own union although some of the unskilled did join the general unions which developed in the 1889-90 boom.

On aspects of the labour force other than quality and earnings little can be said. Some evidence on conditions of employment will be given in the following chapter on lead poisoning. Here it may be said that some employees, especially women employed in white lead departments, received a few marginal fringe benefits in the form of free food provided by their employers (usually in the form of breakfast to avoid their starting work on an empty stomach – supposedly a precaution against lead poisoning) and a room or canteen where they (or sometimes a woman employed by the firm) could prepare their lunches. Several of the larger employers, notably Cooksons, Walkers, Parker and Locke, Lancaster & Johnson, were, by the late-nineteenth century running annual works trips for their employees, but they do not appear to have given any other holidays (except the usual bank holidays) to their employees before 1914. Dependent upon the paternalism of their employers some employees received other benefits. In 1912 Cooksons leased extra land at Hayhole 'to be used for gardens for our work-people' and in the following year Mersey entered the yearly tenancy of a bowling green 'for the benefit of the workmen of the Company'. By the late-nineteenth century Cooksons were paying a bonus, usually £25-£30, to employees completing 50 years service – little enough in the way of recognition of a lifetime's service but more than was done by most firms and, importantly, a help towards keeping employees out of the workhouse when retirement came. Cooksons appear to have been more generous than most firms in respect of *ex gratia* payments and occasionally even pensions to old employees. Most firms did something in this area, even if it was as little as Locke, Blackett's resolution in July 1892 'that a contribution of £1 be paid towards the funeral expenses of old servants of the Company in meritorious cases'. Economy measures led Walkers, Parker to an even more stringent attitude in the early-twentieth century. A request for a pension for an ex-employee of 40 years' service received the stony reply 'that the Board have no fund from which to grant pensions to any of the employees', while in July 1910 the Walkers, Parker board resolved to give three Lambeth employees a month's notice because of the heavy cost of insurance under the Workmen's Compensation Act. They were aged between 65 and 67 and had 29, 33 and 37 years of service with the company and the annual saving to the company at 2 per cent of their wages (the cost of the insurance) would have been £4 2s 6d. No pensions were given and in October when one of the men, unable to obtain work and close to destitution, applied again to the board for assistance, he was refused.

Paternalistic attitudes did exist within the lead firms but they were always more common within the partnerships than the limited companies where shareholders had to be considered and especially in such limited companies as Walkers, Parker which had a woeful profit record between 1890 and 1914. Ultimately it depended on those at the top as to whether and to what extent charitable payments were made and partners had both more scope for making such payments and more to gain from doing so than did boards of directors. For this reason alone paternalism may have declined during the later-nineteenth century with the replacement of partnerships by limited companies. Overall, wages and conditions of employment in the lead industry were probably a microcosm of the industrial economy as a whole, showing considerable ranges but reflecting the overall picture. The one distinctive feature was lead poisoning to which we must now turn.

Notes

1. The figures have been calculated from B.R. Mitchell and P. Deane, *Abstract of British Historical Statistics* (Cambridge, 1962), pp. 160 and 169-71, where they are given as an annual series. They have been grouped here to give annual averages from quinquennial totals in order both to save space and to reduce the significance of annual fluctuations. The final figures for UK lead consumption in column 5 do suggest a considerable hiatus from Burt's eighteenth-century estimates. This is largely a result of his estimates for UK lead output, which may be too high. He estimates (Burt, Lead Industry of England and Wales, pp. 351-4 and 'Lead Production', especially pp. 264-5) that UK lead output was 50,000-60,000 tons p.a. in the late 1760s-1770s, declined slightly in the 1780s, saw little change 1790s-1800s, declined in the 1810s and was perhaps 45,000 tons in the 1830s, from which it grew to the known estimates of Hunt from 1845 onwards. Given that total exports of lead averaged less than 20,000 tons in the 1760s and 1770s, Burt's figures would imply that UK lead consumption in those decades was more than 30,000 tons p.a. and perhaps more than 40,000, not far short of my estimates for the late 1840s, although the latter may be underestimates because of the unreliability of Hunt's figures for output of UK pig lead before 1850. It seems highly unlikely that three-quarters of a century of industrialisation had not seen a rising use of lead, especially since Burt himself points to a rapid growth in UK lead consumption in the first three-quarters of the eighteenth century. It is therefore likely that one of the estimates is inaccurate.

Perhaps the biggest unknown in the problem is the lack of evidence on the availability of scrap lead for reuse. If this were considerable and increasing throughout the period from 1770-1850 it would account for the failure of the supply of new lead to expand and the hiatus would be explained. Although local plumbers etc. probably used scrap materials there is no evidence that the large companies, whose growth we have charted, did so and they must have accounted for the major growth of the industry in the first half of the nineteenth century. Although there is some disagreement between our figures and in our views on the importance of the contribution of secondary lead to total consumption, I should like to express my gratitude to Dr Burt for taking considerable trouble to answer my queries on several points and for correcting and clarifying my ideas.

2. Pulsifer, *History of Lead*, p. 60, 'The argentiferous lead of Spain was formerly desilverised in Great Britain; but the art of refining has lately been fostered by the Spanish Government, and English smelters have lost much of the business.'

3. Ibid., pp. 315-18.

4. This is a highly improbable figure. There are no figures available for British exports of lead chemicals for 1810 but red and white lead exports combined were only 1,000 tons in the peak year of 1805, from which figure they declined to only about 150 tons in 1808 (the last year for which statistics are available), which suggests that most of the exports had previously gone to Europe and had been reduced by Napoleon's economic warfare. See Schumpeter, *English Overseas Trade Statistics*, p. 28. When new figures commence in 1816 white lead exports were only 527 tons and did not reach 1,000 tons until 1827. *BPP*, 1828 [90] XIX, 405.

5. University of Durham, 'Cookson' Box III, 48, pp. 49-50.

6. *Oil and Colourman's Journal*, 25 March 1905, pp. 846-7.

7. Metallgesellschaft A.G., *Metal Statistics* (Frankfurt, various years).

8. This might be put into perspective a little by pointing out that consumption of lead was the second highest of any metal, ahead of copper but a very long way behind iron. UK output of pig iron rose from about 200,000 tons in 1800 to 2.5m in 1850 and 9m in 1900, while copper output rose from about 10,000 tons in 1800, through 30,000 in 1850 to 80,000 in 1900 and that of tin (the fourth metal) from about 3,000 through 12,000 to 45,000. In each of these cases imports became important from about 1850 and the figures given are approximations to the total available supply of the metals in the UK, without deduction for exports. The figures are based on Mitchell & Deane, *Abstract of Historical Statistics*, pp. 131-2, 154-5, 157-9, 163 and 166-7, and on *Records of the School of Mines*, vol. 1, part IV, Statistics of the Produce of Copper, Tin, Lead . . . (1853), pp. 477-9, and *BPP*, 1902 [Cd 818] cxvi part 2, 377, Annual Return of Mineral Statistics of the UK for 1900.

9. Both Roger Burt and Lynn Willies have discussed the significance of secondary lead in the nineteenth century with me. They both suspect that it was more significant than I allow. The first figures for which I am aware for the supply of secondary lead relate to the USA from 1916 onwards. They appear in the annual publication, G.A. Roush (ed.), *The Mineral Industry. Its Statistics, Technology and Trade* (New York). Vol. XXVII (New York, 1919), pp. 413-14 gives a supply of rather less than 100,000 short tons of secondary lead in the years 1916 and 1917 against about 600,000 tons of primary lead. By 1930 it had risen rapidly to 311,000 against 693,000 tons of primary metal. It was specifically noted that the supply of secondary lead came particularly from batteries and it seems likely that consumption of secondary metal would be expanded rapidly in war time. If secondary metal accounted for less than 15 per cent of USA consumption in 1916 it seems unlikely to have been of much significance in the

UK before the late-nineteenth century.

10. Pulsifer, *History of Lead*, p. 291.

11. The 1907 Census of Production, after endeavouring to deal with the problem created by the fact that a number of firms producing white lead were in fact classified as paint manufacturers, who made returns not of white lead but gallons of paint, also concluded that output was about 50,000 tons valued at £1,060,000, while the gross value of the total output of the lead industry was £4,270,000. *BPP* 1912 [Cd 6320] , Census of Production (1907), Final Report, p. 252.

12. Pulsifer, *History of Lead*, p. 271 and Holley, *Lead and Zinc Pigments*, p. 19.

13. The statistics of housebuilding are presented in Mitchell & Deane, *Abstract of Historical Statistics*, p. 239. They show a peak in 1876 of 130,000 houses constructed, a long period of stability at about 80,000 p.a. from 1879-91, then a rise to twin peaks of above 150,000 in 1898-9 and 1902-3, followed by a catastrophic slump to 54,000 in 1912-13 with a negligible output during the war years. Output of sheet lead nationally must have followed this pattern fairly closely, although there were considerable regional variations in housebuilding which would have made differences in demand to particular lead works. In south London, for instance, housebuilding peaked in 1879, H.J. Dyos, *Victorian Suburb* (Leicester, 1961), p. 81.

14. See, for instance, J. Burnett, *A Social History of Housing, 1815-1970* (1978), especially pp. 210-11.

15. The ratio of 55/45 for pipe and sheet may not hold before 1914. It is probable, as a result of the large-scale installation of water supply to existing houses in the late-nineteenth century, that output of pipe rose more rapidly than that of sheet. Conversely in the slump in housebuilding down to 1914 demand for pipe was likely to fall more severely than that for sheet which was less dependent on housebuilding demand.

16. From 1911 the figures are for the output of the members of the UK Lead Manufacturers Association, a cartel of the major sheet and pipe manufacturers in the country. For 1901-10 the figures are for the reported output of the same firms, on which the quotas for the convention were subsequently based. There is no precise evidence on the proportion of total output of sheet and pipe accounted for by these firms but it was probably about 90 per cent. The books of the Association, back to its formation in 1911, are held by the British Lead Manufacturers Association, 68 High Street, Weybridge, Surrey, KT13 8BL. I am particularly grateful to Mr B.R. Lewis-Smith for permission to consult and quote from the records.

17. The figures are roughly as follows:

| | Average annual output (tons) | |
Product	1855-9	1905-9
White lead	20-25,000	50,000
Red lead	8,000	13,000
Sheet	25-30,000 }	123,000
Pipe	10,000 }	
Shot	6,000	10,000
Miscellaneous	?	?
	74,000	196,000
	(70,000)[1]	(183,000)[1]
Pig lead available for home consumption (from Table 6.1)	54,500	181,600

[1] Equivalent pig lead after allowing for weight gain in white lead (the red lead figures are so speculative and the weight gain in that commodity small, 10 per cent, that no allowance has been made in them). It might be thought that the 1907 Census of Production would give suitable figures but there is no way in which the element of double-counting in its figures for pig lead and manufactured lead output can be eliminated.

18. The figures for 1862 and 1873 are taken from B. Plummer, *Newcastle upon Tyne: its Trade and Manufactures* (Newcastle, 1874), p. 121; remaining figures from annual *Abstract of the Tyne Improvement Commissioners' Accounts* (Newcastle, various dates).

19. There is not a great deal of information on company profitability generally in the nineteenth century and what there is suffers from lack of comparability because of the problem of varying and unknown accounting practices. The subject is, however, discussed in some depth in R.A. Church (ed.), *The Dynamics of Victorian Business. Problems and Perspectives to the 1870s* (1980), especially pp. 25-44.

20. M. Girouard, *The Victorian Country House* (1979), pp. 6-7.

21. Ibid., p. 13. The Blacketts of Wylam also appear in the first edition of Burke.

22. Ibid., p. 10.

23. F.C. Hill to J.H. Stewart, 18 Jan. 1938. Hill joined the Johnson firm in 1873 and eventually became managing director.

24. The original signatories were James & Co. Ltd, Foster, Blackett & Wilson Ltd, and Locke, Blackett & Co. Ltd (all of Tyneside), W.W. & R. Johnson & Sons Ltd, Champion Druce & Co., Millwall Lead Co. Ltd, Henry Grace & Co., and Brandram Bros & Co. (all of London), Lewis Berger & Sons Ltd (of Sheffield), and Walkers, Parker & Co. Ltd, while John Hare & Co. of Bristol was added in 1893.

25. A Walkers, Parker board minute of 30 Aug. 1911 noted that Mersey 'was still an unknown quantity' but early in September a meeting took place at Smedley's Hydro, Matlock, where F.C. Hill was on holiday, with W. Sloane, managing director of Mersey, and that firm was brought in. It was not the only 'accommodation' which took place. C.F. Hill was with his father on holiday, while Sloane was accompanied by his daughter, Winifred. The children were subsequently married in 1914.

26. FB & W was given a quota of 2.3 per cent and James & Co. 2.7 per cent, which would give them annual outputs of little over 1,000 tons each, implying a considerable reduction from their probable outputs in the late-nineteenth century.

27. Cooksons appear to have been a particular bother in this respect and in 1909 there were meetings of all the UK regional sheet and pipe associations at which it was agreed that Cooksons would cease price competition. In return Cooksons were to be given an increased quota of sheet and pipe which was to be purchased by the Newcastle and London associations. They, in their turn, were to be recompensed by the payment of £2,100 p.a. from the other regional associations (at the rate of £108 levy per roller and £16 per pipe manufacturer). Price stability was clearly worth complex negotiation. These details come from the minute books of Alexander, Fergusson and I am grateful to Mr J.I.M. Barr for permission to consult them.

28. One fascinating result was that a representative from Krupps was said to have arrived at the works of James, with plans and specifications, the day after the advertisement appeared in the papers.

29. Readers who are interested in discussion of different smelting processes are recommended to consult H.O. Hofman, *The Metallurgy of Lead* (1899),

pp. 83-407, Percy, *Metallurgy of Lead*, pp. 213-492, Tylecote, 'Lead Smelting and Refining', and Raistrick & Jennings, *History of Lead Mining*, pp. 78-85 and 121-2.

30. In Wigham Richardson, *Visit of British Association to Newcastle*, p. 115. It is clear from the context that Cookson meant shaft blast furnaces of the Spanish type, although in the later-nineteenth century ore hearths were also water-jacketed to enable them to work continuously.

31. It is sometimes stated that Pattinson discovered the process as a result of the accidental dropping of a crucible. Since this would have caused the lead to spread on the floor, when it would have been very difficult to separate the first to crystallise, the story seems unlikely. The source for Pattinson's dropping the crucible is Percy, *Metallurgy of Lead*, p. 123. For Pattinson's own account of the process see H.L. Pattinson, 'On a New Process for the Extraction of Silver from Lead', *British Association Eighth Report* (1838), pp. 50-5.

32. Accounts of cupellation, the Pattinson process and the Parkes process, to be mentioned shortly, may be found in Singer, *History of Technology*, vol. IV, pp. 137-8, Hofman, *Metallurgy of Lead*, pp. 412-535, Percy, *Metallurgy of Lead*, pp. 121-74, Smythe, *Lead*, pp. 146-73, and Tylecote, 'Lead Smelting and Refining', pp. 106-8.

33. N.C. Cookson, in Wigham Richardson, *Visit of British Association to Newcastle*, p. 115.

34. E.L. Collis [Medical Inspector of Factories], Report on Dangerous or Injurious Processes in Lead Smelting . . ., *BPP*, 1910 [Cd 5152] XXIX, 51, p. 12 and Smythe, *Lead*, p. 162.

35. Plummer, *Newcastle upon Tyne*, p. 121.

36. See the comments by N.C. Cookson in the *Journal of the Institute of Metals*, vol. 1 (1909), p. 293. That this was not an isolated example is shown by the fact that in 1900 Locke, Blackett paid for one of their employees, Andrew Harbottle, to 'attend a Chemistry Class at the Durham College of Science'.

37. He was to become Hunterian Professor of the Royal College of Surgeons in 1911 and was knighted in 1918. See *Who Was Who 1951-1960*.

38. See, for instance, Pulsifer, *History of Lead*, pp. 283-90 and 317-34.

39. Pattinson's product was probably a mixture of basic lead chlorides. Its apparent success, as compared to other patents for the manufacture of lead chloride, was almost certainly a result of its physical properties, particle size, refractive index, etc. Basic lead chlorides vary considerably in physical properties and one product might be satisfactory as a pigment while another would not. For Pattinson's process see Percy, *Metallurgy of Lead*, pp. 81-4. For the range of lead oxychlorides see Greninger, *Lead Chemicals*, pp. 160-1 and 188-9 where the formula $Pb(OH)_2.PbCl_2$ is given for Pattinson's white lead.

40. In 1882 there was, for instance, an Innocuous White Lead Manufacturing Co. — but it was also innocuous in the commercial sense and soon disappeared.

41. *The Times*, 26 Jan. 1893. Such articles were common. Under the heading 'A new white lead process' and commencing 'The evils attending the manufacture of white lead by the Dutch or stack process are too well'known to require insisting on', *The Times* (12 July 1890) had commented favourably on the MacIvor process.

42. PRO, HO 45/9848/B12393A/4 and 8, James Henderson, HM Superintending Inspector of Factories for the Glasgow Area to R.E. Sprague-Oram, HM Chief Inspector of Factories, 29 Dec. 1892 and 17 Feb. 1893.

43. *Oil & Colourman's Journal*, 1 Feb. 1893, pp. 265-6.

44. In Wigham Richardson, *Visit of British Association to Newcastle*, p. 117. See also Pulsifer, *History of Lead*, pp. 280 and 290.

45. Holley, *Lead and Zinc Pigments*, p. 2.

46. A number of firms, apart from Cooksons, put in chamber plant but none

had lasting success. Of these the most significant was S. Tudor & Co. Ltd, whose works was started in the 1880s but was up for sale in 1906 (*Oil & Colourman's Gazette*, 10 Nov. 1906). Other firms with chamber plant included Wilkinson, Heywood & Clark of West Drayton, Middlesex (*BPP*, 1893-4 [c.7239-I] XVII, p. 380) and H.J. Wright of Ponders End, Middlesex (T.M. Legge, *Report on White Lead Works* (1898), p. 3).

47. The processes are discussed in Holley, *Lead and Zinc Pigments*, pp. 74-107 and Smythe, *Lead*, pp. 272-3. Notes in a Cookson laboratory notebook on a sample of Carter white lead tested in 1912 state, 'very poor colour' and 'We could not meet any of our usual specifications with a lead of the above composition.'

48. The processes at work in Derbyshire in the mid-eighteenth century as described by Gabriel Jars, and at FB & W on Tyneside in 1895, were virtually identical. See Willies, 'Gabriel Jars', and *Chemical Trade Journal*, 19 Jan. 1895.

49. T.M. Legge & K.W. Goadby, *Lead Poisoning and Lead Absorption* (1912), p. 250.

50. Holley, *Lead and Zinc Pigments*, p. 209.

51. The following account is drawn chiefly from 'Petition for prolongation', High Court of Justice, Chancery Division, 1915, B No.060 (31 July 1915) and *The Illustrated Official Journal (Patents)*, Supplement, 5 July 1916, pp. 201-14 [vol. XXXIII], Reports of Patent, Design and Trade Mark Cases [no. 10].

52. Matthews' delusions about the potential of his process were on a grander scale than those of most of the other patentees of white lead processes. He certainly set up Matthews' Newcastle on Tyne District White Lead Co. Ltd and Matthews' Kentish White Lead Co. and possibly other companies as well. The process supposedly began from lead oxide to be manufactured by a patent process invented by James Noad (1895, 21,175).

53. There was probably a number of people working on the problem and apart from the Eckford pot, described below, there was also an Innes pot (patented by William Innes, works manager of Rowe Bros at Runcorn, patents 1912 no. 15,664 and 1913 nos. 10,387 and 16,700). Barton also made further developments in patents 1909 no. 4,465 and 1910 nos. 13,600 and 21,662.

54. I am grateful to Arnold Judson for making available to me photocopies of this correspondence with the American firms, which had been preserved by National Lead Co. and came into the hands of ALM's American subsidiary in 1979.

55. *BPP*, 1912 [Cd 6320], Census of Production (1907), Final report, p. 252. White lead was the only section of the industry for which the printed tables gave employment figures.

56. PRO, HO 45/9938/B 27900, Walkers, Parker & Co. Ltd to the Home Office, 31 Oct. 1898.

57. Ibid., T.M. Legge to E. Gould, 13 Sept. 1898.

58. It is difficult to explain how the tradition of female labour grew up in this employment, which for most of the nineteenth century was entirely dominated by women, but not in others in which there was, presumably, manual work which could have been done equally well (but more cheaply) by women than men.

59. Gateshead Public Library, Minutes of meetings of Gateshead Board of Poor Law Guardians, 4 Nov. 1884. When female employment in white lead was banned in 1898, some firms had difficulty in obtaining male labour, even though the wage for the job rose. Locke, Lancaster & Johnson solved the problem at Millwall by using immigrant labour and a considerable Italian community built up on the Isle of Dogs.

60. PRO, MH12/3096, Sir William Chaytor, Inspector of Factories for the Tyneside district, to the Local Government Board, 11 Nov. 1887. I owe this reference to F.W. Manders.

61. C. Booth, *Life and Labour of the People in London* (1895), vol. VI, p. 103.

62. Legge, *Report on White Lead Works*, p. 5.

63. *BPP*, 1842 [382] XVII, Childrens Employment Commission, Appendix to First Report of Commissioners (Mines) Part II, p. 447 and 1843 [432] XV, p. 145.

64. PRO, HO 45/9938/B27900, T.M. Legge to E. Gould, 13 Sept. 1898.

65. Gateshead Public Library, Minutes, 4 Nov. 1884. PRO, HO 45/9848/B12393A/4, James Henderson, HM Superintending Inspector of Factories for Scotland and the North of England, to HM Chief Inspector of Factories, 29 Dec. 1892.

66. See E.H. Hunt, *Regional Wage Variations in Britain 1850-1914* (Oxford, 1973).

67. Booth, *Life and Labour*, vol. V, p. 379. These figures are more or less confirmed for Tyneside for the same period by the evidence given to the Royal Commission on Labour, *BPP*, 1893-4 [c.6894-IX] XXXIV, Minutes of Evidence before Group C, vol. III. The desilverising workers, who were the best paid, received 33s a week and the most unskilled labourers 20s to 22s per week.

68. Booth, *Life and Labour*, vol. V, p. 379.

7 INDUSTRIAL LEAD POISONING AND THE MEASURES TAKEN TO PREVENT IT

> Some of them gets lead-pisoned soon, and some of them gets lead-pisoned later, and some, but not many, niver; and 'tis all according to the constitooshun, sur, and some constitooshuns is strong and some is weak.
>
> Charles Dickens, *The Uncommercial Traveller*, Ch. 34.

Extent of lead poisoning and efforts to control it from the nineteenth century to the present. Evidence of government enquiries from the 1840s, reports of factory inspectors from the 1870s, Poor Law Guardians' complaints; from 1883 Special Rules established to control conditions of employment in white lead works, their extension, the banning of the employment of women in white lead and notification of lead poisoning as an industrial disease. From 1900 the number of cases of and deaths from lead poisoning fall rapidly as precautionary expenditure on hygiene increases. Since the 1960s industrial lead poisoning has become insignificant and attention has turned to lead emissions from vehicle exhausts.

That those who worked in the manufacture of lead products were at risk from lead poisoning[1] had been known since classical times but in centuries when there were so many factors which could cut life short and when life expectancy, especially for working men, was low, lead poisoning remained just another problem. It was accepted as a natural hazard of particular types of work. In 1747, *The London Tradesman*, a widely used guide for those who were contemplating apprenticing children, stated

> There are works at *Whitechapel*, and some other of the Suburbs, for making of White and Red Lead ... the Work is performed by ... Labourers, who are sure in a few Years to become paralytic by the Mercurial Fumes of the Lead; and seldom live a dozen Years in the Business.[2]

The time of the Industrial Revolution was not one in which the value of life generally was held in high esteem or in which the labour force was seen as anything but a factor of production whose relationship with the

209

employer ended with the financial one — a wage in return for labour provided. Nor was it a period in which the State laid down controls over conditions of works — indeed the period was one in which the view became increasingly held that the State would do positive damage by interfering with the productive process. This being the case it is not to be wondered at that lead poisoning went on unchecked and largely unnoticed.[3] It was a hazard of employment which was similar to the dangers faced by pitmen or by seamen. If employment was a financial contract then *caveat emptor*, with good fortune going to those workers who found employment under someone whose paternalistic feelings were stronger than the norm.

The problem of conditions of employment was, therefore, one throughout society and it was inevitable that public attention would be drawn to other, large-scale occupations before one so insignificant as lead manufacturing. From very early in the nineteenth century, the State had taken an interest in textile factories and by an Act of Parliament in 1833 had appointed the first factory inspectors to begin the process of control of employment conditions. There followed in 1842 legislation to ban the employment of women and young children underground in coal mines and the next two decades saw further Acts which developed the detail of control within textiles and coal mines. It was in the 1860s, following the reports of the Commission on the Employment of Children, that intervention was extended to a wide range of industries and trades.

The earlier Commission on Children's Employment, which had led to the passing of the Mines Act of 1842, had investigated a wide range of factories, including lead works but little attention was paid to them since they offered few opportunities for child employment. The subject of lead poisoning was raised in passing and it was shown that the companies paid some attention to the problem. In answer to a question as to what steps were taken to counteract 'the unhealthy tendency of the work', William Sloane, a foreman at Chester, replied,

We supply those employed with medicine and medical advice, we supply soap, water, and towels for washing, and we provide clothes for them to work in, which they put on when they go to work, and take off as soon as work is over. We also have these clothes washed. There is a room for the use of the women, in which there are water-closets. We caution them not to eat with soiled hands, and use our best endeavours to keep them in health; and when we find any of them losing health we try to put them to other employ till they get better.

George Burnett, managing partner at Locke, Blackett, gave very similar evidence and argued 'that severe cases of illness arising from the effects of lead in the system are comparatively rare'.[4] It does, however, seem unlikely that the incidence of lead poisoning was falling — indeed as output and employment rose the number of cases almost certainly increased even if incidence remained unchanged. Some new processes increased the risk. In the late-eighteenth century it was common for white lead to be 'air-dried' in open-sided sheds but the desire to speed production and reduce costs had led to the widespread introduction of drying stoves — in effect large ovens with tiers of shelves on which bowls of white lead paste were placed. It appears that more lead dust was raised and breathed in by employees, especially in the emptying of these stoves, than had been with the air-drying sheds. Moreover some of the developments, which contemporaries thought had reduced lead poisoning, were not particularly effective. The grinding of white lead under water, in order to reduce dust, proved unsatisfactory and the introduction of fans to blow the dust away is unlikely to have been beneficial. There is also doubt as to the extent to which employees took even primitive care and used the facilities for cleanliness available to them, since they did not understand the problems involved. It is impossible to assess the prevalence of lead poisoning in the early- and mid-nineteenth century because of the lack of evidence but the extent of the problem in later decades implies that it was not a new phenomenon.

Despite awareness of the existence of lead poisoning,[5] neither of the Children's Employment Commissions made recommendations on the subject and all that was done was that the 1864 Factory Act noted the existence of lead poisoning in the Potteries and prohibited the eating of meals in the drying rooms of earthenware factories, where pottery was given its lead glaze. This was a period in which the State was beginning to recognise the problem of dangerous trades but was initially more concerned with externalities and attempts to reduce the impact on third parties. Hence an 1862 report on 'Noxious Vapours' led to the 1863 Alkali Act imposing regulations on chemical factories and an inspector to enforce them but even here

The information which the Committee have received in respect to the injury arising from lead smelting . . . is, however, in some respects contradictory, and does not appear sufficiently conclusive to warrant the Committee in founding upon it any specific recommendation.[6]

If the impact of lead smelting on the local environment was dismissed as *non proven*, the lead manufacturing firms were not even mentioned for their external impact, let alone their impact internally on their employees.

It was not until 1878 that further developments came, with a twin-pronged attack. Parliament's concern about the protection of private property was followed up by another report on the impact of 'Noxious Vapours', to which both A.O. Walker and Norman Cookson gave evidence,[7] which recommended that lead works be placed under the Alkali Acts for inspection purposes. While this regulation was for the prevention of external damage, the second prong of the attack was concerned with employment in the works. From about 1875, the half-yearly reports of the Factory Inspectors[8] contained regular comments on the employment conditions of women in white lead works. Although under the 1864 Factory Act the inspectors had the power to inspect such factories they were not empowered to take any action on the question of lead poisoning. Nevertheless they were in a position to obtain information and draw attention to the problem and in an 1875 report Alexander Redgrave, Chief Factory Inspector, commented that, 'A short time since my attention was called to the subject of lead poisoning, and to the conditions of persons employed in whitelead works . . .'.[9] As a result he had instructed the sub-inspectors to report on the London works and a considerable section of his report is devoted to the evidence they obtained. He stressed that 'No general rules are laid down or carried out in white-lead works. All depends upon the extent of the precautions thought necessary by the manufacturer himself' and while a variety of precautions was taken, 'there is nevertheless much suffering among these people'. Several of the preventive measures were ineffectual since many of the workpeople interviewed objected to wearing gloves because they could not handle the implements freely when doing so and would not wear a respirator, or 'muzzle' as they called it, because they became too hot. Redgrave concluded his report with a call for the legislation of rules for all white lead works but this was too far in front of general opinion to be put to Parliament. Government listened to professionals, such as the factory inspectors, but usually required evidence of strong public outcry before it would venture into a new area of intervention.

The initiative made by the factory inspectors, for there is no evidence of any other significant pressure, did bring about one major development affecting the white lead industry in the 1878 Factory Act.[10] In that Act the employment of children and young persons (under the age of 18) was prohibited in any process in the manufacture of white lead which took place after the stacks had been filled. The immediate effect on employers was negligible since few, if any, children and not many girls under 18 were employed in such heavy work as emptying stacks and filling the drying stoves. The significant points were that the legislature had specifically recognised that the major problem in lead poisoning was caused by dust raised in the work (i.e. there was no prohibition of work in setting blue lead into the stacks) and secondly a principle had been established. Increasingly throughout the century Parliament had legislated on the hours and conditions of work of children, young persons and women, since such groups were not regarded as free agents capable of making a rational bargain with their employers. Now, however, it made the huge step forward of deciding that it should intervene to prohibit certain categories of employment in a manufacturing industry. This was a step which was pregnant with potential extension, both to women in white lead manufacture, and, since lead manufacture was far from being the only occupation deleterious to health, to other industries. It is, however, obvious that the provisions of the Act were going to do nothing to stem the tide of lead poisoning but a different tide, that of the public outcry which was needed to give urgency to the factory inspectors' case, was just beginning to flow.

It appears to have commenced with the Poor Law Guardians on Tyneside and in London. Since many women took work in white lead manufacture only because they had no alternative support, the result of a bout of lead poisoning would lead to their dependence on the Poor Law and its medical apparatus. Undoubtedly this had been happening for some time but by the early 1880s, under the pressure of the rising cost of maintaining the sufferers, the Guardians were complaining and endeavouring to obtain action. The immediate cause of their success appears to have been the fact that they enlisted the support of two working-class MPs, Thomas Burt and Henry Broadhurst. They had asked questions in the House and forwarded letters on the subject of lead poisoning to the Home Office.[11] The Gateshead Guardians took a special interest in the subject, since they suffered from having a large working-class residential district and no doubt felt aggrieved that Walkers, Parker, across the river at Elswick, paid rates to Newcastle while Gateshead had to support its lead poisoning victims.[12] The minutes of the meetings of the Guardians contain regular reports of women, usually in their early twenties, who had become dependent on poor relief because of lead poisoning. These would be the worst cases; previous histories of illness are frequently mentioned, some of the women were blind as a result of poisoning and were to be maintained in the Blind Asylum, while there is a steady trickle of reports of deaths. Gradually the local authorities moved from considering the problem themselves to passing it on to central government. After the death of Elizabeth Boyle (aged 26) from lead poisoning, the Gateshead Guardians, on 23 May 1882, considered a resolution to refer the problem to a committee but eventually passed an amendment:

That as the attention of the Board had repeatedly been called to the great fatality attendant upon the employment of females in the White Lead Works in the neighbourhood and that Decrepitude, Palsy, blindness and often death are the frequent results of Lead Poisoning — 'That the Board memorialise the Home Secretary to establish an enquiry into the method of such employment with the view to the introduction of machinery to do the most dangerous part of the work.'

Meanwhile the death in Shoreditch infirmary of Hannah M'Carthy, who had worked at the Islington white lead works, had been followed by a verdict by the coroner's jury of death from lead poisoning, together with a rider giving 'an expression of their opinion that some measure should be passed by the Legislature compelling the proprietors of lead factories to look after their workpeople, and take such measures as would put a stop to the present wholesale poisoning by lead'.[13] Little different from many other straws in the past, this was the additional weight which brought action. Questions were asked in the House and in April 1882 the Home Secretary instructed the Chief Factory Inspector to prepare a report on white lead factories.[14] This appears to have been the signal for several of the local authorities, which had white lead factories in their area, to send details of lead poisoning cases to the Home Office. The campaign gives the appearance of having been orchestrated and Redgrave may well have written to the local authorities to obtain evidence in order to buttress his case. The Gateshead Guardians showed that 15 cases of lead poisoning had come on to poor relief in their area in the three months to 31 October 1882. At St Leonards, Shoreditch, 23 people had been admitted to the infirmary suffering from lead poisoning in the 18 months to May 1882, of whom three had died, while in the Poplar Union there had been 26 cases during 1881, all but three of them women. Among other instances, the Holborn Union mentioned 54 cases of lead poisoning in the previous twelve months.[15]

Redgrave presented this evidence in his report in 1883 and commented that the visits he had made to all the white lead works in the country, together with the enquiries he had instituted, have 'shown me that the temporary illnesses and permanent disabilities which affect those working in whitelead far exceed anything that has come before the public'. He had examined the regulations laid down and the care taken to enforce them in the various works and had

> found the greatest possible diversity, and having regard to the indifference exhibited by some of the whitelead manufacturers whose works I have visited, to the serious unhealthiness of the occupation, I have come to the conclusion that it is necessary that those manufacturers who have neglected that which is a moral duty, should now be compelled to place their works in at least as good a condition as those of most of their competitors in the trade.

Redgrave therefore reverted to his 1876 call for special rules to be imposed in white lead factories but was not prepared to advocate the more radical options which were being mentioned to him. 'It would', he commented, 'be opposed to all the principles of the Factory Acts to forbid the employment of adult women in an occupation in which danger may be reduced to a minimum' and 'it would be impossible to make it compulsory that machinery should be used for the manufacture of white lead'. Like politics, effective factory legislation was the art of the possible and the possible depended on the state of society's attitudes on any topic at a particular time. Regulation, Redgrave decided, was possible and he showed that the manufacturers would in general support this development provided that it was 'impartially enforced in all whitelead works'. That Redgrave was anxious to push regulation only as far as would be acceptable to most employers (for the sake of having rules which might actually work) may be seen from his comments on medical inspection of employees, which had been adopted 'with such good effect' by two firms which he would 'gladly see . . . followed in every whitelead work'. However, he continued

> it would be carrying legislative regulations to too extreme a point to insist upon this, especially as it has been urged upon me in the strongest terms that English firms have already to strain their utmost to contend with foreign competition, and that great care must be taken not to burden our manufacturers with restrictions which would still further hamper the trade.

The evidence of pragmatic empiricism in government intervention could hardly be clearer.

The result of Redgrave's report was the passage of the Factory and Workshop Act, 1883 (46 & 47 Vict. c.53) White Lead Factories, which stipulated that manufacturers should produce and enforce special rules from 1 January 1884. Although such legislation already affected coal and metalliferous mines, this was the first application to a manufacturing industry and was to provide the precedent for extension to many more. Although the rules were to be proposed by the employers they had to conform to a schedule of regulations in the Act and be approved by the Home Office.[16] The rules were mainly concerned with cleanliness: firms had to provide baths, water, soap, brushes and towels and ensure that their workers used them; acidulated drink[17] was to be made available as were overalls and respirators; smoking was not to be allowed on the premises; white lead beds were to be saturated with water (to reduce dust) before they were stripped and anyone unwell was to be sent to a doctor.

Although such rules might do something to reduce lead poisoning in the less well-run works they would not do much to affect the basic problems. The local authorities, which had started the enquiry, were well aware of this. Before the rules were even in force the Gateshead Guardians had complained that they would be of little use and that what was needed was the introduction of mechanical drying stoves. While such a response may not be considered surprising, the severe criticism which Redgrave's report received from a leading article in *The Times* is notable and suggests that Redgrave had misjudged the state of public opinion and might have obtained considerable support had he suggested stronger measures.

> He [Redgrave] does not seem to realise the enormous, the practically insuperable, difficulty of making poor and ignorant people observe a variety of minute precautions, especially such as take the irksome form of compulsory and unaccustomed cleanliness. It is impossible to doubt that such precautions, if ordered by Act of Parliament as a condition of the carrying on of lead works, would be systematically evaded by those for whose benefit they were designed . . .[18]

The Times was not, however, making a plea for non-intervention and it went on to state its position, that

An Act of Parliament to forbid poor and ignorant men and women from being employed in an occupation dangerous to health and life, and to compel the substitution of machinery, would commend itself to all those in whom benevolence assumes a practical form. It would certainly be more useful and more easy to be enforced than one which directed the inhabitants of Poplar or of Gateshead to clean their teeth before eating, and to be careful to brush away all dirt from under their fingernails.

Whether or not the opportunity had been missed to propose more stringent measures which would have been passed into legislation, the 1883 Act was in existence and was unlikely to be altered until it had been given some trial. The Gateshead Guardians, having been informed of two cases of girls who had become blind after working in white lead works, memorialised the Local Government Board that action be taken on the Guardians' pet topic, the provision of 'mechanical means' to fill and empty the drying stoves. The response they received was that the 1883 Act had only just come into force and that it was not in force when the girls contracted lead poisoning.[19] In other words 'wait and see' was the official position, if not the even stronger view of Sir William Chaytor, factory inspector for the north-east district, 'I am perfectly certain nothing more can be done on the part of the employers.'[20]

Restrictions to deal with lead poisoning fit classically into the pattern of much nineteenth-century government intervention in employment and other areas. A few individuals got together to campaign about a particular problem, their case received some press support, Parliament instituted an enquiry and on its report did nothing or passed permissive legislation. The whole process was then gone through again, the outcry being more persistent, and Parliament passed stronger legislation with moderately effective enforcement procedure. Those employers affected by the legislation (for example, shipowners in the case of regulation of merchant shipping or chemical manufacturers forced to adopt new techniques by the Alkali Acts), who would often have campaigned against the legislation as ruinous, then learned to live with it. They would claim that everything in the garden was now lovely and that no further improvements were possible, or if they were that to introduce them would only lead to ruin. As a result Parliament felt that it had done its duty and turned to other matters but the problem did not go away since reformers and philanthropists kept raising the problem and ultimately raised sufficient new evidence and influenced sufficient people to force the institution of a new enquiry and so the process goes on. Intervention was a slippery slope, on which one might stand still but could not move upwards and any movement was likely to send one downwards, deeper into legislative involvement. Often, therefore, within a decade of one piece of legislation which gave more or less general satisfaction, another would be passed enforcing restrictions rejected or not even considered by the former. Thus did society progress. As with poverty, there was, and is, no perfection, only relative progress as standards change over time.

While there was some evidence that the introduction of special rules was beneficial, any feelings of complacency should have been destroyed by the continuing evidence of regular deaths from lead poisoning[21] and complaints from Poor Law authorities. However, with the publication in 1891 of a book on lead poisoning by Dr Thomas Oliver,[22] complacency was no longer realistic. Oliver was Professor of Physiology at the University of Durham and Physician to the Royal Victoria Infirmary in Newcastle, to which many of the Tyneside lead poisoning cases were admitted. Together with K.W. Goadby and Dr T.M. Legge, Oliver represented a new input to the debate over lead poisoning. If the 1870s and 1880s were the period in which the factory inspectors called the tune, that after 1890 was dominated by the doctors, although, as we shall see, their findings were transformed into public outcry by the newspapers.

Apart from providing a highly technical pathology of lead poisoning, Oliver showed the extent of the disease subsequent to the passing of the 1883 Act. He had been much involved with treating lead poisoning cases and showed that no fewer than 135 had been treated in the Royal Victoria Infirmary alone, in the five years to June 1889. Whether or not Oliver's revelations were a causative factor, the Home Office enforced new special rules for white lead works from 1892. The only noteworthy addition which they contained was the provision of compulsory weekly medical examination, which in 1883 Redgrave had considered 'would be carrying legislative regulations to too extreme a point'. Having listed the manufacture of white lead (along with arsenic extraction and some others) as a dangerous trade, the Home Office in April 1893 appointed a Departmental Committee, which was to make much the most thorough enquiry yet into lead poisoning.[23] Among other things it was appointed to enquire 'whether the special rules ... are sufficient' and it is likely that it was established following the general concern which occurred after the publication in the *Daily Chronicle* of two articles on conditions in white lead works.[24]

The writer of those articles began by noting that ' "white cemeteries" is a name by which the white-lead works are known amongst the women here [Newcastle] whose fate it is to spend their lives and very often to meet their death, in the manufacture of one of the most deadly of poisons'. He claimed that there were 600 women on Tyneside in such works which were for them 'what the dock gates are for men. Society metes out stern measures for its weak ones and its failures. To the women of Newcastle who cannot get bread, it offers white-lead.' On the day on which he had arrived in Newcastle the local papers reported an inquest on a girl of only 17 years of age who had died from lead poisoning after working for only five months at the Elswick works. The *Daily Chronicle* reporter went to see her parents and was told, 'She used to come home and clean herself thoroughly, and shake out the dust from her underclothes. They were like "a poke of flour", said the father.' To his credit the reporter visited the Elswick works, around which he was shown with 'every facility', and he commented, 'Everything seemed to be well provided for, and an exquisitely clean dining room was being got ready for the midday meal of soup which the firm provides at its own cost.' He was not, however, to be seduced by the evidence that all that was reasonable was being done and wanted 'unreasonable' action.

His second article, after pointing out that there had been ten inquests at Newcastle in the previous two years on victims of lead poisoning, went on to deal with the special rules. 'They grant immunity to the capitalist for his murderous process, and throw the onus of defence upon the workpeople.' He went on

You want to muzzle the process, not the people [a reference to the use of respirators]. The Minister who let mad dogs run about the streets and enforced the use of greaves upon all citizens, and instituted municipal caustic supplies at the street corners would be handed over to the proper authorities. But wherein does our State policy, in the case of dangerous trades in general and the white-lead trade in particular, differ from this dangerous dog policy?

Highly critical of employers, factory inspectorate and government, he concluded that the only answer lay in the banning of the manufacture of basic lead carbonate and its replacement by lead sulphate.

The *Daily Chronicle* had reported on a number of industrial evils and the white lead articles were splendid examples of the tradition of Victorian investigative journalism. Much exaggerated and written in the language of righteous indignation, they contained enough truth (quoting, for instance, Dr Oliver) to compel belief and further investigation. Moreover they were backed by a leading article in the newspaper, which fitted them into a topic of great contemporary concern — British attempts to bring religion, morality and civilisation to Africa at a time when the shortages of all three were obvious at home: [25]

> [were] some Congo tribe . . . to adapt certain of our secular observances to its sacrificial ritual; if for instance, it made its women besmear themselves with poison, or condemned them to work in the fashion described by our Commissioner, treading the powdered white death underfoot, breathing it, swallowing it, carrying it about on their heads, mounting ladders with it, no amount of consideration for the deity in whose honour the rights were practised would check us from equipping a missionary expedition to teach humanity to the horrible black. The great god Trade is another matter. If we do not treat him with proper respect disaster and eclipse must follow.

James Henderson, Superintending Inspector of Factories for Scotland and the North of England, a man with considerable knowledge of industrial lead poisoning, endeavoured to calm the disturbance which obviously followed at the Home Office. He informed the Home Secretary that 'The picture drawn by the writer [of the *Daily Chronicle* articles] . . . is greatly exaggerated', quoted the example of Alexander, Fergusson where there had been no fatal case for ten years and only a few non-fatal ones and concluded that if the employees 'would only avail themselves of the means which are there provided for their protection' under the Special Rules, there might be 'complete immunity' from lead poisoning.[26] Public interest had, however, been raised and now had to be assuaged. This could not be done by issuing a statement that all would be well if white lead workers cleaned their teeth regularly and it was decided that a committee of the Home Office would be set up to inquire into the conditions of work in the manufacture of all lead products. Home Office notes show that civil servants were aware that much was at stake — 'This is a most important enquiry, and it is very desirable that the composition of the committee should be such as commands the confidence of the Manufacturers' and they went on to comment, 'There is such a strong feeling existing at present between the Manufacturers of Sulphate of Lead and the ordinary White Lead that it is desirable neither party should have any reasonable ground of complaint.'[27]

The membership of the Departmental Committee, chaired by James Henderson, and including two other factory inspectors and Dr Oliver, could hardly have been criticised by the manufacturers. The Committee visited 46 works throughout the country and took evidence from nearly 200 witnesses, including owners and managers of both lead carbonate and lead sulphate works, medical men and employees. Inevitably the Committee heard the old cry from some manufacturers that all was being done that could be done and it is the variation in attitude between employers which is perhaps the most notable feature of the evidence. E.A. Walker of the Millwall Lead Co. argued that they could not afford to replace female with male labour and noted that they were 'dependent for our labour on fresh hands turning up every morning'. He concluded that 'white lead has come to such a state now with all this expense, that you have to observe regarding your workpeople, that you might as well have your money in consols, as far as the profit goes'.[28] He was immediately followed, however, by George Whiteley, manager of the Johnson white lead works in nearby Limehouse, who told the Committee that his firm had not had women working in white lead manufacture for ten years. Norman Cookson and his general manager, W.M. Hutchings, showed that they had developed a mechanical drying stove into which it was not necessary for the women to go to place and remove the dishes of white lead paste and that this had reduced the incidence of poisoning. They also argued that it was necessary for the packing of white lead to be done under a hood with an updraught to take away the dust. By contrast James Leathart (of Locke, Blackett) was asked, 'Do you have a hood where you fill your casks with dry white lead?' and replied 'No, we have not. It is a capital place where we pack. It is quite in the open, and there is very little dust made, a mere nothing; and there is such a fine current of air that it all goes off.' In general Leathart gave the impression of much less concern than did Cooksons over the health of employees. There was no check on women before they started work or if they were off for a day or two whereas at Cooksons, in order to prevent evasion of the medical inspection prescribed by the 1892 Special Rules, women had to see the works doctor before they began work and when they returned after being off — even if only for a day. Other firms also showed conflicting evidence. While Mersey had recently ceased employing women in any part of the works (regarding them as more susceptible to lead poisoning), the managing director, William Sloane, was adamant that the Special Rules 'are as sufficient as it is possible they can be, and admit of no improvement'. In his defence it must be said that lead poisoning cases were less significant at Mersey than in any other works. Similarly at Walkers, Parker's Chester works only men were employed in white lead but at Elswick women were employed.

From the confusion of the evidence the committee had to produce recommendations and in time-honoured practice they selected what they regarded as the best practices of individual firms rather than adopting a more radical approach. They came 'to the conclusion that there is at present no substitute that can take the place of carbonate of lead made by the old Dutch process' and they recommended its continuation with precautions, 'rather than its replacement by other processes which, though less dangerous, produce an article for which there is no constant demand either at home or abroad'. While this conclusion was justified by the evidence it is doubtful that this could be said for the Committee's rejection of the mechanical stove — 'No perfectly satisfactory mechanical stove has yet been devised, and their very high price is also a bar to their general introduction.' One cannot help but suspect that the Committee was influenced by the opposition of other established manufacturers to Cooksons who, as we have seen, were standing outside the price association set up in white lead in 1892. Since mechanical stoves worked effectively at Cooksons, where lead poisoning cases were second lowest to those at Mersey, their general enforcement might greatly have reduced the incidence of the disease.

What the Committee did was to tighten up the Special Rules by, for instance, insisting on medical inspection before a woman commenced or recommenced work and the compulsory fitting of hoses to stacks which were to be thoroughly watered before the white lead was stripped. The major recommendations were, however, reserved for the question of the employment of women. The Committee started from the assumption that women were more susceptible to lead poisoning than men and recommended that the age at which women should be allowed to work in lead should be raised from 18 to 20 and that from 1 January 1896 no woman should be employed in contact with white lead. Finally the Committee recommended that Special Rules be extended to the manufacture of red lead and other lead products and that cases of lead poisoning be compulsorily notified to the factory inspectors.

The report of the Departmental Committee can only be regarded as a very conservative reaction to the evidence and one which could not be expected to do a great deal to remedy the problem. The response to the report was inevitably mixed. Radicals saw that it would do little to deal with the problem. Henry Labouchere, proprietor of the magazine *Truth*, commented

The report on white lead is, that it is impossible that it can be used in manufacture without poisoning the user . . . Precautions cannot prevent the poisoning. On reading this I naturally expected that these admissions would be followed by a recommendation on the part of the Commission to prohibit the use of this dangerous stuff in manufactures. I was mistaken. The Commission limits itself to the recommendation that girls alone, under twenty years of age, should not be poisoned.[29]

The alternative view, an excellent example of the changing attitude towards government intervention in industry, was put by the journal *Engineering*, in a leading article:[30]

The Manchester school of politicians, which preached the doctrine that individuals and classes were quite able to safeguard their own interests, both personal and financial, without aid from the State, is now pretty nearly extinct, for experience has clearly shown the fallaciousness of its views . . . The sense of a civilised community now demands that the State should endeavour to exercise a paternal care over those that are not competent to take care of themselves.

There were many improvements which might be made, the journal continued, but habit prevented their implementation:

Evidently the owners need some pressure brought to bear upon them in order to enable them to recognise that they live at the close of the nineteenth century, and to see that not only is it indecent to allow ignorant women to continue in such tasks, but also that it is uneconomical. The disinclination that we should otherwise feel in concurring in the recommendation to abolish female labour at two years' notice is a good deal reduced by the knowledge that it is now used, in part, for such unnecessary purposes.

Between two such positions there was a gulf, not unbridgable but one which would ensure that alterations in the official attitude towards lead poisoning would come only slowly and in the light of the build-up of new evidence and increasing public pressure.

Some of this was already coming, since at the same time that the Departmental Committee was sitting, there was a Royal Commission on Labour and to that Miss May Abraham, one of the first two female factory inspectors who had been appointed in 1893, made a report on conditions in the manufacture of white lead.[31] This produced further evidence of considerable slackness in enforcement of the existing Special Rules in some works, with twelve deaths from lead poisoning in the previous four years of employees at the Sheffield works of Lewis Berger & Sons Ltd. Newspaper reports of inquests on lead poisoning victims continued to be common in the mid-1890s and it is certain that reported cases were only the tip of an iceberg. Moreover there was clear evidence that the existing Special Rules of 1892 were not being effectively enforced. The verdict at one inquest was that 'the deceased died from lead poisoning contracted at the lead works in consequence of the neglect of the special rules by the management'.[32] The factory inspector subsequently proceeded against the firm, Walkers, Parker at Chester, and obtained a conviction.

In 1896 there came another of the pieces of extravagant journalism, along the lines of the *Daily Chronicle* articles, which was to keep the subject in the public eye. *Pearson's Magazine* ran a series of articles under the general title 'The White Slaves of England', of which one explored the Newcastle lead works.[33] The article contained an example of a regular Victorian standby, the death-bed account — given by an old white lead worker; went on to imply prison-like conditions at the lead works, referred to the miscarriages which were common with women working at such places and described the ex-lead worker who, because of wrist-drop, had 'to eat like an animal, with his mouth to his plate'. Add to this the authenticity of some quotations from Dr Oliver and the reader was dripping with sympathy for the girls whom the writer described as lined up for the regular medical inspection anxious 'to look at ease, in spite of the horrid pains that were gripping them' and desirous to convince the doctor that they were not ill, since to be laid off work meant no income and eventually the workhouse.

It is easy to point to the contrived nature of such writing but the article hit the right note in keeping the controversy alive. The first major step, following the report of the 1893 Departmental Committee, was that notification of cases of lead poisoning was made compulsory on medical practitioners and factory owners by the 1895 Factory Act, with effect from 1 January 1896. Once again the lead industry was the guinea pig for a new area of intervention and one of crucial importance, since it was only from accurate statistics that it would be possible to tell whether or not lead poisoning was under control. It is certain that the first statistics are inaccurate — one source gives them as 357 cases of lead poisoning in 1896, 370 in 1897 and 480 in 1898[34] — and that reasonably reliable statistics do not commence until 1900, but a large step forward had been achieved.

The next development was the banning of women from working in white lead, proposed by the 1893 Committee to be effective from 1 January 1896 but delayed as a result of opposition from the employers. Walkers, Parker, Locke, Lancaster and several other firms tried to obtain the rejection of the proposal but it was made clear by the Home Office that the abolition of female employment was going to be enforced and the opposition merely caused the implementation of this regulation to be delayed until 1 June 1898.[35] In addition from 1 January 1900 new Special Rules were enforced which, among other amendments, provided for the provision of dust extractors over white and red lead packing plant and from 1901 the abolition of the old drying stoves and their replacement by stoves into which the dishes were pushed with poles, which avoided the necessity for workers actually to enter the stoves.

All of this added to a plethora of regulation such as was not imposed on any other manufacturing industry. This did not mean that lead manufacture was the most dangerous of trades but once an industry had experienced intervention it was liable to further enquiry and improvement. The Special Rules of 1900, for instance, resulted from the report on white lead works by T.M. Legge, HM Medical Inspector of Factories (an appointment which was an innovation resulting from the introduction of compulsory medical examination of employees). Among other things Legge drew attention to two significant points. Firstly, lead poisoning was more severe in those works which packed dry white lead for sale rather than grinding it with oil, which was another factor suggesting the importance of dust in causing lead poisoning. Secondly, he stressed that casual employment was a major factor in leading to high incidence of lead poisoning, presumably because casual workers were less careful about observing the Special Rules than were regular employees in white lead factories. The statistics he quoted made this point very clearly:

	No. of factories	Average no. employed	Approximate no. employed during 1898	No. of lead poisoning notifications in 1898
Regular employment	13	822	1,000	50
Casual employment	9	641	3,000	250

Cooksons and Mersey (with the largest white lead works in the country and the latter with no case of lead poisoning in the three years since notification had commenced) were once again singled out as offering best practice techniques.[36] Compared with conditions in these works, those at some other firms were appalling, as was made clear in the Chief Factory Inspector's report, on which a Home Office minute had commented:[37]

> It is remarkable that three firms, employing together 432 persons are accountable for 242 cases of lead poisoning – while the remaining fifteen firms – employing about 1600, have only 136 cases.
> The three chief offenders are all Newcastle people: *viz.*
>
> | | Employees | Cases |
> |--------------------------|-----------|-------|
> | Foster, Blackett & Wilson | 285 | 88 |
> | Locke, Blackett & Co. | 181 | 118 |
> | James & Co. | 66 | 32 |

Certainly from 1900 there are statistics of lead poisoning cases and deaths which have been generally taken as being reliable. Table 7.1 gives the detailed figures for the major industrial sectors involved.

Table 7.1: Notified Cases of Industrial Lead Poisoning, 1900-10 (deaths in brackets)[38]

Year	White lead	Red lead	Smelting	Sheet & pipe	China & earthenware	Coach building	All industries
1900	358(6)	19	34(1)	17(1)	200(8)	70(5)	1,058(38)
1901	189(7)	14	54(3)	17	106(5)	65(4)	863(34)
1902	143(1)	13	28	12	87(4)	63(1)	629(14)
1903	109(2)	6	37(2)	11	97(3)	74(5)	614(19)
1904	116(2)	11	33(1)	7	106(4)	49(4)	597(26)
1905	90(1)	10	24(1)	9	84(3)	56(3)	592(23)
1906	108(7)	6	38(1)	7	107(4)	85(7)	632(33)
1907	71	7	28(2)	6	103(9)	70(3)	578(26)
1908	79(3)	12	70(2)	14	117(12)	70(3)	646(32)
1909	32(2)	10	66(5)	9(2)	58(5)	95(6)	553(30)
1910	34(1)	10	34(5)	4	77(11)	70(6)	505(38)

The figures show clearly that within the lead manufacturing trades employment in the manufacture of red lead and lead sheet and pipe was not a significant cause of lead poisoning and that it was in white lead manufacture that the most serious problems existed. However it is also clear that, as a result of the enormous level of attention given to the trade, the problem was being brought under control in the first decade of the twentieth century, by the end of which lead smelting (employing fewer workers), to which little legislative attention had been paid, was equally as important a source of lead poisoning. It seems certain that for the preceding decade or two, for which no adequate figures of lead poisoning are available but in which total employment in the lead industries was little lower, the incidence of lead poisoning must have been considerably higher.

Table 7.2: Annual Average Numbers of Cases of (Deaths from) Lead Poisoning, 1900-19[39]

	1900-4	1905-9	1910-14	1915-19
Total	753(27)	599(28)	548(34)	279(21)
Smelting	37(2)	45(2)	40(4)	34(2)
White lead	183(4)	76(3)	31(1)	17
Red lead	13	9	8	11
Pottery	119(5)	94(7)	67(10)	19(6)
Paints & colours	46(1)	39(1)	20	12
Industrial use of paints	139(7)	144(7)	157(10)	55(6)

Subsequent statistics, presented in Table 7.2, imply a continued decline in the incidence of lead poisoning and were certainly used to argue that case during the 1920s. It is noticeable that there was very little decline between the quinquennia 1905-9 and 1910-14 and the considerable decline which came in the final quinquennium was less the result of declining incidence than of declining employment as a result of the limited availability of lead during the War. It is, therefore, somewhat doubtful as to whether there was a declining trend in the incidence of lead poisoning, once the major impact of the agitation over white lead and the imposition of more stringent special rules in that industry in 1900 had taken their effect.

During the period for which figures have been given agitation and enquiry over lead poisoning continued apace,[40] although after 1900 the enquiries concerned those occupations which had been overshadowed by the white lead agitation but which had been shown by the notifications to have significant numbers of cases of lead poisoning. They are of interest here, partly because they kept the question of lead poisoning before the public eye but ultimately because they introduced a campaign for the abolition of the use of lead in such occupations, which threatened employment in the lead industry itself.

The occurrence of lead poisoning as a result of the use of lead rather than in the manufacture of lead we have seen to have been a subject of debate for centuries. But the shift of attention from lead works specifically to the use of lead generally, which came from the late 1890s, might be attributed to the sober, objective and seemingly reliable attempt to outline the facts on lead poisoning in the Potteries by a special correspondent of *The Times* in 1898.[41] Once again this shows the significance of the media and public outcry in stirring response from government, since it is unlikely to be coincidence that in the following year Oliver and Prof. T.E. Thorpe, Principal of the Government Laboratory, produced an official report on the impact on health of the use of lead in the manufacture of pottery.[42] They concluded that 'leadless glazes, of sufficient brilliancy, covering power, and durability, and adapted to all kinds of table, domestic and sanitary ware, are now within the reach of the manufacturer' but, recognising that conservatism was likely to lead to the continued use of lead, they recommended the replacement of white lead with lead frit.[43] In the years before 1914 this recommendation was gradually adopted and in a note for the Walkers, Parker board in May 1913 the Chester manager commented that 'it would be necessary for the Company, sooner or later, to undertake this class of business' since 'lead frit was now being made by several White Lead Manufacturers'.

It was, however, to the unregulated trades (those which were not under the Factory Acts) and especially to painting that the greatest attention was to be paid because of the growing evidence that the number of deaths from lead poisoning in them exceeded those in all the regulated trades combined. The figures are given in Table 7.3. It is almost certain that the figures for plumbers and painters understate the total numbers of deaths since there was no mechanism to ensure notification in such fragmented trades but it is also important to note that although the absolute numbers were high the incidence was low since there were approximately 15,000 house painters at the time.

In 1911 the Home Office set up departmental committees to enquire into the dangers attendant on the use of lead paints in the painting of buildings and in the painting of coaches and carriages.[45] There had been a good deal of advance discussion that an enquiry might recommend the abolition of lead in paint and the white lead manufacturers marshalled their forces accordingly. By November 1912 they had already spent more than £3,000 on legal advice and expert witnesses, according to a minute of the Walkers, Parker board. Apart from arguing that any restriction on the use of lead paint would lead to serious loss of employment and a considerable increase in imports of leadless paints (which were hardly made in this country), they stressed the fact that regulation had reduced lead poisoning in the lead manufacturing trades and argued

Table 7.3: Deaths from Lead Poisoning, 1900-13[44]

	1900	1901	1902	1903	1904	1905	1906	1907	1908	1909	1910	1911	1912	1913
House painters	31	30	26	32	30	19	30	32	29	34	31	35	37	31
Plumbers	11	11	6	7	9	9	6	7	15	13	6	13	10	6
All regulated occupations	38	34	14	19	26	23	33	26	32	30	38	37	44	27

that it should be applied to painting. Despite this evidence the committee on the use of lead in the painting of buildings recommended that
'a law should be introduced prohibiting in this country the importation,
sale or use of any paint material which contains more than 5 per cent
of its dry weight of a soluble lead compound . . .'.[46] This was a very
unpleasant blow for the white lead manufacturers, although nothing
was done about the recommendation because of the impact of the
First World War.

It was, however, inevitable that the question of lead poisoning would
again arise as soon as the War was over, although the statistical evidence
does not suggest that the number of cases rose in proportion as lead
output increased after wartime shortages. In 1920 the Report of the
Departmental Committee on the painting of coaches and carriages was
published, making a similar recommendation to that of its predecessor
on house painting. This helped to fuel the discussion but perhaps the
most significant factor was that the question of lead poisoning was
taken up by the International Labour Office (ILO), a body set up under
the auspices of the League of Nations. The first British response was the
passing in 1920 of the Women and Young Persons (Employment in
Lead Processes) Act, which embodied the recommendations of the
November 1919 ILO conference in Washington, but which added little
to previous legislation. Secondly, more stringent regulations, concentrating on the avoidance of dust, were introduced for the white lead
industry on 1 October 1921. These were not steps which were likely to
satisfy the ILO, which at its Geneva conference in 1921 had approved a
draft convention calling for the prohibition of white lead in interior
painting, with the support of the British delegate, Sir Montagu Barlow,
the Minister of Labour.[47]

This was the signal for renewed pressure from the white lead manufacturers in an attempt to prevent reduced demand for their product.
Some of the evidence they produced was objective, pointing, for
instance, to the fact that there was no accurate information on lead
poisoning in an unregulated trade such as painting and, therefore, that
the statistics which were being bandied about were unreliable. On other
occasions, however, their campaign smacked much more of the subjec-
tive, with biased presentation of evidence. In an apparently scientific
study, printed in a highly reputable journal, C.A. Klein, chief chemist
of the Brimsdown Lead Co. Ltd and Prof. H.E. Armstrong, FRS,
produced a peculiarly unscientific paper, continually appealing to the
emotions and concluding by adapting to white lead paint George
Saintsbury's comment on wine, 'On those who would deprive us of it,

let the curse of nature rest.'[48]

The campaign was aimed chiefly at the reliability of the evidence taken by the two Home Office committees. W.G. Sutherland, Secretary of the National Association of Master Painters and Decorators, had dissented from the painting committee's report and produced a minority report in which he challenged the evidence of HM Office of Works that zinc oxide paints were a suitable replacement for lead paints for exterior work. He also argued for regulation of the painting trades (by the factory inspectors) rather than abolition of lead paint. His criticism of non-lead paints was reinforced by experience of their use during the war years, when lead paint was in limited supply, and the Home Secretary, therefore, decided to appoint a new committee 'To re-examine the danger of lead paints to workers in the painting trades . . .'. It looked especially at the question of substitutes for white lead paints and made much of the fact that HM Office of Works was no longer impressed with zinc oxide paints for outside use or for inside use where heavy wear and tear occurred.[49] Overall the report of the committee, which was published in 1923, presents an amazing contrast to its predecessors. Where the earlier reports had appeared anxious to ban lead paints, the 1923 report went out of its way to ignore evidence which might have encouraged a ban. Much attention was paid to those users who had found zinc oxide paints wanting but the evidence from local authorities, railway companies and tramway operators who were satisfied with their use of leadless paints was largely ignored and the report concluded, 'We are satisfied that for outside painting, and for certain kinds of internal painting, there is at present no efficient substitute for lead paint.'[50] It is not surprising that a letter from the White Lead Makers Section of the London Chamber of Commerce, sent to all the white lead manufacturers, could comment that it considered 'the nature of the Report and the recommendations to be even more satisfactory than was expected'.

The manufacturers were not entirely out of the wood since the 1923 report recommended the ratification of the ILO Convention with regard to most types of indoor painting, and the Labour Government brought forward a bill in 1924 to legislate for this. The resultant opposition brought together an unlikely collection of bedfellows. Both Sir Thomas Legge and Sir Thomas Oliver opposed the prohibition of white lead paint and called for the imposition of regulations.[51] Their arguments were supported by workmen in the white lead industry, in close collaboration with their employers, and orchestrated by Thomas Kennedy, manager of Cooksons' Howdon Works. Kennedy spoke at lead works throughout the country and arranged letters to MPs and trade union leaders and the drawing up of petitions praying for the introduction of regulations not prohibition. Eventually all the activity proved unnecessary since the Bill was lost with the fall of the Labour Government, but a favourable climate of opinion had been created and when the Conservative Government came to act on the problem it introduced a bill based on the principle of regulation of the painting trades rather than prohibition of white lead paints for internal use and this became effective from 1 January 1927.

The passage of the Act marked the end of the significant campaign for the control of lead poisoning. The 1923 Report was the last in a long series of government inquiries over a 40-year period into the causes of industrial lead poisoning. The chief reason for this must be the general decline in notifications of the disease despite the increased attention it was getting (and therefore the reduced likelihood that cases went unrecorded). This decline must be attributed to the growing effectiveness of regulation. Although the increased use of lead which occurred with the return to peace after 1918 brought a rise in the number of cases, the total remained below the pre-war level. The quinquennium 1920-4 saw an average of 309 notifications and 26 deaths against 548 and 35 respectively for 1910-14.[52] In an annual report at the end of the inter-war period the Chief Factory Inspector felt able to comment:

> a feature of the year worth referring to here is the further reduction in the cases of lead poisoning. Since the disease became notifiable, the cases occurring in a year have fallen from well over one thousand to below one hundred. Such a reduction implies not merely that a fewer number of workers has been actually affected but that also the ill-effects of lead which may ultimately produce invalidity and death have been correspondingly reduced.[53]

In addition to regulation, such factors as the introduction of wet rubbing down of old paintwork, as a result of the development of wet-or-dry sandpaper (in which the lead manufacturers played a part), also helped to reduce the level of lead poisoning by reducing the level of lead dust in the air breathed by painters.

Although the lead manufacturers had managed to ride out the major storm of controversy and threat to their level of production they still had reason for concern and caution, largely as a result of the development of problems in new and unregulated areas of lead use. In the

1920s the motor vehicle was beginning to offer a major new demand for lead products, especially for batteries (for which lead plates and lead oxide pastes were required) and Sir Thomas Legge pointed out that there were more cases of lead poisoning in 'the motor and wireless accumulator industry than in any other trade'.[54] This was a problem which could be dealt with by the extension of regulation but, coming as it did shortly after the report of a Departmental Committee of the Ministry of Health on the use of lead in petrol, it still served to disturb the rather fragile crust on the volcano of public outcry. Tetra-ethyl lead had only become available in the UK in the mid-1920s and, following an American enquiry into its dangers, the Ministry of Health had established its committee, whose report stated that 'we are of the opinion that there are no reasons for prohibiting the use of Ethyl Petrol in the U.K., and we do not recommend any legislative action'.[55] It was to be another 40 years before the lead (and petrol) companies were to be seriously bothered with this question again. Nevertheless the manufacturers were anxious to keep a very low profile on the subject of lead poisoning, as may be seen from the alterations made to a 1930 ALM draft typescript for an advertising booklet. As a result of the introduction of Lead Fritts in the glazing of pottery, the booklet commented that potters could use, 'the beautiful lead glaze without any of the hygienic difficulties associated with the use of white lead'.

By the 1930s a great deal had been achieved and it was no longer in the lead manufacturing but in the lead using industries that the problem of lead poisoning was most severe. The lead manufacturers had moved from the late-nineteenth-century position where some, paternalistically, did all of which they were aware to improve the working environment of their employees, while others did a minimum and actively opposed legislative intervention. By the 1930s they had come to terms with the dangers which the manufacture of their product brought and were anxious, for their own future security, to minimise them. Hence in 1934 regular testing of blood-lead levels, an important advance indicator of other symptoms of lead poisoning, was commenced at the Millwall works of ALM, after experimentation over the previous five years. The pioneering work was done largely by the chief chemist, L.G. Jones, and on the basis of his experience William Carrott, the Millwall works manager, presented a report to the company on 28 May 1937 in which he wrote:

Medical inspection directed merely to comply with the Home Office Regulations is insufficient by itself and should include a genuine endeavour to safeguard the health of the workmen. Just keeping employees above the standard of health below which they can be regarded as suffering from slight lead poisoning should not be tolerated . . .

Blood-lead testing became instituted throughout the ALM merger, although it did not become a requirement of the statutory regulations for lead works until 1964.

The period since the Second World War has seen a continual tightening of the standards which the lead manufacturing (and using) industries have to meet, as a result of which the incidence of industrial lead poisoning has become almost insignificant. For the years 1930-7 the average annual numbers of notifications of and deaths from lead poisoning (in all industries) were 182 and 21 respectively (and in 1937, the last year, the figures were still high at 141 and 19). In the years 1971-80 the annual average of notifications was 41, with only a single death from lead poisoning in the decade, despite a higher consumption of lead and considerably more stringent definitions of the notifiable level of lead absorption. In so far as lead manufacturing was concerned the reduction was a result of co-operation between the companies and the factory inspectors, although there have been isolated instances where the inspectors have prosecuted as a result of failure to meet standards of hygiene. Increasingly severe restrictions have been placed, and met, on the permissible blood-lead level,[56] the amount of lead in air in the factories and the amount of fume allowed to escape into the outside atmosphere. Manufacturers have been assisted in meeting these standards by the development of more efficient filtration units, the first of whose primitive predecessors had been introduced in the early 1900s as, in effect, internal flues, in which dust and fume from manufacturing processes were trapped.[57] Initially, as in so many other areas where legislation has forced the consumption of waste products, the development was beneficial, since the companies soon paid for the cost of fume collection in the value of lead saved. In more recent decades, however, the extra cost of increasingly efficient dust control has far outweighed the saving of lead involved and the amount now wasted into internal works' and external atmospheres is very small and a far cry from the nineteenth-century farmer's outcry about pollution of his crops and poisoning of his cattle.

Nevertheless lead poisoning refuses to go away as a problem and the important question remains as to whether, as a country, we have become paranoic on the subject and are enforcing massive expenditure (which might more valuably be employed elsewhere) for little return. The major problem lies in the use of lead in petrol, which is only of significance here in so far as it accounted for about 60,000 tons of lead or one-fifth of the total used in the UK in 1980. As recently as 1966 one study (admittedly with vested interests) could conclude:

> a great deal of attention is being directed nowadays towards reducing the emission of any harmful substances into the air and although it has been suggested that lead anti-knock compounds in petrol might constitute a health hazard, the evidence accumulated so far indicates that there is no significant risk.[58]

The subsequent decade and a half, with increasing research activity and measuring against increasingly stringent standards, has shown that there is risk and, although blood-lead levels for urban dwellers are much below those at which lead poisoning becomes feasible, wisdom dictates that the risk should be minimised. In the United States legislation has begun to control the level of lead in petrol and this pattern is to be followed in the UK with a consequent reduction in the volume of lead consumed. In view of the facts that lead builds up in the body over time and of the sheer volume of lead going into the atmosphere and being deposited in soils, this seems sensible. The problem is no longer one of the health of specific groups of workers and where industrial employees do show higher than usual blood-lead levels the problem is easily overcome and the old symptoms no longer occur. The problem is now a more general social one as a result of the fact that lead pollution levels are certainly higher now than they were at the time of the late-nineteenth century outcry on lead poisoning. Moreover while such levels may have minimal impact on adult health, they may put children at risk of brain damage.[59] Inevitably such a possibility has led to a recent outcry in the media and to a campaign for the abolition of lead in petrol. Much of the debate has been emotive in the extreme and has only obfuscated the issue. The facts are still uncertain and dogmatic statements on either side do nothing of use. It is in the interests of all parties to collaborate in further research and in ensuring a lowering of the level of risk.

While attention to the reduction of the lead content of petrol is desirable, the outcry which has occurred as a result of media publicity on the supposed dangers of lead poisoning from lead works seems less justified. In 1972, for instance, there were extravagant claims about damage to the environment as a result of lead emissions from the

Millwall works of ALM.[60] Subsequently there have been several scares about high blood-lead levels in children of lead works' employees or those living near lead works.[61] There is, however, no serious evidence that such people are at noticeable risk from lead emissions. It would, however, be to fly in the face of the evidence of the last 100 years and of this chapter not to note that standards of judgement change over time. There can be little doubt that more stringent regulations with lower permissible blood-lead levels and further research into human physiology will ensure that lead poisoning remains a subject of active debate in the future. Nevertheless economic factors are important and society has to recognise that there is a trade-off between the levels of expenditure on ameliorating various social ills. The returns from expenditure on improving hygiene in lead works are becoming progressively smaller and it is necessary to bear in mind that illness and death are the woods, of which lead poisoning is the bark on a very small tree.

Notes

1. Lead poisoning occurred chiefly as the result of lead getting into the blood stream. This arose mainly from absorption via the lungs due to inhalation of lead dust in the air of the workplace but also through the digestive system from the ingestion of lead particles. Clinical evidence of lead absorption included symptoms of tiredness, headache, constipation and anaemia. More prolonged exposure could give rise to the presence of a blue line in the gums although this did not necessarily indicate lead poisoning since many employees had such a line for years without any further symptoms. Severe symptoms of lead poisoning were colic, wrist and/or foot drop (with paralysis of the extensor muscles), encephalopathy, blindness and, in extreme cases, death. See also *British Medical Journal*, 23 Nov. 1968, 4, p. 501.

2. R. Campbell, *The London Tradesman* (1747, reprinted Newton Abbot, 1969), p. 107.

3. There is a school in historical study which is inclined to argue that things should have been different in the past, largely, it seems to me, because we now regard those things as wrong and no longer allow them to happen. Quite apart from the question of the availability of resources in, say, 1830 to do what we would do in 1980, it is anachronistic to apply modern values to an earlier period.

4. *BPP*, 1842 [382] XVII, Royal Commission on the Employment of Children, Appendix to First Report of Commissioners (Mines), part II, p. 447 and 1843 [432] XV, Appendix to Second Report of Commissioners (Trade and Manufactures), part II, Report and Evidence of Sub-Commissioner J.R. Leifchild, p. 145.

5. In his report for 1863 (*BPP*, 1864 [3416] XXVIII, 1, 6th Report of the Medical Officer of the Privy Council, pp. 21-2), John Simon had drawn attention to the question of lead poisoning and posed the question as to whether processes which diffuse lead dust 'are not processes which ought to be discontinued or modified'. His comments were based on a report by Dr George Whitley on 'the occurrence of lead poisoning among persons who work with lead and its

preparations' (ibid., pp. 350-7). I am grateful to Mr K. Edwards of the Dept of Pharmacy at the University of Aston for drawing my attention to this report.

6. *BPP*, 1862 [486] XIV, Report of the Select Committee of the House of Lords on Injury from Noxious Vapours, p.v.

7. *BPP*, 1878 [c.2159] XLIV, Report of the Royal Commission on Noxious Vapours.

8. Almost every report up to 1914 contains some detailed evidence on conditions in white lead factories and cases of lead poisoning. From the 1920s, as the incidence of poisoning fell, mention becomes less common but the reports remain an invaluable source.

9. *BPP*, 1876 [c.1434] XVI, 17, p. 11.

10. The Act was largely a codifying one, which brought together and tidied up and extended the regulations contained in a number of preceding Acts. For the provisions of the 1878 Act see H.A. Mess, *Factory Legislation and its Administration* (1926), p. 12 and B.L. Hutchins and A. Harrison, *A History of Factory Legislation* (2nd edn 1911), p. 202. These two books are invaluable for the detail of the development of factory legislation.

11. PRO, HO 45/9620/A15330/3.

12. For instance, it was resolved 'That the girl Sarah Kennedy be maintained in the Blind Asylum at the cost of the Guardians. The Committee desire to express their regret that Messrs. Walker, Parker & Coy. do not see their way to discharge their moral responsibilities in this case.' Gateshead Public Library, Minutes of the Poor Law Guardians, 18 May 1885.

13. PRO, HO 45/9620/A15330/2.

14. Report by Alexander Redgrave, Esq. CB, Her Majesty's Chief Inspector of Factories, upon the Precautions which can be Enforced under the Factory Act, and as to the Need of Further Powers for the Protection of Persons Employed in Whitelead Works, *BPP*, 1882 [c.3263] XVIII, 957.

15. Reports from local Poor Law authorities are in *BPP*, 1883 [c.3516] XVIII, 929, Communications to the Secretary of State with Report by Alexander Redgrave, CB.

16. PRO, HO 45/9620/A15330/56, 58 and 59 contain the special rules approved for 23 white lead firms. Since one firm, Condy's White Lead Co., Church Road, Battersea, only employed men (ibid., A15330/5, Redgrave to Home Office, 24 Sept. 1889) and yet had to establish special rules, it is clear that the 1883 Act applied to adult males. Previous legislation on manufacturing industries had been concerned about the hours and conditions of employment of children, young persons and women and had affected adult males only incidentally. It may be that the 1883 Act broke this principle, albeit only tentatively. It was probably also the first occasion on which the Factory Acts had, effectively, enforced the provision of a meal room, see Mess, *Factory Legislation*, p. 18. In practice this had largely occurred already. On 29 January 1880 the Home Secretary had issued an order banning the taking of meals by children, young persons and women in white lead factories except in rooms solely used for meals. See Redgrave's first report, *BPP*, 1882, p. 2.

17. Known as 'treacle beer' in one firm, the principle was that basic lead carbonate would be precipitated as basic lead sulphate, which it was thought was insoluble in gastric juices and would pass through the body. The following recipe was given to Walkers, Parker at Elswick by their Chester works.

| No. 1 | { Strong sulphuric acid | 5 oz. |
| | Water | 32 oz. |

No. 2	{ 6 lb of loaf sugar
	32 ounces of water
	1 oz. essence of lemon

Mix with No. 1

1 oz. of the mixture to 2 quarts of water.
Quantity for each person up to ½ pint.

There were also anti-lead poisoning tablets and Cooksons, and no doubt other firms, issued their workers with a booklet, *Lead Poisoning and How to Prevent it*. One of its comments, that the authorities 'are most emphatic in stating as the result of experiment that to take alcoholic drinks is the worst thing possible for lead workers', is most unlikely to have gone down well.

18. *The Times*, 17 April 1883. In this prognostication the paper was correct, since evasion of the rules was to be a major problem. The paper was unjust in accusing Redgrave of being unaware of the difficulties of enforcing rules – his first report (*BPP*, 1882) had discussed the problem.

19. Gateshead Public Library, Minutes of the Poor Law Guardians, 4 Nov. and 16 Dec. 1884. I am grateful to F.W. Manders for providing me with the evidence on the Gateshead agitation.

20. PRO, MH 12/3096, Sir Wm Chaytor to W.E. Knollys, Local Government Board, 11 and 12 Nov. 1887.

21. These were noted even by the Chief Factory Inspector in his reports. Although he claimed that 'The fatal cases which have come under my notice have diminished noticeably, and I attribute the result to the general provisions of that [1883] Act', a year later he was to note that 'during the past three months three men have been certified by the medical men as having died from lead-poisoning contracted at the Chester Lead Works'. *BPP*, 1890-1 [c.6330] XIX, 443, p. 33 and 1892 [c.6720] XX, 463, p. 2.

22. T. Oliver, *Lead Poisoning in its Acute and Chronic Forms* (1891).

23. The Committee was small (consisting of only six members, including three factory inspectors, and Dr Oliver). Its report was rapidly published as *BPP*, 1893-4 [c.7239] XVII, 717, Report from the Departmental Committee on the Various Lead Industries.

24. The articles, written by Vaughan Nash, are in *Daily Chronicle*, 15 and 21 Dec. 1892. These articles did for lead poisoning the same job that, ten years earlier, W.T. Stead had done in the *Pall Mall Gazette* for working-class housing conditions. They transformed a problem which was known to be of significance into one on which serious action had to be taken. At the time the *Daily Chronicle* claimed the largest circulation of any London daily.

25. *Daily Chronicle*, 15 Dec. 1892, p. 5. General Booth's *In Darkest England and the Way Out* had been published in 1890.

26. PRO, HO 45/9848/B12393A/4, 29 Dec. 1892.

27. Ibid., B12393A/34.

28. *BPP*, 1893-4 [c.7239-I] XVII. The minutes of evidence are indexed by name of witness.

29. *Truth*, 21 Dec. 1892, p. 1322.

30. *Engineering*, 22 Dec. 1893, p. 764.

31. *BPP*, 1893-4 [c.6894 – XXIII] XXXVII, part 1, pp. 151-5, Reports from the Lady Assistant Commissioners on the Employment of Women to the Royal Commission on Labour; Report by Miss May E. Abraham on the Conditions of Work in the White Lead Industry. It is an interesting anomaly that while the factory legislation of the nineteenth century appertained only to female adults it was enforced by male factory inspectors for the first 60 years.

32. *BPP*, 1896 [c.8067] XIX, 89, Report of Chief Factory Inspector.

33. R.M. Sherard, 'The White-lead Workers of Newcastle', *Pearson's Magazine*, vol. II (1896), pp. 523-30.

34. Holley, *Lead and Zinc Pigments*, p. 146.

35. The result of the change to male employment was an increase in the number of male cases of lead poisoning, which suggested that the belief that women were more susceptible was incorrect.

Period	Notification of lead poisoning cases at Newcastle	
	male	female
1/12/1897 – 31/5/1898	19	66
1/6/1898 – 30/11/1898	82	12

Quoted in Legge, *White Lead Works*, p. 5.

36. See also PRO, HO 45/9938/B27900, T.M. Legge to E. Gould, HM Superintending Inspector of Factories, 13 Sept. 1898, in which the contrast is drawn between Cooksons and FB & W where there had been four deaths in a year, where dust was prevalent and exhaust hoods ineffective for lack of draught.

37. Ibid., 9939/B27900/28. It should be noted that the three firms mentioned employed 532 not 432 as stated in the minute, while the number of cases was 238 according to the table and not 242. Numbers of employees and notifications for each firm are given in the original document, while figures for four Newcastle firms for the years 1894 to 1895 are given in ibid., 9856/B12393 AC/1. They are all quoted in Rowe, History of Lead Manufacturing, pp. 452-3.

38. *BPP*, 1920 [Cmd 631] XX, 49, Appendices to the Reports of the Departmental Committees on the Use of Lead in Painting, p. 3. The figures relate only to industries under the Factory Acts and exclude the large number of cases and deaths in housepainting and plumbing. There is a more detailed breakdown of cases by trades in Legge and Goadby, *Lead Poisoning*, p. 47 and Mess, *Factory Legislation*, p. 46 carries the figures on to 1924.

39. Home Office, Factory Department, Form 324, Jan. 1921, Memorandum on Industrial Lead Poisoning, pp. 3-4.

40. There appear to have been at least 13 parliamentary, or Home Office departmental, enquiries into lead-based industries between 1882 and 1920. They are all listed in the bibliography at the end of this book.

41. *The Times*, 27 Sept. and 8 and 18 Oct. 1898.

42. *BPP*, 1899 [c.9207] XII, 277, Report by Prof. T.E. Thorpe and Prof. T. Oliver on the Employment of Compounds of Lead in the Manufacture of Pottery, their Influence upon the Health of the Workpeople.

43. Lead frits were composed of lead oxide and silica and were much less dusty than white lead. By 1918 common specifications were 75 per cent PbO and 25 per cent SiO_2 for lead monosilicate and 65 per cent PbO and 35 per cent SiO_2 for bi-silicate. These specifications would be chemically inaccurate since frits contained small amounts of aluminium oxide. Modern specifications would be: lead bisilicate – PbO 63.5 per cent-65.5 per cent, Al_2O_3 2.8 per cent-3.2 per cent and SiO_2 balance; lead monosilicate – PbO 83.5 per cent -85.5 per cent, Al_2O_3 from less than 0.2 per cent up to 2.2 per cent (dependent upon grade), SiO_2 balance. The subject is discussed in detail in F. Singer, *Low Solubility Glazes* (1948). I am grateful to Mr L. Williams, ALM's chief chemist at Chester for clarifying this subject for me.

44. Mess, *Factory Legislation*, p. 59.

45. *BPP*, 1914-16 [Cd 7882] XXIV, 901; 1920 [Cmd 630] XX, 1, [Cmd 631] XX, 49 and [Cmd 632] XX, 125, Reports of the Departmental Committees on the Use of Lead Compounds in Painting (1, report on painting of buildings; 2, report on painting of coaches and carriages; 3, appendices; 4, minutes of evidence of both committees).

46. Ibid., [Cd 7882], p. 102.

47. Most of the ILO debate was concerned with painting and is clearly

outlined for other countries as well as Britain in International Labour Office, *Studies and Reports Series F (Industrial Hygiene) No. 11 White Lead* (Geneva, 1927).

48. H.E. Armstrong & C.A. Klein, 'Paints, Painting and Painters, with Special Reference to Technical Problems, Public Interests and Health', *Journal of the Royal Society of Arts*, no. 3588 (26 Aug. 1921), pp. 655-85.

49. *Report of the Departmental Committee on Industrial Paints* (1923), pp. 27-8.

50. Ibid., p. 31.

51. *Newcastle Chronicle*, 1 Dec. 1923 and *The Times*, 22 May 1924.

52. Mess, *Factory Legislation*, p. 46.

53. *BPP*, 1938-9 [Cmd 6081] XI, 39, pp. 62 and 64.

54. *The Times*, 27 Feb. 1929.

55. Quoted in ALM, Southern Area News Bulletin, Dec. 1970, p. 5. See also the leading article in *The Times*, 28 July 1928.

56. Between 1972 and 1981, for instance, the maximum blood-lead level for employees was reduced from 120 to 80 microgrammes per 100 millilitres.

57. In addition to the use of bag-houses many other developments have reduced the level of lead dust in the atmosphere. Until the end of the 1950s scrap lead was frequently moved at the works by hand shovelling and barrowing to furnaces but since then the introduction of mechanical shovels and conveyor feed have reduced the risk. Automatic water sprays have been introduced to keep down dust on roads and at scrap lead stores and mechanical sweepers and automatic vacuum cleaners have replaced hand sweeping which had served largely only to raise dust.

58. International Lead and Zinc Study Group, *Lead and Zinc. Factors Affecting Consumption* (New York, 1966), p. 18.

59. The problems are much too complex for discussion here and the reader is referred to DHSS, *Lead and Health. The [Lawther] Report of a DHSS Working Party on Lead in the Environment* (1980), which contains both a summary of the results of research into the extent of lead pollution and its impact on the human body and a detailed bibliography of recent publications on the subject. A more recent article (W. Yule, R. Lansdown, I.B. Miller and M.A. Urbanowicz, 'The Relationship between Blood Lead Concentrations, Intelligence and Attainment in a School Population: a Pilot Study', *Development Medicine and Child Neurology*, 3 (1981), pp. 567-76) has raised a further outcry in the media. The *Guardian* (1 Oct. 1981, p. 23), for instance, devoted a full page to articles, arguing that lead emissions from vehicle exhausts cause brain damage in children, under the heading 'Still pumping poison into our children' and with an illustration of a little girl, looking apprehensive because of the petrol pump being pressed to her head. By contrast the Medical Correspondent of *The Times* on the same date drew attention to the fact that the article by Yule *et al.* had not corrected its sample of children for age. The children who had high blood-lead levels were younger than those with low blood-lead levels and the Correspondent of *The Times* commented, 'Need anyone be surprised that nine-year-olds are better at spelling and reading than seven-year-olds from the same background' and concluded that 'the scientific evidence does not at present support any connection between the amounts of lead in the blood of healthy city children and their performance at school'. Preliminary results of recent American research do, however, suggest that the major contribution to blood-lead levels comes from lead in petrol. Center for Disease Control, Atlanta, Georgia, 'Blood Lead Levels in the US Population', *Morbidity and Mortality Report*, vol. 31, no. 10 (1982).

60. As a result the Medical Officer of Health for the London Borough of Camden instituted measurement of the fall-out of lead from the atmosphere and

blood-lead tests on local children. Three children, all of Millwall works employees, showed blood-lead counts above 40 micrograms per 100 millilitres which of course attracted the interest of the press and television. Samples had, however, been taken by fingerprick in their homes where the risk of contamination of the sample was high and subsequent hospital tests showed the levels to be considerably lower. The fall-out tests showed, over a considerable period, that lead in air levels were safe by existing standards and that fall-out from the lead works was less significant in causing lead pollution than exhaust gases from motor vehicles.

61. See, for example, *Sunday Times*, 10 Feb. 1980, a report on the Chester works of ALM.

8 THE TRANSITION TO LIMITED COMPANIES, 1889-1914

I shoot the hippopotamus
With bullets made of platinum
Because if I use leaden ones
His hide is sure to flatten 'em.
                    'The Hippopotamus', Hilaire Belloc

The conversion to limited liability of existing lead manufacturing companies and the formation of new companies in the late-nineteenth and early-twentieth centuries. Levels of output and profitability and general assessment of each company's performance from about 1890 to 1914. Walkers, Parker & Co. Ltd; Mersey White Lead Co. Ltd; Locke, Blackett & Co. Ltd; Locke, Lancaster and W.W. & R. Johnson & Sons Ltd; Brimsdown Lead Co. Ltd; Cookson & Co. Ltd; other companies.

In Chapters 4 and 5 the development of the major lead companies was discussed up to the late-nineteenth century, when many of them converted from partnership to limited company organisation. In one instance, Walkers, Parker, this took the form of selling the partners' assets to a limited company in which there were large numbers of share-holders. This was not difficult for a large and well-known partnership but it was certainly more difficult for the smaller companies, although Locke, Blackett managed to do it relatively successfully, as did Mersey, a new company set up from scratch in 1889. More commonly the smaller companies converted to limited liability status but were to remain what were later to be called 'private' or 'close' companies in which the old partners retained an over-riding proportion of the shares, while small numbers were sold to members of the families involved and to senior managers. As such the conversion to limited status was less concerned with access to the capital market than with legal security.[1]

It is less than coincidence that so many conversions to limited liability took place from the late 1880s to the early 1900s and it was certainly not unique to the lead manufacturing industry. Although there are many examples of old-established concerns becoming limited companies in the first 30 years after limited liability legislation was first intro-duced in 1855, it remains true that it was from the economic boom of the late 1880s that this became common.[2] For some the sight of rising prosperity after a decade or more of low profits no doubt encouraged a sale, while for others the experience of the depression encouraged a move to a more financially secure form of organisation and yet more were probably persuaded to convert in order to follow the fashion of the moment.

It is common to lament the lateness of corporate development in the British economy and to blame the continued existence of family firms for the relative industrial decline of that economy *vis-à-vis* other western economies in the late-nineteenth century. It sometimes seems that this is done purely from the late-twentieth-century evidence that large companies are usually limited companies while family firms are small, therefore partnership organisation prevented the economy from growing. This is clearly nonsensical, for many firms had grown to a large size under partnership control in the nineteenth century and, although not large by the standards of engineering and shipbuilding firms, Walkers, Parker had successfully reached an employment approaching 1,000. The view that partnership organisation stifled initiative, based on the idea that the third generation of the family frequently lost interest in the firm, ignores the fact that such firms often introduced paid management, with a share in profits and sometimes the eventual offer of a partnership. Finally, while the rise of the limited company is frequently applauded in this kind of analysis, it is rarely recognised that power often remained in the hands of the old partners (who were unlikely to change their spots with their firm's title) and, even where it did not, there has been little attempt to produce evidence to show that limited companies were more dynamic than their predecessors. As we look at the development of the lead companies up to the First World War it will be interesting to see what light they throw on received opinion.

Walkers, Parker & Co. Ltd

This was the first lead company of any significance to take joint stock status and this was done purely as a means to get the firm out of the partnership dispute which had dragged on from 1881 to 1888. Walkers, Parker & Co. Ltd was incorporated on 21 January 1889 and a prospectus issued. After stressing the long and successful history of the partnership comprising 'the largest and most Important Lead Manufacturing Establishments in the world', and listing the recent profits, the prospectus concluded that:

The business is capable of expansion under a more centralised and economical system of management, and it is believed that last year's results [net profit £58,362] will be improved upon; but, even if only maintained, there would, after paying the interest on the Debentures and Preference Shares and providing for the management of the Company, remain a surplus of about £25,000.

The prospectus called for subscriptions for part of the share capital — 20,000 6% Cumulative Preference £10 shares and 3,500 4½% Irredeemable Mortgage Debentures of £100 each — the remaining capital, £300,000, in the form of 30,000 £10 ordinary shares, being taken by the vendor. It will be noted that the company, which had been purchased on 27 October 1888 from the partnership by E.J. Walker for £500,000, was being sold to the public some three months later, capitalised at £850,000.

Not surprisingly this called forth some fairly sharp comment from the financial press. Under the heading, 'The reputation of an old business undertaking in peril owing to the greed of the company promoter', the *Lighthouse* commented, 'Never in the history of company promotion have we heard of a more glaring attempt to palm off an undertaking on the public under questionable circumstances and equally questionable auspices'[3] and concluded:

Mr. E.J. Walker must either take the customers of Walker, Parker & Co. to be a pack of bigger fools than they are, or he must flatter himself that they place unbounded confidence in him . . . we have no hesitation in saying that he is a very greedy man, to put it mildly.

It seems unlikely that E.J. Walker was the initiator of the promotion and it is probable that someone else was promoting the company, perhaps Arthur Burr, who was subsequently criticised for his influence over the company and especially the appointment of his brother, Walter, as accountant and a member of his staff, J. Barrett-Lennard, as secretary. Much subsequent criticism complained that these two, although they knew nothing of the lead industry, dominated the company's policy and ignored the opinions of the experienced managers of the various works. There is certainly some suggestion that Arthur Burr was involved in financial transactions with E.J. Walker before the sale to the limited company and he had felt so touchy on the subject as to bring a libel action against the *Daily News* for certain comments that paper had made.[4] Burr certainly had a large holding in the company, since on 30 April 1890 he transferred it to J.C. Stogdon in payment of a debt of £60,000.

The board of directors was probably unaware of the financial activities which had gone on, since its members appear to have been selected as respectable business and professional men who would make suitable figureheads. The chairman, W. Wentworth Walker (no relation to the lead manufacturing family), belonged to a Wolverhampton firm of lead merchants, T.W. & J. Walker, and provided the nearest link to knowledge of lead manufacturing.

Despite the unfavourable financial press which the prospectus received, the limited company managed to get off the ground, largely with financial support from Liverpool and London. Current economic conditions should have been favourable to its development but in 1889 the net profit was £38,000, some £20,000 less than in the previous year and this was followed by a profit of only £2,000 in 1890. As a result a committee of two directors, the secretary and the accountant was set up to examine the works and its report expressed the 'opinion that the late Managing Director [E.J. Walker] is to a very large extent responsible for the bad results shown'. This report was made available to shareholders at the annual general meeting for 1891, at which a net loss of £16,000 was shown, and it may well be that E.J. Walker was being used as a scapegoat to protect the directors and the rest of the company structure from shareholders' criticism.

Walker had apparently been forced to resign as managing director since he had proved as unacceptable to his colleagues in the limited company as he had done to his relations in the partnership. Possibly his major sin had been to improve and furnish the old partner's house at Chester, in which he was to live, passing the cost, of almost £3,000, through the Chester accounts in 1890. While this incensed most of his fellow directors, one, J.A. Carson, supported him in a series of criticisms of the management and of Arthur Burr's influence upon it. It is, therefore, improbable that Walker was entirely responsible for bad management and the company's financial plight but at the 1892 AGM his earlier reputation was against him and both he and Carson were voted off the board.[5]

Quite apart from the ability or otherwise of E.J. Walker as managing director, the firm was suffering from a number of problems. In the north-west there was a wholesale loss of management when the limited company was formed. W.M. Hutchings, the Bagillt works manager, resigned to become Cooksons' manager, while William Sloane and C.J. Day, Chester works and office managers respectively, resigned and set up the Mersey White Lead Co. Ltd in 1889, taking with them several of the Chester workmen. Not only did this mean that new managers had to be found but it is clear that Mersey took over many of Chester's contracts as a result of Sloane's personal contacts. Moreover it was suggested that Sloane and Day deliberately ran down the contents of the white lead stacks at Chester, before they left in March 1889. There is some evidence to support this in the Chester stock books, which show falls in the stocks of pig lead from £61,000 to £35,000 between October 1888 and April 1889 and from £9,000 to £5,000 in stocks of white lead — falls which exceed those which might have been expected as a result of rising sales during a boom year. Day subsequently claimed that he was instructed to reduce stocks and the decision may have been made by E.J. Walker. It seems likely, therefore, that the limited company lost out on the rising trade in its first year or so of existence as a result of lack of capacity and increasing competition, and it is interesting to note that Mersey's first sale of white lead, made in May 1889, was to Walkers, Parker at Chester.

Walkers, Parker also suffered from the financial speculation which was flying around. On 28 May 1889 Hutchings wrote from his new Tyneside base to Day at Mersey that Walkers, Parker & Co. Ltd 'appear to have very poor credit and to be very much mis-trusted by everybody and to have difficulty in financing their requirements. I hear of firms who will not sell them anything except for cash in advance of delivery.' While the inability to obtain credit would raise financial charges and reduce profits, it was a problem which might relatively rapidly have been overcome by establishing a reputation for financial viability but it was followed by a more serious problem with the fall in price of lead. Having recovered from a low of £11 per ton in 1884 to around £13 10s in the late 1880s the price began to dip in the early 1890s and fell below £11 in 1892. It seems likely that Walkers, Parker began to build up their lead reserves, in response to the earlier run-down position, not only when prices were falling but also as demand for their products was turning down with the trade cycle after 1890. To a considerable extent this policy could be blamed on inexperienced management by Burr and Barrett-Lennard and lack of control by a directorate which did not include a lead manufacturer. The Chairman told the annual general meeting in 1891

> The almost universal custom in the lead trade was to sell, not as in most other trades, according to the cost of the article, having regard to the first cost, the price of fuel, of wages, etc. but to sell the

manufactured article based upon the price of pig-lead on the day on which the sale was made. The result was that in a falling market they sustained a loss over which they had no control, and in a rising market they enjoyed an additional profit which they had not earned by any labour of their own. He maintained that this was a pernicious system of selling, and as long as it existed it placed their trade in all its ramifications, at the mercy of speculators.[6]

This showed recognition of the problem, to which the answers should have been careful raw materials purchasing policies and the creation of a reserve fund to offset losses from falling prices out of gains when prices rose, but little was done.

There were, however, some attempts to improve the company's position. The Dee Bank[7] works was put up for sale in 1889 but although a sale was negotiated it fell through because of the purchaser's inability to raise the money. Secondly, new premises at Neptune Street, Liverpool were taken at a cost of £7,000 to replace the Paisley Street premises on which the lease was due to run out in 1892. Even this venture was ill-fated since it was soon decided that it was unnecessary to have manufacturing premises (for lead sheet and pipe) at Liverpool, when Chester and Dee Bank were nearby. The Neptune Street works was then closed for manufacturing about 1895, subsequently reopened in 1897, following another change of mind, and finally closed after further financial problems in the early-twentieth century and eventually sold to J. Bibby & Sons in 1906 for £8,000. It would, therefore, appear that some of the positive steps taken by the company were unsuccessful. There were, however, some signs of positive response to the need to increase profits. The managers of the main works were initially appointed on fixed salaries of £450 p.a., which was low for their responsibilities, but in March 1890 these were adjusted and A. Billyeald, the Lambeth manager, was, for example, moved to a salary of £600 p.a. with 2½ per cent of his works' net profit over £5,000 p.a. and 5 per cent over £15,000. In addition, early in 1891 a general manager, C.E. Bainbridge, was appointed on a salary of £800 p.a. plus profit-sharing incentives, the Chester manager was sacked after less than two years in the job and offices and plant were closed and reorganised in order to produce financial savings.

Measures such as these appear to have had little impact since there were net losses in five of the first thirteen years of the company's existence, during which period the total balance was a net profit of less than £10,000 on an original capital employed of £850,000.[8] In the first year of operation, 1889, a 5 per cent dividend was paid on the ordinary shares and the 6 per cent preference shares received their dividend in the first two years. Thereafter there was no dividend of any kind until 1903. This was a performance of disastrous proportions which it is difficult to believe could not have been bettered if the company had remained in the hands of the Walker family.

Inevitably there was considerable dissatisfaction among directors and shareholders. In January 1892 an unofficial meeting of the Liverpool shareholders was held to which the chairman of the company was refused admission. There was strong criticism of the existing board but this was temporarily overcome by the expedient of appointing a director from Liverpool, J.M. Somerville, in the place of another director of short-standing who conveniently resigned. This did little to stem the overall tide of shareholder discontent, which led to a turbulent AGM in March, but the directors managed to ride out the storm and get the accounts adopted and their report, blaming E.J. Walker for the problems, was accepted. The board now had a breathing space but economic conditions, with the price of lead falling, were against a return to profitability. The board minute on the 1892 accounts noted that:

> The question of certain items in the balance sheet, including losses, unavailable assets, and depreciation was discussed at great length, and it was agreed to deal with these items by reducing the Ordinary Share capital from the sum of £300,000 in £10 shares to £75,000 in shares of £2.10.0. each.[9]

A capital reduction had been inevitable since the gross overcapitalisation when the company was formed in 1889, but a reduction of £225,000 ought clearly to have been seen as insufficient since the company was not even making sufficient profit to pay the preference let alone an ordinary dividend. The reduction needed to be of at least £300,000, the value of the ordinary shares, which it will be remembered were additional capital to the price paid for the partnership's assets, and probably a further £100,000 to allow for such things as the fall in lead price. The company's assets had been grossly inflated, the chairman telling the 1890 AGM, 'With regard to the way in which they arrived at the goodwill, it was the difference between the assets of the company and the capital, £850,000 . . .',[10] and £143,000 was written off goodwill in the capital reduction of 1893. Smaller sums were written off plant, buildings and investments but this still left considerable overcapitalisation.

From the point of view of the preference shareholders the writing down of the vendor's shares was no doubt seen as an acceptable way of dealing with the problems. It may, therefore, have encouraged them to moderate their criticism and the 1893 AGM went off quietly. Trouble, however, was never far away from the company in these times. There were still outstanding legal actions over the distribution of the purchase money in 1889 and E.J. Walker's debt to the company and the secretary and general manager (Barrett-Lennard and Bainbridge) were forced to resign when the board discovered that their names had appeared on the prospectus of a company to manufacture white lead by a new patented process.

The chapter of accidents, which is the history of the company's first years of operation, could hardly be put down to misfortune. The board was lacking in knowledge of the trade and took insufficient control over the management. As a result the company's financial standing fell disastrously, bankers would not increase overdraft facilities and lenders pressed for loan repayments. With a net profit of only £4,000 in 1893 there was serious unrest at the AGM with an attempt to reject the accounts and another to set up a shareholders' committee of inquiry. These were withdrawn on the board agreeing to the appointment of two new directors, R.S. Mason and William Newall, and this was the beginning of total reorganisation. When the results for 1894 showed a net loss of nearly £10,000 Newall replaced W.W. Walker as chairman and all the remaining original directors resigned and were replaced by appointees of the AGM.

For those who believed that it was the composition of the board of directors which was responsible for the company's performance this was both a triumph and the opportunity for their opinions to be tested. In the years 1895 to 1900, however, while there was a net loss in only one, the total net profit was only £39,000, despite a rise of nearly 80 per cent in the price of pig lead and an improvement in the trade cycle, circumstances which should have enabled the company to make a net profit of £40,000 p.a. Once again the board did not contain anyone with knowledge of lead manufacturing (at least two members were solicitors and the chairman owned a boat-building yard at Chester). Numerous attempts were made to tighten up on works efficiency and ensure that production costs were lowered but the board does not appear to have achieved any effective control. Each of the works was inspected to see what improvements could be made but it is unlikely that the board could judge effectively and the minute of 2 April 1895 that 'instructions be given to the several managers that it would be necessary in future to bring everything requiring attention before the Board', although understandable, is more likely to have promoted dissatisfaction than efficiency. The Liverpool, Chester and Dee Bank managers were sacked and all three works controlled from Chester by a new manager, while a clerk and a traveller were dismissed when investigations of accounts showed that embezzlement had taken place. The Chester manager, H. Davison, was given a roving commission to examine the various works for possible economies, and one result was the introduction of piecework payment at Newcastle. These steps did not get to the root of the problem and some of the economy measures were probably the opposite of beneficial. The purchase of lead was centralised and when Billyeald, the Lambeth manager, as he had done in the past, made an unauthorised transaction, which led to a loss of £300, he was ordered by the board to repay the sum out of his salary. It is unlikely that this produced a contented manager, obviously an important requirement when the managers were the senior men with knowledge of the trade, and although the principle of centralisation of purchase was sound it no doubt frustrated local managers, since the necessity to obtain board authorisation caused delays.

It was apparent that the new board had failed to pull the company round and it came under pressure early in 1898 from both Brown, Janson & Co. and the Bank of Liverpool for the reduction of overdrafts. Moreover shareholder unrest, stilled by the appointment of the new board, soon broke out again, with the demand for a shareholders' committee of inquiry at the AGM in 1898, following a net loss of over £1,000 in the previous year's accounts. The board was able to reject this demand but the financial problems were becoming so serious as to affect the company's performance. Both banks had taken control of the company's supply of pig lead (insisting that it be held on their accounts until the overdrafts were reduced) and this caused the works to be stopped on several occasions for lack of raw materials. Output and efficiency were consequently reduced and, in addition, the banks on several occasions insisted that part shipments of lead were sold, sometimes at a loss, in order to reduce overdrafts. All sorts of expedients were adopted to find sufficient money to obtain pig lead and pay debenture interest in the late 1890s, including the raiding of a deposit account containing the insurance money paid to effect repairs following a fire at the Chester works. In such circumstances it is not surprising that repairs and renewals were neglected and lack of investment meant that low productivity was certainly a cause of the company's problems.

The regular rejection of expenditure proposals from works managers is an instance of the problem and when they did spend money they were criticised. A fire at the Elswick works in 1900 caused the manager to buy 1,000 yards of fire hose – a step which the board criticised because he had not obtained written estimates and submitted them. He replied that 'it was necessary to act at once to replace the old hose, which had become utterly useless'.

Consequent upon the inability to get the problems under control, even the peak of the cyclical upswing of the late 1890s produced profits of only £19,000 and there was no dividend 'for the simple reason', the chairman explained at the AGM, 'that every penny was required for working the business'.[11] While the directors succeeded in getting their report accepted and the rejection of motions for their replacement and the setting up of a shareholders' committee of inquiry, they could have had little long-term hope. A drop in profit to less than £3,000 in 1901 ensured that the problems surfaced again and David Thomson, in moving a resolution for a shareholders' committee of inquiry, commented that the directors' report 'was of the character that they were all so familiar with – namely, only a long string of excuses'.[12] The committee was approved and four shareholders appointed, two from Liverpool and two from Carlisle (including Thomson, chairman of a plumbers' merchants), reflecting the strong shareholding in the north-west.

The immediate consequence was further serious trouble with the banks, which were much disturbed by the setting up of the share-holders' committee and wanted the directors' personal guarantee of the overdrafts. The banks refused to release pig lead and the Newcastle works was stopped for several days on at least two occasions. In May 1901 the board agreed to co-opt two members of the shareholders' committee, including Thomson, and the banks were temporarily subdued by the issue of mortgage debentures as security for the over-drafts. From this point the company's position began to improve, apart from the fact that it immediately had to face by far its biggest net loss, £44,000 in 1901. Since this could be easily attributed to a fall in the price of pig lead from £17 per ton in 1900 to less than £13, it was accepted with relative equanimity by the shareholders. Action was taken to improve the management. In July 1901 W. Eckford came from United Alkali Co. Ltd to become sub-manager at Chester and, as we have seen, was to make major technical contributions, while the exist-ing managers were given clear indications that they would be sacked if improved productivity did not occur and they were given profit-sharing targets which were realistic for the company's existing position. One example, from the board's communication to the Chester manager, who was subsequently sacked, will suffice to show that the position was made very clear:

> [the directors] are very dissatisfied with the management of the business and works in your district. The increase of expenses, though the sales have so seriously declined, the absence of information as to its cause, the want of close acquaintance and attention on your part to the gravity of the situation and of constant examination of its causes, day by day, week by week, indicates in the judgment of the directors that you have failed to realise the responsibilities and duties of your position. The Board defers taking any action with reference to the matter, as concerning yourself, until they see the effect of this communication upon the management of the business in the near future.

Apart from tightening up on management, with the aim of reducing expenses (in which the directors took a lead by reducing their own fees), measures were taken to reduce the company's indebtedness by selling surplus assets, especially land. Both at Chester and Elswick the company had inherited more land than was likely to be needed for future expansion and it was wise to capitalise on it.

Together financial economies and capital realisations began to improve the company's position and after 1901 the board minutes are no longer full of frenetic concern over the state of the bank accounts. As cash built up it was gradually used to cancel the debentures. Between 1901 and 1906 the outstanding value of first and second debentures was reduced from £325,000 to £50,000, with a consequent annual saving in interest payments of more than £12,000. That alone is almost equal to the difference between the average annual profit performances for the periods 1890 to 1901 and 1902 to 1913, respectively losses of £2,400 and profits of £10,700, which would suggest that any other improvements made after 1901 were cancelled out in profit terms by deteriorations in other factors. The company was clearly not buoyant. On the accounts for 1902, which showed the largest profit since 1889, *Leading Opinion*, a financial weekly, commented on 'the extremely critical condition' of the company and, noting the 'directors' pleasure in presenting the accounts', wrote:

We would like to add our congratulations to the directors upon the extraordinary facilities which they evidently possess for the enjoyment of pleasurable sensations; for to the merely ordinary and commonplace individual there appears to be little enough in these documents to excite rapture.[13]

As the article pointed out, the preference dividend had not been paid since 1890 and was in arrears by the sum of £144,000, while reserves were £2,372, and £26,000 p.a. was required for interest payments alone. While considerable success had met the attempts to deal with the latter problem by 1906 and the nominal capital employed had been much reduced, a long article in the *Financial Times* could still comment on the company's unfavourable position.[14] Although dividends on preference shares had begun to be paid again from 1903, this had been only at 3 per cent, one-half of the nominal level, and arrears had risen to £174,000.

After half-a-dozen years the new directors had failed to bring the company to prosperity and it was inevitable that they would be challenged as they had challenged their predecessors. The opportunity came with the proposal to pay no preference dividend for 1906 and after some months of negotiation the major preference shareholders succeeded in obtaining a capital reorganisation. The preference shares were enfranchised as 'A' ordinary shares in return for which the whole of the arrears of preference dividends were written off. This removed a burden of debt which appeared irreducible and offered the ex-preference holders not only voting power but the hope of a return on their investments. For the year 1907 this hope was fulfilled with a dividend of 6 per cent and although it fell to only 3 per cent in the years 1909 to 1911, it recovered to 5 per cent in the years immediately preceding the War. The results for these years, with profits averaging a little under £11,000 for the years 1909 to 1913, were hardly satisfactory, reflecting as they did a net return of less than 5 per cent on total capital employed.

Some practical achievements had been made in the first decade or so of the twentieth century. Of these electrification was the most notable, both for lighting and power purposes, although it was done on a piecemeal and not always successful basis. There was also some modernisation of plant. At Chester in 1902 the company installed its first overhead electric travelling crane to fill and empty the white lead stacks and the example was followed at Elswick in 1909, while a new rolling mill was installed at Lambeth in 1904 at a cost of £1,700.[15] There was also reorganisation of the works. Neptune Street, Liverpool, was at last sold in 1906 and the company was then left with only a warehouse at Liverpool, at the Duke's Dock, a sensible arrangement in view of the nearness of the Chester premises. Also in 1906 the rolling mill and pipe presses were transferred from Dee Bank (where part of the works was closed) to Chester in a further piece of rationalisation. There were, indeed, suggestions that Dee Bank should be closed entirely. Its only rationale lay in the availability of local supplies of ore but these were in decline and for much of the time one of the two blast furnaces at the works was out of blast and it is clear that Dee Bank was an inefficient works, partly responsible for the drain on company finances.

While some works were being closed and others reduced, the company was actually opening a new one at Hull. This decision was less paradoxical than it might seem on the surface. There was a considerable lead trade in the Yorkshire area and none of the major producers was established there. In 1905 premises were taken at Charlotte Street and a pipe press transferred from Newcastle. This move, however, provoked a reaction from one of the more considerable of the smaller blue lead manufacturers, R. Fell & Sons of Skipton, which within months had established a works at Hull, although most certainly with no intention of working it. In November 1905 Fell arranged a meeting with Walkers, Parker and offered three options: (1) open competition; (2) price agreement with a share of the Hull trade, which Walkers, Parker estimated would 'mean a loss of about half the Company's pipe trade in Hull'; (3) 'Mr. Fell might be prepared to consider the sale of his works to the Company at a high price.' Despite its potential strength the company settled for the third option, apparently on the ground that it could not afford the cost of fighting even such a small fry. Although he asked £6,500, Fell accepted £4,000 (and no doubt still made a considerable profit).

Once again even a positive step taken by Walkers, Parker had been less than satisfactory and costs ended up higher than were expected. It is, therefore, hardly surprising that the sum of a series of uninspired decisions was the still poor profit performance of the years before the War. Relatively, Walkers, Parker had continued to decline in significance within the lead industry, except in red lead where the Eckford equipment had maintained it in the van of progress. Just before the War Walkers, Parker's share of the Convention quotas (which certainly accounted for more than 90 per cent of UK output in red and white lead and probably not far short of that figure in blue lead) was about one-sixth for both blue and white lead and one-third for red lead. It remained by far the largest company if rather a tired giant.

Plate 1. River Thames with both Lambeth shot towers (Belvedere Road left centre and Commercial Road further to the left) and the Red Lion Brewery on the right. View taken from Brayley's *History of Surrey*, published in 1850.

Plate 2 *(above).* Windmills for grinding white lead at the Islington works, circa 1800.

Plate 3 *(below).* Share certificate of Brimsdown Lead Co. Ltd.

Plate 4 (above). Shot tower at Morledge works, Derby, 1885.
Courtesy of Derbyshire Record Office

Plate 5 (above). Elswick shot tower, circa 1955.

Plate 6 (below). Morledge lead works, Derby, early nineteenth century.

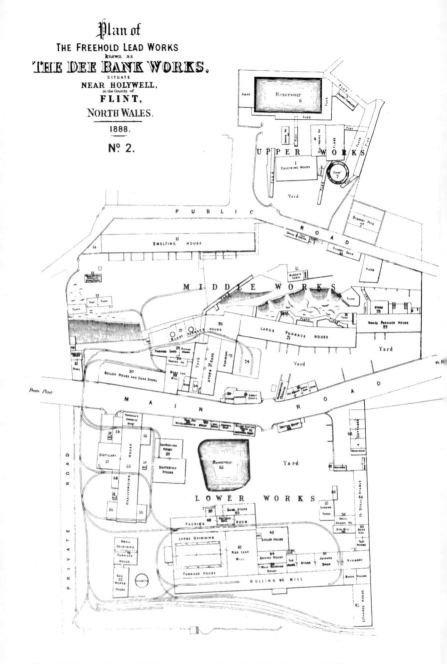

Plate 7 *(above).* Plan of Walkers, Parker's Dee Bank lead works, 1888.

Plate 9 *(opposite page, right).* Blast furnace for smelting lead at Chester works, circa 1950.

Plate 8 *(above)*. Bootle lead works with barges on the Leeds and Liverpool canal, 1948.

Plate 10 *(above).* Tapping molten lead from the furnace into pigs at Millwall works, 1949.

Plate 11 *(below).* Tilting furnace for conversion of copper matte from lead blast furnace at Elswick works, 1949.

Plate 12 *(above)*. Skimming dross from 12-ton molten lead block after casting, Millwall, 1949.

Plate 13 *(below)*. Block of lead on mill before rolling at Millwall works, 1947.

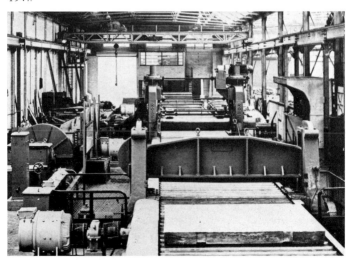

Plate 14 *(above)*. A pair of finished sheets of lead on mill, preparatory to coiling, Millwall, 1949.

Plate 15 *(left)*. Pouring solder into moulds and removal of finished sticks, Millwall, 1949.

Plate 16 *(above).* Stocks of extruded lead pipe at the Millwall works, photographed in 1949.

Plate 17 *(above)*. Coiling lead pipe after extrusion, Millwall, 1949.

Plate 18 *(below)*. Tapping lead frit furnace at Chester works, circa 1955.

Plate 19 *(below)*. Testing samples of silver lead at Millwall works, circa 1950.

Plate 20. Pouring lead shot at the top of the Lambeth tower — an early twentieth-century photograph.

Plate 21 *(above).* Elswick lead works, Newcastle upon Tyne, circa 1957.

Shot production at Elswick, circa 1960

Plate 22 *(left).* Pouring molten lead into the shotrunner's card.

Plate 23 *(right).* Drops of lead falling through the card.

Plate 24 *(above left)*. View down the tower, with the water tub below. Plate 25 *(above right)*. Shot falling into tub of water at the base of the tower. Plate 26 *(below)*. Sorting shot on inclined plane.

Plate 27 *(above)*. A white lead stack being filled at Millwall works in the early twentieth century. In the foreground one layer, or 'height', with blue lead on top, has been completed and is partly covered with boards before the next layer. The fork was used for spreading the tan bark and the workman is filling the pots with acetic acid.

Plate 28 *(below)*. Drying stoves for white lead at Millwall, circa 1935.

Plate 29 *(right)*. Handbill advertising one of the pills sold around 1900 as a lead poisoning preventive.

BARRINGTON'S
Lead Poisoning Preventive

To workers amongst Metallic Lead, Red Lead, White Lead, Litharge, Plumbers, Potters, and Paint Grinders, this mixture is especially adapted, both as a Preventive and a Cure for Lead Poisoning.

For all persons employed in handling Lead in any form or subject to Lead fumes or Lead dust it is advisable to take this mixture, which has been used by a worker in Litharge (Yellow Lead) for five years, and who never once suffered from Lead Poisoning or Colic

This mixture contains no injurious minerals, and is entirely free from Mercury, in fact is highly recommended as a cure or preventive of Mercury Poisoning

Price 1/1½ per bottle, Post Free, in United Kingdom

J. BARRINGTON
Frimley, Carlyle Road,
SOUTH EALING, LONDON

Northern Printing Co., 69, Liverpool Rd., Islington, London, N.

Plate 30 *(below)*. Labour force at Millwall white lead works, photographed in the early years of the twentieth century.

Plate 31 *(top)*. Group of long-service employees (ranging from 41 to 53 years' service) at Millwall in the late nineteenth century.

Plate 32 *(centre)*. Employees at Burdett Road, Limehouse, lead works, circa 1900.

Plate 33 *(left)*. Employees in solder department at Millwall lead works, during the Second World War.

Mersey White Lead Co. Ltd

It is appropriate to deal next with this company, both since it was set up shortly after Walkers, Parker[16] and because it was commenced by ex-employees of Chester. William Sloane and Charles Day would undoubtedly have sided with A.O. Walker, who had been managing partner at Chester, in the partnership dispute and they obviously decided that they would not get on with E.J. Walker, who, as managing director of Walkers, Parker & Co. Ltd, was going to live in the manager's house at Chester. They purchased for £3,500 the premises and wharf of the by then moribund Sankey White Lead Co. Ltd at Warrington and began to manufacture white lead by the stack process. Mersey was registered in February 1889 with an initial capital of £25,000 in £10 shares and this nominal figure was doubled in November 1889, the capital being provided almost entirely by Chester residents who knew Sloane and Day.[17] In addition borrowing powers were increased from an initial £10,000 limit to £75,000 by 1898. Unlike Walkers, Parker, however, where borrowing had got out of hand and was a sign of weakness, Mersey was borrowing from a position of strength on the basis of expanding production. The company was able to raise debenture capital readily from local friends and acquaintances and in so doing was strengthening its position by improving its gearing ratio, since ordinary share capital remained unchanged.

Mersey initially benefited from the wide experience and contacts of its management. Day was questioned in the continuing saga of *Walker* v. *Walker* before the Court of Chancery:

- Q. Your late position with the Firm had of course given you information as to who the customers were.
- A. Of course naturally I had been there 28 years.
- Q. And you used that information for the purpose of advancing your interests in the new Coy.
- A. Yes decidedly. I should have been of no use to the Coy if I did not know anything about the business . . .
- Q. Have you been getting on pretty well do you think in transferring the custom to your new business.
- A. Yes very fairly I think.[18]

Not only did Sloane and Day have direct contacts with the old customers of Walkers, Parker (which that firm's new management and directorate lacked) they also had links with the employees. At least three senior workmen, George Thompson, who became Mersey foreman (after 32 years at Chester), William Jackson, foreman cooper and Martin Jordan (27 years at Chester) followed their managers to Mersey. The new company was, therefore, well placed for success, although it was only starting on a small scale and Sloane and Day were not initially to do very well out of it, being paid £400 p.a. each plus (jointly) 10 per cent of net profits after all expenses and a 10 per cent dividend had been paid.

The early experience in the cyclical upswing of 1889 was, however, beneficial and by the end of 1890 the firm had increased its capacity to 38 stacks. Moreover in several fields it was to show itself more dynamic than the large, long-established companies. In 1891 tramways to all the stacks were constructed and all internal freight movement was subsequently done by rail, which compares interestingly with Chester where horse and cart was a major form of internal transport until 1938. We have also seen that Mersey was in the van of progress in dealing with prevention of lead poisoning.[19] In the early 1890s cash-flow problems, common to many expanding companies, occurred but were overcome without great difficulty, usually as a result of temporary loans by directors. Sales were gradually increased from around £30,000 p.a. to £60,000 in the mid-1890s and £100,000 by the turn of the century and the company became financially stable. Even in the first partial year of trading a net profit of more than £1,000 had been achieved and the figure rose steadily to £10,000 by 1899. This was a considerable achievement in the light of Walkers, Parker's performance and it is noticeable that, apart from a slight dip in 1892, profits continued to rise during the period of major price fall, which the directors of Walkers, Parker considered the major reason for their own company's lack of success. Mersey's success, by contrast, must be credited to Sloane's knowledge and management (Day having died in 1892). Undoubtedly the company's concentration on a single product, white lead, was in some senses an advantage for a small firm, since management's attention was concentrated. On the other hand white lead manufacture tied up capital, in the form of work in progress, for a longer period than any other lead product and should have made the company more vulnerable to the price fall. There was a stage when Mersey considered backward integration when it made an offer for the Llanerchymor lead-smelting works in Flintshire but fortunately for the company, in view of the declining profitability of the north Wales industry, the offer was not accepted and Mersey stuck to white lead alone.

By the mid-1890s 'The business of the Company having increased beyond the capacity of the works . . .' the board decided on further expansion and eight more stacks were built. As a result, with an annual output of over 5,000 tons of white lead, Mersey became the largest producer of that commodity from a single works, although Walkers, Parker's total output was greater and Cooksons' was soon to become so with the introduction of its chamber process. This was the last major expansion which the Company was to make and, with the exception of the building of stacks to fill odd spaces in the works, capacity remained constant until the First World War with a total of 55 stacks. To some extent this resulted from recognition that Mersey would find it difficult to increase sales further in a market which was expanding only slowly if at all but it must also have had something to do with the inertia effect of age on Sloane, who had provided the initial dynamism. There was some justification for sitting back to enjoy the fruits of earlier efforts. By 1900 a considerable reserve fund had been built up, although part of it was needed in the following year to meet a loss of £13,000 caused by the fall in the price of pig lead, on which the directors noted, 'so rapid a fall has not occurred since the year 1816'. Nevertheless the company was still able to pay a dividend of 10 per cent from reserves and used the warning to create a further reserve fund to equalise dividends, in the event of future profit fluctuations resulting from changes in the price of pig lead. Apart from this hiccup the company's reserves rose and, on the advice of its solicitors, were used to reduce its debenture capital by repurchase whenever possible. Net profit figures which only averaged around £4,000 p.a. up to 1913 were a justification for a policy of non-expansion, and investment of surplus cash in government and railway stock, although it offered a lower return, required little effort. Finally, in 1911, Mersey rejected its early philosophy and agreed to join the White Lead Convention – its conversion to middle-aged conformism was complete.

Locke, Blackett & Co. Ltd

The prospectus for the conversion of the old partnership to a limited company was issued in April 1891, offering for sale 10,000 £5 preference shares and 10,000 £5 ordinary shares together with £30,000 in mortgage debentures. We have seen that the likely reason for the conversion was an attempt to restore the fortunes of an ailing company and the new board certainly comprised a fairly powerful Tyneside directorate.[20] The prospectus forecast the hope of a 10 per cent dividend on the ordinary shares and in each of the first three years of the company's existence 7½ per cent (net of tax) was actually paid. This must have provided some encouragement to the shareholders, who, although coming from a wider geographical background than did the Mersey holders (as was to be expected from the conversion to joint-stock status of a long-established and well-known partnership) were heavily concentrated in the north-east.[21] In 1894, however, consequent upon the fall in price of pig lead there was a sharp dip in net profit which necessitated a transfer from reserves in order to meet the preference dividend. There was no ordinary dividend and, although 2½ per cent was paid in the following year, there was again no dividend in 1896 and a long period of financial difficulty had begun. It seems likely that the board had not been able to establish the effective management which the company with its three diversified works required if it was to succeed in an increasingly competitive environment. The position may have deteriorated after the death of James Leathart in 1895, since his son, T.H. Leathart, who succeeded him as managing director[22] does not seem to have been a success in the job.

In 1897 there was a net loss and the board obviously decided that the company could not be resuscitated and proposed that the shareholders accept an offer for the company which entailed its liquidation. The offer came from a syndicate composed of N.C. Cookson, C.F. Forster, and A.H. Lancaster, respectively directors of two of the largest Tyneside and the largest London lead manufacturers. There was some opposition from shareholders to the offer and they eventually imposed conditions which led to its withdrawal. Faced with having to carry on an unprofitable company the directors now resigned *en masse* and were replaced among others by members of a shareholders' committee. Initially the new board had apparently no intention of continuing the business since it re-entered negotiations with the syndicate. However, its demands were too high and N.C. Cookson wrote on 18 April 1898, 'I regret that the price you name for the whole of your undertaking is quite out of the question from our point of view and we beg to decline it with thanks.'

There was now little for the new board to do but endeavour to follow a more effective policy of rationalisation than its predecessor, which had already taken some steps in the right direction. In 1894 it had been resolved to sell the lease of the Wallsend smelting works and, after some negotiations with Cooksons who proved uninterested, it was disposed of to the Wallsend and Hebburn Coal Co. This enabled concentration of production and some reduction of expenses and there was also an attempt to modernise some of the plant. In 1892 for instance a water-jacketed blast furnace had been ordered from the USA at a cost of £1,700 and new white lead drying chambers had been installed. The new board adopted a stronger policy of modernising plant, expending £3,500 in its first two years on replacing the Pattinson desilverising plant with a Parkes' zincage and on introducing machinery for filling and emptying stacks. Such a policy was continued up to 1914, the most notable but unsuccessful example being the introduction in 1905 of the Betts electrolytic process for refining lead.

In addition to modernisation there were continued attempts at rationalisation mainly related to the old Gallowgate works which was something of a liability, being on the edge of Newcastle and dependent on road transport for all its freight movement. Several attempts were made to sell the site but despite their failure the board of directors was ever hopeful and noted in its interim report for 1908:

> During the last ten years no offer for this valuable site, or any portion of it, has been received; but, as the neighbourhood has greatly improved, the development of this property can only be a question of time, and by the sale of this land a sufficient sum should be realised not only to pay off all the Debentures but also to effect the concentration of the works at St. Anthony's, which would result in important economy in manufacture.

As late as the 1930s, after manufacturing had ceased at Gallowgate, there were still to be hopes of selling the site. There was, however, one successful site disposal when the premises in the Close, which had been used as a warehouse and wharf, were sold in 1903. There was also some product rationalisation. The manufacture of lead shot and orange lead were given up and red lead production ceased at Gallowgate in 1905 and was concentrated at St Anthony's, which was the works which the company was most concerned to modernise and develop.

None of these steps was sufficiently radical to offset the underlying inefficiency of the firm, which can probably be ascribed to an inappropriate works structure and poor management. Irrespective of the ability to sell the site it would probably have paid the company to close the Gallowgate works with its high costs of production and transport and concentrate output entirely at St Anthony's where there was room for expansion. That this was not done must to some extent be put down to the less than dynamic management of T.H. Leathart and his brother, J.G., who became company secretary in 1904 and a director from

1910. The board of directors seems to have shown little in the way of active direction of policy and the company returned to being largely a private fiefdom of the Leatharts. After some moderately successful years from 1898 to 1902 (including 1901 when the fall in the price of pig lead had produced losses for most companies), the company paid no dividend from 1903 to 1911, a period which included losses in 1907 and 1908 and there was a capital reduction in 1910. In the immediately following years profitability improved and, like most of the lead manufacturing companies, Locke, Blackett was able to benefit from wartime prosperity.

Locke, Lancaster and W.W. & R. Johnson & Sons Ltd (LL & J)

That unwieldy title (reflecting the need to maintain the goodwill attached to the families' names) was the result of the merger in 1894 of the two major London lead manufacturing companies with which we have been concerned. In 1892 the Johnsons had converted their partnership to joint stock status[23] and had been responsible for initiating the White Lead Conference. It is easy to see that the advantages of amalgamation would not be lost on a firm which was already coming to grips with the disadvantages of competition, especially in the 1890s when a major merger wave was underway in British industry.[24] It was not, however, a case of one firm gobbling up another merely with the intent of eliminating competition. Locke, Lancaster was a firm of some significance, which had expanded and adopted up-to-date technology in the 1880s and the merger was one in which vertical integration was the most significant factor. While the two firms overlapped in the production of sheet and pipe, they complemented each other in the remaining major areas of lead production, Locke, Lancaster having smelting facilities and Johnsons producing chemicals. The merger also made sense geographically since the companies' three works, Commercial Road and Burdett Road, Limehouse and Bridge Road, Millwall, were concentrated in a small area of east London.

In 1892 the Johnson company was registered with a capital of £70,000 (made up of 6,800 ordinary and 200 'B' shares, each of £10), plus a debenture issue of £25,000. This was a considerable increase on the last known figure of the Johnsons' partnership capital of £27,000 in 1885. It may be, however, that the latter figure involved undervaluation of land and buildings and there was considerable expansion of the firm in the late 1880s. In 1894, to accommodate the merger, the capital was increased to £115,000 and an additional £15,000 of debentures issued, while in 1895 and 1896 there were further increases to give a capital of £160,000 together with £50,000 of debentures (the latter being entirely paid off by 1914). While the Johnsons had numerical superiority on the board and supplied the chairman of directors until 1927, both families continued to make major contributions to the firm, although each kept largely to its original works. A.H. Lancaster (who left a fortune of more than £½m when he died in 1928) and after him his nephew, Harry Charles Lancaster, were mainly concerned with the running of the Bridge Road smelting and desilverising works. The Johnsons also remained involved in their original works as active managers. R.B. Johnson, for instance, who had been educated at Oxford, joined the family firm and came to manage the Burdett Road works, where, with the assistance of a local engineer, he developed a mechanical stove (patent no. 22, 391, 1891), 'for the purpose of conveying the white lead to be dried into the stove and removing it therefrom after drying without the necessity of handling, thereby preventing the inhaling and absorbing of the dust and fume by the workmen'.

The first major step made by the new company was one of expansion, causing the increases in capital of 1895-6. In 1895 the works of the Millwall Lead Co. Ltd[25] was purchased, a site on the Isle of Dogs opposite Greenwich Hospital, where both white and blue lead were manufactured. Not only did this move offer the opportunity for expansion of output, especially of white lead since there was insufficient space for this development at Burdett Road, but it also provided the scope for a new development. In December 1896, in conjunction with the London lead-smelting firm of H.J. Enthoven & Sons, LL & J formed the London Lead Smelting Co. Ltd, to work a new lead smelting process on the Millwall site.[26] This proved unsatisfactory, since it failed to eliminate sulphur in the lead ores and subsequent smelting proved difficult. The process was subsequently replaced by American Dwight Lloyd sintering plant and the London Lead Smelting Co. Ltd was liquidated. Although the company was unsuccessful it had brought two of the major London lead smelters together and this contact was to lead to more significant links during the First World War.

The Millwall works did, however, provide suitable scope for the development and modernisation of existing processes and it was claimed that the first electric-powered rolling mill in the country was installed there. This may well have been the lead foil rolling mill which was transferred from the Commercial Road premises when these were closed in 1903. This decision was an eminently sensible piece of rationalisation since the old premises, which the Johnsons had taken in the 1820s, had no scope for expansion and had only road access. In one sense the rationalisation was even more far-sighted, since the second lead foil rolling mill was shipped to Ceylon where LL & J, in association with Harrisons & Crosfield, had set up a company called Colombo Lead Mills Ltd. This made good commercial sense since production costs were lower in Ceylon but it was also to be of advantage after the First World War when Empire countries began to raise tariff barriers against imported manufactured goods. LL & J was for a long period to reap a very worthwhile reward from this investment. Total tea lead output built up to around 6,000 tons p.a. in the years before 1914, of which Colombo accounted for more than one-half.

There is little further evidence about LL & J before 1914, with the exception of some annual output and profit figures. White lead output between 1902 and 1914 averaged about 6,500-7,000 tons p.a., which would place the firm third in size of production to Cooksons and Walkers, Parker (the three of them jointly producing around 60 per cent of UK output). In the production of lead sheet and pipe the firm was of less significance with peak outputs in the early 1900s respectively of 7,000 and 2,000 tons p.a., although these figures fell in the years before 1914, as the building cycle turned down, to only 2,500 and 1,500 tons. Nevertheless the firm's financial position was very sound and the fact that the £50,000 debenture issue had been paid off by 1914 would suggest that surplus cash had been generated over most of the preceding two decades. The first available set of accounts, for 1913, shows the firm to have been awash with cash assets. The General Reserve Fund had reached £75,000 which the board decided should be a maximum (although a further £25,000 was added in the following year), in addition to which there were several other reserve funds including £13,500 against price fluctuations in pig lead and £65,000 in a Dividend Guarantee Fund. These figures were struck after the declaration of a dividend of 25 per cent on the ordinary share capital of £125,000. The net profit for 1913 was £56,000 and for the following year £73,000. These are figures which compare favourably with the best returns on capital which the Johnsons made in the period to 1885, as shown in Table 5.7. Coming from a totally different source they confirm the very high profitability of the company, although in the absence of any details it is not possible to analyse the cause of this success. It may well be that the figure for capital employed was unrealistically low and that capital development had been financed out of revenue over the years (and this is especially probable in the case of

Colombo Lead Mills Ltd). Nevertheless this would still imply a profit-earning capacity which was totally out of the reach of firms such as Walkers, Parker and Locke, Blackett.

The distinction, which most obviously offers an explanation, is that between joint stock companies in which the board of directors had no knowledge of lead manufacturing and failed to appoint competent managers and those companies in which the old families with a detailed knowledge of the industry retained control of their firms. This argument is, of course, the complete antithesis of that which is frequently used for this period – that family firms lacked dynamism and that British industry needed an injection of professional management.[27] Providing that it was of good quality there was nothing wrong with professional management but too often it was not and had little or no knowledge of the industrial process with which it was involved. In such circumstances the well-run family firm (of which there were many) could run rings around the joint stock companies with professional management. This point is further illustrated and supported by the evidence of the next two case studies in the period to 1914.

Brimsdown Lead Co. Ltd

Brimsdown was one of the many companies set up to develop one of the new white lead processes of the late-nineteenth century. What made it different from all the others was that it was backed technologically by some highly competent research chemists and financially by Ludwig Mond. It had, therefore, attributes which are often postulated as the potential basis for successful industrial growth in the late-nineteenth century.

The company was originally set up in 1893 as Bischof's White Lead Syndicate Ltd[28] to develop the patented process of Professor Gustav Bischof. At some stage in the later 1890s Ludwig Mond became interested in Bischof's solution to the problem of the rapid manufacture of basic lead carbonate.[29] Contemporary accounts described the process as commencing by converting pig lead into litharge, which was then reduced to a sub-oxide of lead by means of water gas (the latter being a product of another company in which Mond was interested). Water was then added, which created lead hydrate and this was treated with acetic acid and carbon dioxide in order to produce basic lead carbonate by the normal chemical route. The white lead slurry was passed through a filter press in order to remove excess water and then mixed with linseed oil in a pug-mill, thus removing the remaining water and producing ground white lead ready for market without any dust or direct human handling. The resultant material was chemically identical to stack white lead but had a much smaller and more regular particle size, which gave it greater covering power as a paint (Brimsdown claimed 50 per cent greater), but which may not have appealed to the conservative painters. The firm stressed the hygienic nature of its process in order to benefit from the wave of publicity on lead poisoning and advertised that 'the Home Secretary has recognised this claim to such an extent as to exempt the Bischof Syndicate's works from all the onerous special rules which are enforced with regard to other white lead works'.

Mond became involved financially and was chairman of the vendor syndicate which in 1900 sold the company to the public as the Bischof White Lead Corporation (1900) Ltd.[30] This company was set up with a nominal capital of £350,000 (£200,000 in preference and £150,000 in ordinary shares) of which £100,000 (equally divided between preference and ordinary) was to be paid to the vendors together with £50,000 in cash. As was his way, Mond was obviously thinking in terms of a very large-scale enterprise. More immediately he was thinking of a large-scale profit, since the vendors had little to offer except the Bischof patent and the drawing-power of their names in return for their £150,000. If, as was to prove the case, the process was less than highly successful, the company was going to carry a huge burden of capital, the charges on which it would not be able to meet.

Mond did not appear on the board of the new company but was represented by his son, Robert.[31] Also on the board were J.F.L. Brunner (son of Mond's partner in Brunner, Mond) and E.L. and J.F. Pease of the Darlington Quaker family. This was obviously intended to be a directorate which would draw on a wealthy investing public but not even the names on the board could encourage the public to take up shares to the value of £¼m in an untried process and only 64,672 of the 150,000 Cumulative 7% Preference Shares and 72,505 of the 100,000 Ordinary Shares offered were taken up. The remaining shares were taken up by the Commercial Development Corporation Ltd, in which both Mond and S.A. Bartlett, who had been appointed secretary of Brimsdown, were involved. In 1902 this Corporation went bankrupt and Mond had to step in and guarantee the financial support which it had provided.

The company took over plant and land at Hythe Road, Willesden Junction in London, at which Bischof had been doing experimental work. It seems unlikely that much white lead had been produced although there must have been some of good quality. There is a well-attested story that Holman Hunt's painting the 'Light of the World' had deteriorated as a result of defective priming and being hung too close to hot water pipes. Hunt decided to paint a replica and his friend Mond recommended him to use Bischof's white lead as a base for his paints.[32] Nevertheless production could only have been on a very small scale since in February 1901 Bischof was authorised to purchase 25 tons of pig lead and in April experimental plant was still being erected. It is clear that the company, although formed, was far from being in a position in which commercial production could be commenced. There appear to have been two immediate priorities, to find a site for commercial production (the Willesden site being too small) and to strengthen the management and directorate (this was necessary since it soon became clear that Bischof had no commercial or organisational sense and the directors had too many other interests to be able to give Brimsdown much attention). In July 1901 W.A. Humfrey was appointed to manage the Willesden works at a salary of £500 p.a. but this was inevitably to cause conflict with Bischof, who had been retained to develop his process, and in July 1902 he was given three months' notice. The directorate was strengthened with the appointment of Bertram Hopkinson in July 1901 (who resigned in 1904 when he was appointed to a chair of mechanical engineering at Cambridge University) and E. Speyer (who was to become chairman) in October 1901.

The search for a new works site was a long one fraught with numerous problems. Negotiations were concentrated on Thames-side sites in Kent and Essex during 1901 but failure to find a suitable site at an acceptable price led to a switch to sites on the Lee Navigation and Grand Junction Canal in Hertfordshire. It was not, however, until May 1902 that a site on the Lee Navigation at Millmarsh Lane, Brimsdown, was purchased. The long saga of the company's difficulties in obtaining a works' site is an early symptom of its many subsequent problems. It seems unlikely that a well-run company could find it difficult to obtain a suitable site in under 18 months. In particular Bischof was very fussy as to what he regarded as a suitable site and there appears to have been a lack of decisive direction.

This was clearly the view of a number of shareholders, since a great deal of criticism was voiced at the AGM in June 1902 and an unsuccessful attempt made to reject the payment of the directors' fees. In October a group of shareholders (which included an obviously disgruntled Bischof, who had been sacked, and Dr Stephen Miall[33]) endeavoured, again unsuccessfully, to appoint a shareholders' committee to enquire into the running of the company.[34] The influence of Mond and his friends and supporters was too great for the dissidents to achieve any marked success but the latter appear to have had reason for their concern, since Mond appears to have interfered in an extraordinary way in the running of the company. H.C. Bainbridge tells the story of how he was introduced to Mond, who appointed him as works chemist at Willesden even though he had no experience.[35] At this time Mond was probably suffering from overwork and in December 1902 he experienced a heart attack. Coming shortly after the collapse of the Commercial Development Corporation Ltd, this did nothing to encourage continued support for Brimsdown. Between 1901 and 1903 six directors resigned (including the chairman) and there were three more resignations by 1906, by which time all the original directors had left the board.

As a result of the failure to speed up the erection of the Brimsdown works the company's losses were considerable. The Willesden experimental plant was capable of producing only two to three tons of white lead per day and it is doubtful if it produced that quantity on a regular basis. Between 1901 and 1905 the company averaged an annual trading loss of £5,000 and the heavy overheads for such a small turnover pushed the net loss up to an average of nearer £10,000 p.a. It is unlikely that the shareholders appreciated being informed by the directors' report of 19 March 1903 that 'a considerable time must elapse before the Company can reasonably be expected to pay dividends'. The directors, however, took advantage of the death of Bischof, which had occurred in January, to direct the shareholders' attention away from the company's failures:

Owing to the fact that more than one-third of the consumption of the U.K. is imported White Lead of an inferior quality, the trade have looked with suspicion on the origin of the Corporation's White Lead because of its apparently foreign title. The Directors have consequently decided to ask the Shareholders to change the name of the Company to 'The Brimsdown Lead Company'.

A change of name did nothing to improve the company's fortunes. The Willesden works was closed in January 1904 but there were delays in the installation of plant at Brimsdown and further problems even when the works came into production in April 1904.[36] There were continual financial pressures which were only relieved by Mond's agreement to take up extra shares, guarantee the bank overdraft and advance cash in return for debentures. In their annual report for the year 1903 the directors had repeated the comment that the payment of an ordinary dividend would take some time but could 'see no reason to doubt that the corner will soon be turned, and that the position of the Company will shortly be much improved'. The annual report for 1904, however, presented many of the shareholders with a *fait accompli*. In February 1905 the directors invited a number of large shareholders to a meeting to discuss the company's financial problems. The decision of the meeting, conveyed to the AGM in May, was that the company should go into voluntary liquidation and be reconstructed. The capital write-down, which totalled £141,000, was to be used largely to write off the accumulated loss of the previous years and part of the inflated value of the goodwill and patent rights which the company had purchased in 1900. The directors claimed that 'This would make the value of the Willesden Lease, the patent rights and goodwill of the business to be taken over by the reconstructed Company, about £40,000, which the meeting considered to be a fair and reasonable figure.' However, the opportunity was taken to recommend that the preference share dividend be made non-cumulative, a wise decision since the new company, trading as The Brimsdown Lead Co. Ltd and Reduced, was to prove little more successful than its predecessor.

Even after its reconstruction the company was still overcapitalised and its continued losses required borrowings (which were difficult to obtain and were probably made at high interest rates), which considerably increased expenses. Secondly there were initial difficulties in selling the increased output from the Brimsdown plant. The mid-1900s was a period of severe competition in the trade, when Cooksons, in particular, were expanding their output considerably. That Brimsdown's white lead was different visually from that of its competitors may have been a problem and there were certainly many complaints from customers about 'quality', which were probably to do with visual appearance. That the product was in fact sound is attested by the achievement of regular orders from government departments which also purchased white lead from long-established firms. By 1910 the firm had built up to annual sales which, it claimed, involved it in considerable reduction in order to meet the convention quota of 4,000 tons p.a. which it received from the following year. Ultimately it seems clear that although the Bischof process was one of the few new processes which was satisfactory from chemical and technological viewpoints it was uncommercial. It simply could not produce white lead at a price which enabled it to compete successfully with chamber and stack white lead. It provides an example of a case where a scientist (even one who had been so commercially successful in other fields as had Mond) had let his enthusiasm for a scientific solution to a problem to run unchecked by economic reality.

After 1904 the net loss began to fall and from 1906 the company began to make trading profits for the first time, although between 1905 and 1912 the new company made a cumulative net loss of more than £50,000. When a net profit was finally achieved, in 1913, it was only £246, giving a return of 0.3 per cent on capital employed. Even this was achieved as a result of making no allowance for depreciation and the accounts show no depreciation charge at all between 1900 and 1915. In the palmy years during the War we shall see that the company was actually able to make profits, although they were insufficient to pay a dividend, and were pitiful besides those of other lead companies. In 1914 Brimsdown's only hope for the future lay in the fact that it had begun to build up a strong research department. In R.S. Brown the company had a general manager of genuine ability and in C.A. Klein a chemist who was to make major contributions to the industry. In the merger-conscious era after the War Brimsdown was to be of interest for its research potential even though its white lead production remained uncompetitive.

Cookson & Co. Ltd

Among the large firms Cooksons was notable for the strength of the family involvement in the firm's development and it is not surprising that it was the last to give up partnership organisation. During the 1890s Norman Cookson was in supreme control, the only active director, although towards the end of the decade two of his sons, Harold and Clive, came into the business and were to become directors in the 1900s. Perhaps to their influence and to the growing age of their father should be attributed the conversion of the partnership to a limited company with effect from 1 January 1904. The company was set up with a nominal capital of £250,000 in 25,000 ordinary shares of £10, although all of the issued shares went to the three partners in the old firm, Norman received 10,883, his brother G.J., 5,727 and Clive 572. A few other members of the family (especially the women) received sizeable holdings in an issue of part of a nominal £125,000 of 4¼% debenture stock. In taking over the assets of the partnership the limited company appears to have paid £236,000, of which £113,000 was for stocks, £110,000 for the works and £13,000 for investments.

The firm had grown rapidly in the previous decade or so but it had been able to do so largely from family resources. In 1905 the outside financing was limited to bank overdrafts totalling £44,000 from Lambton & Co. and Brown, Shipley & Co. and a loan of £25,000 against debentures from the North British & Mercantile Insurance Co. Without any of the 'padding' for goodwill in its capital, which was the unfortunate lot of firms such as Walkers, Parker, Cooksons had made a very comfortable return of 9 per cent on capital employed in the period 1889 to 1905. The years to 1914 were less satisfactory, largely as a result of losses in 1907 and 1908 totalling £33,000.[37] This produced some temporary financial difficulty and the accounts for 1908 show that bank overdrafts had risen to £117,000, while reserves, which had been £72,000 in 1906, had fallen to only £11,000. The chief reason for this setback appears to have been that the company had built up a huge level of stocks, which reached a peak of £337,000 in 1907, at a time of high prices and had to sell them on a falling market. Recovery in 1909 and 1910 was slow, with low profits, but in the remaining years to 1914 a return of around 7 per cent on capital employed was achieved.

To what did Cooksons owe this satisfactory performance in a difficult period when several of the lead companies were writing their capital down rather than increasing it as did Cooksons? Primarily the success must be put down to the entrepreneurial abilities of members of the family. Norman Cookson had proved his capabilities in the late-nineteenth century and his son Clive, after Harrow and a period of gold prospecting, was to do so in the twentieth. Possibly their most important trait was the capacity to recognise the need to bring in qualified outsiders. Norman Cookson had employed trained chemists and he recounted that

As a manager, they selected a man of the very highest scientific education and skill, and they took him, although he had not seen anything of two of their largest branches of manufacture, red lead and antimony refining and so on; but he thought if he got a metallurgist of high education and skill, the ordinary rule-of-thumb work would be absolutely nothing to him.[38]

Under Clive Cookson this policy was extended, science graduates were gradually brought into the company's management and non-family directors were appointed. In 1909, for instance, H.S. Tasker, a Cambridge double first in Natural Sciences, was taken on and was eventually to follow Clive Cookson as the second chairman of ALM. In

1911 A.J. Hugh Smith[39] was employed from Hay's Wharf Ltd and became a director in the following year, along with Frank Reid who had joined the company in 1878 on a salary of £35 p.a. In order to obtain and keep good men adequate salaries with regular increases were paid. Smith was to be paid £750 in the first year, £1,000 in the second and £1,250 in the third, together with 10 per cent of any net profit over a 10 per cent return on capital. He was also to have the option to subscribe for shares at par. In 1913-14 he received commission payments on profits averaging £1,400 p.a. and during the War these averaged £6,000 p.a. Managers were also treated generously. W.M. Hutchings was paid a salary of £800 p.a. when he came to Cooksons in 1889 as manager from Walkers, Parker at Bagillt (the Walkers, Parker managers by contrast received about £500 p.a.). In addition he received a bonus related to profits (which Cooksons paid to all their salaried employees from 1889) and around the turn of the century he was receiving almost £2,000 p.a. The Cooksons were therefore prepared to pay for good leadership and management and they appeared to be remarkably successful in selecting it.[40]

Secondly the firm's geographical structure may have been of significance in its performance. Although it had a wide range of products these were concentrated at two works in close proximity at Hayhole and Howdon on the Tyne.[41] This made for ease of management control, as compared for instance with Walkers, Parker, where the directors were for ever having to travel long distances to deal with problems at particular works. The disadvantages of having a single centre of production were overcome by having sales offices in major towns, including London, Bristol, Birmingham, Manchester, Leeds and Glasgow. The whole organisation was then controlled from a head office in Newcastle, in Bank Chambers, Sandhill until 1905 when larger premises were taken in Milburn House. This was necessitated by a rapid growth in salaried staff from eight in 1875 to 16 in 1890, 28 in 1900, about 50 in 1908 and about 80 (of whom 20 were women) in 1915. Not only did the company expand through regional sales offices it increasingly looked abroad for expansion to the areas from which its raw materials came. In 1903 Cooksons set up a subsidiary called Republican Mining and Metal Co. Ltd, to buy up Mexican mines and smelters from which it was receiving supplies of both lead and antimony. While this was the most important, several other companies[42] were set up to control overseas mining and metal trading operations and metallurgical experts were employed.[43]

Finally it may be said that Cooksons' performance may have owed something to the balance given by its wide range of products and especially its interests in antimony as well as lead. This point was singled out by Norman Cookson at the AGM in 1908 when he commented that 'although lead business for the year [1907] had been extremely unprofitable the antimony business had been very satisfactory'. In that year the profit on antimony operations was £36,000 while the Howdon desilverising operations had experienced a loss of £42,000, and in 1911 when antimony made a profit of only £502 the desilverising profit was almost £12,000. Profits by source are given in Table 8.1.

Table 8.1: Cookson & Co. Ltd, Profitability, 1905-14[44]

Year	Antimony	Howdon	Hayhole	Chamber works	Investment income	Net profit
			Profit/(loss) (£)			
1905	13,645	32,152	(6,951)	(9,159)	4,135	29,012
1906	32,073	4,748	5,412	1,193	12,266	42,473
1907	36,007	(42,389)	2,172	(931)	50,351	44,753
1908	9,456	(34,355)	3,764	(4,374)	22,267	(3,241)
1909	6,989	6,164	(3,790)	(2,600)	3,313	5,736
1910	7,086	616	1,947	(2,207)	8,995	17,230
1911	502	11,642	4,446	4,773	6,038	29,012
1912	5,931	12,614	4,269	11,362	838	32,612
1913	(1,759)	(991)	11,819	16,699	3,154	25,579
1914	9,395	(5,934)	12,073	23,714	1,062	33,898

Throughout this period white lead remained Cooksons' chief product with output levels of 12,000-14,000 tons p.a. (although this fell back to around 10,000 tons in 1913-14). Of this the major part came from the chamber works where process costs at £2 10s to £3 per ton averaged about £1 per ton less than the costs of producing white lead by the stack process (and in addition, because of the speed of the chamber process, capital was turned over more quickly). It seems likely that the throughput of stack white lead was to some extent varied according to demand, because of its higher production costs (and this was especially true during the First World War, when, with a shortage of lead, several of the 36 stacks were left empty). However, it is probable, as a result of their experience of the low production costs of the chamber process that Cooksons worked their stacks more efficiently than most firms. By the years before 1914 the stacks were being filled with 108 tons of lead each, much higher than the national norm of perhaps 70 tons, with the result of higher output, since corrosion levels appear not to have been different. As with most firms, an increasing proportion of output was being sold as paint (ground with linseed oil) rather than as dry white lead. As well as white lead the firm built up its output of oxides, largely red lead, from less than 2,000 tons in 1900 to more than 3,000 by 1914 and although lead sheet and pipe sales declined with the building cycle, output of antimony was about 4,000 tons in the years before 1914.

Cooksons thus approached the profitable opportunities offered by the war years in a very sound condition. Of all the companies in the lead industry it had been the most successful in following a policy of expansion in a difficult period. It had made profits during a period of severe competition and was to increase them enormously under the cartel and the disappearance of foreign competitors. It had made the step from family firm to large organisation without losing the advantages of direction by those whose capital was at risk, while it had gained the benefit of competent outsiders. All these were to stand it in good stead and within a decade the company was to be the driving force in an amalgamation of the lead industry which was to engulf many of its much longer established rivals.

Other Companies

There is no major evidence on any of the other companies which came into the ALM merger and in any event they were to be of less significance but the following brief details will outline their development.

James & Co. Ltd

This producer of white and red lead at the Ouseburn on Tyneside was the first to become a joint stock company. It did so in 1884 as a result of an amalgamation with John Ismay & Co., another local white lead manufacturer. The nominal capital of the amalgamated company in 1884 was £100,000 in 1,000 shares of £100, although only 546 were issued and they were held almost entirely by members of the Ismay family, who appear to have been responsible for taking over the James works after the death of Walter James. The shares, which were written down to £40 in 1905 when a capital distribution of £60 was made to the holders as a result of the closure of the old James works and its sale to Newcastle Corporation, came entirely into the hands of the Ismay family by 1914.

No profit figures are available for the company but the fact that it always managed to pay a dividend on its ordinary shares suggests that it was consistently profitable. Throughout the 1890s and 1900s the dividend was never less than 2½ per cent and it was 6½ per cent in the years 1907 and 1908 in which its neighbour, Cooksons, made losses. The company would appear to have been small and modestly profitable and its only claim to being innovative was that in 1895 Charles D. Ismay took out a patent (no. 23,969) to improve the wet grinding process for white lead. There were some initial difficulties with the process since the white lead sometimes hardened and became unusable but the process attracted the attention of the Home Office and was adopted by several companies, including Cooksons in 1903.

Foster, Blackett & Wilson Ltd

We saw in Chapter 5 that this firm became a limited company in 1890, although the shares remained almost entirely in the hands of the previous partners, A.J. Foster and C.F. Forster. In April 1896 the company was voluntarily wound up and a new company of the same name registered.[45] There is no evidence as to why this was done but the opportunity was taken to introduce 5,000 £10 preference shares into the capital structure and offer them to the public, although the ordinary share capital and direction of the business remained in the hands of Foster and Forster. The fall in the price of pig lead in the early 1900s may well have caused the company to return to unprofitability, since in 1905 there was a capital reduction writing £3 10s per share off all the ordinary shares (reducing the nominal value to £6 10s and the issued shares to £4 paid). A balance sheet for 1909 shows the company still to have been in difficulties, with a debit balance of £11,000 to the profit and loss account. The unhappy saga continued with a second voluntary liquidation in 1913 and the reconstruction of a third company of the same name, which purchased the assets of the old company for the sum of £49,031. In this A.J. Foster owned 21,000 of the 23,000 issued ordinary shares and with a net profit of £4,800 in 1914 the War had clearly brought a change in the company's fortunes.

Quirk, Barton & Co. Ltd

Set up in 1876 to smelt and refine lead and antimony ores, this firm had originally been a partnership between Thomas Barton and William H. Quirk, with works at Rotherhithe on the Thames. It had subsequently developed the manufacture of tea lead, lead sheet and pipe, solder and litharge. After the death of Quirk it was converted into a limited company in 1906 at which time a balance sheet suggests that it was in financial difficulties. Apart from liabilities to the old partners totalling

£41,000, creditors totalled £58,000 and bank advances and other loans were £47,000. With debtors of only £27,000 and stock of £50,000 it is clear that short term liabilities were financing fixed capital and the balance was only achieved by the inclusion of £10,000 for goodwill. In 1911 the company was in serious financial difficulties which necessitated a reconstruction and the formation of a new company with the same name. Walkers, Parker had noted at the end of the previous year that the company had liabilities of £160,000 but assets of only £130,000 and was concerned about the sum of £600 owed to it by Quirk, Barton. Eventually Walkers, Parker and other creditors were paid in the form of shares in the new company created in 1912. Although retaining G.H. Quirk as representative of the lead interests, the board of the new company was composed of bankers and industrialists (such as S.D. Lazarus and E.L. de M. Mocatta) who were creditors of the firm. The company was hopelessly in debt (balanced by an item of £92,000 for goodwill) and although it made small net profits from 1911 onwards (and these increased to around £10,000 p.a. during the War) no dividend was paid until 1925.

Of the other companies, such as Cox Bros and Alexander, Fergusson, what little is known of their performance in the period to 1914 has already been given at the end of Chapter 5. The overall impression which one gets of the period 1889-1914 in lead manufacturing is one of considerable financial difficulties. The cyclical depression of the early 1890s together with a major fall in the price of lead caused many companies to incur net losses. The same was true of the years 1901-2 and 1907-8 and a remarkable number of companies experienced capital write-downs and reconstructions during the period. Cyclical fluctuations in the economy as a whole had, however, occurred before this period and there is no reason to believe that they were any more severe in the late-nineteenth century. What was different was the level of competition (both as a result of rising imports and the increasing output of some British companies such as Cooksons and Mersey), which placed severe pressure on the less efficient firms during trade recessions. There is no doubt that conversion to limited liability status was often a defensive reaction to such pressures, as were the attempts at cartelisation.

Some firms, however, managed to prosper during the period and it is notable that these were the family-run companies: Mersey with its limited objective of making and selling a large volume of white lead; Cooksons which managed to expand dramatically and introduce technical innovations; and LL & J which developed a successful amalgamation and rationalised its production. These companies contrast markedly with the limited companies with professional management, such as Walkers, Parker and Brimsdown (with all its technical expertise and backing from Brunner, Mond) which suffered capital reductions and continued unprofitability.

Notes

1. In 1914 some 80 per cent of all limited companies were private rather than public ones, L. Hannah, *The Rise of the Corporate Economy* (1976), p. 25.

2. For the development of limited liability companies see J.H. Clapham, *An Economic History of Modern Britain. Free Trade and Steel 1850-1886* (Cambridge, 1932), pp. 133-44, H.A. Shannon, 'The First Five Thousand Limited Companies and their Duration', *Economic History* (supplement to the *Economic Journal*), vol. II, 7 (1932), pp. 396-424, idem, 'The Limited Companies of 1866-1883', *Economic History Review*, IV (1933), pp. 290-307 and J.B. Jefferys, Trends in Business Organisation in Great Britain since 1856 . . . (University of London, PhD, 1938), Chaps. II and III.

3. There was a brief article in the issue on 12 Jan. 1889, p. 239, followed by a more detailed account on 19 Jan., pp. 257-9. The latter brought forth the threat of legal action from solicitors to E.J. Walker but nothing further appears to have occurred, apart from continued criticism by the paper, see *Lighthouse*, 2 Feb. 1889.

4. *The Times*, 24 Sept. 1889.

5. The fall in profit for 1891 led to several further critical accounts in the *Lighthouse*, 12 Sept. 1891, pp. 1985-6; 26 Sept., p. 2022; and 31 Oct., p. 2106. E.J. Walker does not seem to have made much money out of the flotation of the company, which again suggests that there was someone backing him. Apart from the evidence that he used company money to develop the Chester house, surely not the action of someone who was wealthy, he was, in 1892, to apply for the manager's job at Chester, which carried a salary of about £500, and he died in 1905 leaving an estate valued at only £420 8s 3d.

6. *The Times*, 12 March 1891.

7. The nearby Bagillt works appears not to have been sold to the limited company but to have been sold separately by the partnership, perhaps to Arthur Burr (since a Walkers, Parker minute of 23 March 1891 notes that Burr had offered the Bagillt rolling mill to the company for £350).

8. *Financial Times*, 8 March 1906. The detailed figures are given in Appendix VII.

9. Initially, of course, the ordinary shares represented no-one's hard money and since no share register has survived for the period it is not possible to tell who owned the shares in 1893. It certainly was no longer E.J. Walker since a board minute of 10 March 1891 noted that he had disposed of all his shares.

10. *The Times*, 6 March 1890.

11. Ibid., 7 March 1900.

12. Ibid., 7 March 1901.

13. *Leading Opinion*, 4 March 1903.

14. *Financial Times*, 8 March 1906.

15. This involved the rejection of an interesting offer of co-operation. Locke, Lancaster & Johnson had offered to roll for Walkers, Parker on special terms to

avoid the necessity of the latter's laying down a new mill (and, therefore, reducing the existing over-capacity) but the Walkers, Parker board rejected the offer.

16. The company numbers in the Companies Register are 28,073 for Walkers, Parker (subsequently transferred to ALM) and 28,302 for Mersey.

17. When, in 1899, the original issue of debentures came to be replaced, letters of acceptance from holders are written to Mersey in personal terms, asking for advice, enquiring after members of the secretary's family, etc.

18. Ms. notes on *Walker* v. *Walker*, before Edward Ridley, Official Referee, 27 June 1890. Ridley clearly disapproved and in the examination of Sloane commented, 'it seems to me that this conduct was reprehensible and I tell him so to his face' and later, 'I suspect you have been mismanaging this all the way through. What business had you to leave stacks lying idle [22 of the Chester stacks had been left empty, although Sloane claimed that he was instructed to do this] like this? It is waste of capital.'

19. Sir Thomas Legge wrote on 12 Jan. 1925 to C.F. Hill, chairman of the Mersey directors, 'It is 26½ years since Mr. Sloane took me round the factory and made me realise what intelligent anticipation could do in the way of preventing incidence of poisoning and in your case it has continued ever since. The lesson has not been lost on me.'

20. Apart from Leathart, who was to be managing director for five years, the chairman was Col. F.F. Sheppee of the Birtley Iron Co., and the remaining directors were Armorer Hedley of Thomas Hedley & Co., Capt. G.J.W. Noble of Sir W.G. Armstrong, Mitchell & Co. and J.G. East a Tyneside lead broker. The new company's first step was to write £27,000 off the partnership's valuations of plant, stock and buildings, thus bringing assets more into line with reality.

21. An analysis of the application forms for shares in 1891 gives the following figures:

	Newcastle residents		Residents of Northumberland & Durham		Others	
	No.	% of total	No.	% of total	No.	% of total
Preference shares	5,642	45.6	3,054	24.7	3,675	29.7
Ordinary shares	3,757	47.3	3,438	43.3	755	9.5

There were a few large holders, including Noble and Armstrong on Tyneside and A.H. Lancaster from London, but most shareholdings were small and included a sprinkling of working men (a joiner, a bank porter, a signalman, a caretaker, a gardener, etc.) as well as the Durham County Colliery Enginemen's Mutual Aid Association which took 100 ordinary and 20 preference shares.

22. T.H. Leathart was not in fact appointed until 1897 after an interregnum of two years in which he was on several occasions proposed by Armorer Hedley (who was his grandfather) for the job but rejected by the remaining directors, and a general manager was appointed. Relations between members of the board were unsatisfactory and after the appointment of Leathart there were disagreements between him and the general manager. Such an atmosphere was hardly conducive to the running of an efficient company.

23. The ALM records contain no evidence on the conversion and the Companies House file on the company (no. 35621) has been destroyed.

24. See H. Macrosty, *The Trust Movement in British Industry* (1907) and L. Hannah, 'Mergers in British Manufacturing Industry 1880-1918', *Oxford Economic Papers*, 26 (1974), pp. 1-20.

25. Little is known about the Millwall Lead Co. Ltd, which had taken over the works of Pontifex & Wood in December 1889 but it was clearly a response by

some members of the Walker family to the loss of the Walkers, Parker works. Two of its directors, F.J. and E.A. Walker had previously been partners at the Lambeth and Elswick works. The Walkers had a dominant share of the firm's capital. Of an issued capital of £45,000 in £100 shares, six members of the Walker family owned 300 shares, while interestingly the remaining 150 shares were owned equally by directors of H.J. Enthoven & Co. and Locke, Lancaster & Co. At this stage neither of these companies had a manufacturing capacity in lead chemicals and both were obviously interested in the opportunity to develop in this field. The new company however proved unsuccessful and it was wound up in 1895. The company file is in PRO, BT 31 4631/30409.

26. The company had a capital of £35,000, shares being owned by directors of Enthovens and LL & J, with F.C. Hill as managing director. The company file is in PRO, BT 31 7162/50565. The process had been patented (nos. 8,064, 1896 and 3,795, 1897) by T. Huntington and Ferdinand Heberlein, manager and assistant of a firm with works at Pertusola in Italy.

27. But see, however, the comments of Hannah, *Rise of the Corporate Economy*, especially p. 25, 'Significantly many of the firms which did not retain the services of the former owners (or which retained the services of owners lacking the managerial capacity to exercise overall control in an enlarged company) encountered serious managerial difficulties.' On business organisation, family firms and professional management see also B. Supple (ed.), *Essays in British Business History* (Oxford, 1977), especially pp. 20-6.

28. The company was incorporated on 28 Jan. 1893 with a capital of £30,000 in £10 shares, increased to £50,000 in 1897. PRO, BT 31 5497/38074.

29. J.M. Cohen, *Life of Ludwig Mond* (1956), pp. 211-12. Bischof's patents were numbers 11,602 1890, 12,034 1893, 30,444 1897 and 13,202 1898.

30. The company's file is in PRO, BT 31 31832/68222.

31. Robert (later Sir Robert) Ludwig Mond (1867-1938) was educated at Cambridge, Zurich, Edinburgh and Glasgow (where he studied under Kelvin) and undertook chemical research at Brunner, Mond. He resigned his directorship in 1902 on the ground of ill-health. See *DNB*.

32. Cohen, *Ludwig Mond*, pp. 217-18 and H.C. Bainbridge, *Twice Seven* (1933), pp. 102-6.

33. Miall was invited to become a director in December 1902. He subsequently became managing director and wrote a standard history of the chemical industry, *A History of the British Chemical Industry* (1931). He was a solicitor, with a doctorate of law, of the firm of Winterbotham & Miall.

34. *Financial Times*, 1 July and 17 Oct. 1902 and *Evening Chronicle* (Newcastle), 13 Oct. 1902.

35. Bainbridge, *Twice Seven*, pp. 89-93.

36. Ibid., pp. 149-54.

37. Detailed figures are given in Appendix VII.

38. *Journal of the Institute of Metals*, 1 (1909), p. 245.

39. Educated at Eton and Trinity College, Cambridge, where he took a first in history, he had been Gladstone Prizeman in 1902. He subsequently became a director of ALM, but resigned in 1927 on his appointment as managing director of Hambros Bank Ltd, a position he retained until 1951. See *Who Was Who 1961-70*.

40. Apart from Tasker and Smith they also employed E.P. Reynolds from 1910, another Cambridge graduate, who was to become a director of ALM and then in 1940 managing director of Goodlass Wall & Lead Industries Ltd.

41. It is noticeable, after all the works closures of the twentieth century, that both of the Cooksons works are still in existence, although none of the other companies which came into the ALM merger has retained its works intact.

42. These included in 1911 the Three Castles Lead Co. Ltd, a joint operation with Rowe Bros to develop the Barton lead oxide patents in the USA.

43. F.G. Trobridge, educated at the Universities of Durham and Bonn and the School of Mines at Freiburg, for instance, had a doctorate in metallurgy. He was subsequently to become general manager of Cooksons in 1931.

44. The usefulness of such figures depends upon the pricing policy for intra-firm sales and the allocation of overhead expenses, about neither of which there is any evidence. Between antimony and lead the profit distinction is probably reliable but it is likely that the distinction between lead departments is unfair (it is probable, for instance, that Howdon took most of the impact of the fluctuations in the price of pig lead while the relative profitabilities of Hayhole and Chamber works appear to be unlikely). Investment income came chiefly from subsidiary companies.

45. PRO, BT 31 6809/47898.

9 AMALGAMATION IN THE INDUSTRY AND THE FORMATION AND EXPANSION OF ASSOCIATED LEAD MANUFACTURERS LTD

Something lingering, with boiling oil in it
. . . something humorous, but lingering, with
either boiling oil or melted lead.
          W.S. Gilbert, *The Mikado*

The impact of the First World War on the lead companies — rising profitability, control of lead supply by the Ministry of Munitions, encouragement towards amalgamation. The first Associated Lead Manufacturers Ltd, sponsored by Enthovens, formed in 1919 but collapses because of financial problems. The need to regain overseas markets and the formation of British Australian Lead Manufacturers Pty Ltd. The formation of the second Associated Lead Manufacturers Ltd in 1924 with the merger of Cooksons and LL & J and its takeover in the next five years of most of the major UK lead manufacturers, including Rowe, Brimsdown, Locke, Blackett, Foster, Blackett and Walkers, Parker. Finally in 1930 forward integration into paint manufacture with the merger with Goodlass Wall to form Goodlass Wall and Lead Industries Ltd.

In their report for 1914 the directors of FB & W expressed the hope that 'the volume of business will continue to increase in spite of the war, as it is expected that the English manufacturers will capture the large trade hitherto done by Germany, and it is hoped that this annexation will, to a large extent, be permanent'. The War certainly did reduce imports of lead manufactures and it enabled all the companies to expand their profits considerably, although the volume of output did not increase, partly because of reduced demands for lead sheet and pipe (because of the almost complete cessation of house-building) and partly because of a reduction in the supply of pig lead from abroad because of shortage of shipping space and government control over imports. Nevertheless the War was a very profitable period for all the lead companies, in marked contrast to the previous 20 years which had been very unsatisfactory ones for many of them.

Even Brimsdown managed to make net profits in every war year, although these never exceeded 3 per cent on the total capital employed and did not prevent a capital reconstruction, in which £100,000 was written off to eliminate a £36,000 debit balance to the profit and loss account, reduce goodwill by £54,000 and also effect a reduction in the asset value of plant. This, of course, did much more to improve Brimsdown's balance sheet than did its small wartime profits but if such a company managed to make profits it is hardly to be wondered at that others did rather better. At Cooksons, where net profits had been in the range £20,000-£30,000 p.a. in the years immediately before the War they averaged over £100,000 in the years 1915 to 1917, with rates of return on capital employed in excess of 10 per cent and the firm's dividend, which had never previously exceeded 7½ per cent, was maintained at 10 per cent. Of course excess profits duty made a considerable hole in such profits but it by no means creamed off the whole of the extra profits. For the period 1914-19 Cooksons paid excess profits duty of about £190,000 on declared net profits of £379,000, although they subsequently received a repayment of £60,000.[1] There were obviously ways around the payment of excess profits duty which were acceptable to the government. Cooksons for instance installed plant for the manufacture of bullet rod for the Ministry of Munitions and its share of the cost of the plant was allowed against profits. Moreover the Ministry paid 47 per cent of the cost of the buildings and 24 per cent of the cost of plant involved but in 1918 Cooksons established a net loss of £72,000, partly as a result of an agreement with the Ministry that they should be allowed to write off the value of some buildings. Consequently the company ended up with buildings, presumably of some value, which had been partly paid for by the Ministry and had managed to establish a loss which reduced the liability for excess profits duty.

The story of Locke, Blackett was similar. A previously unprofitable company, which had only begun to pay dividends in 1912 after a gap of a decade, it managed to increase the rate of dividend in 1915 from 5 to 10 per cent, consequent upon profits which more than doubled pre-war levels. To an even greater extent than Cooksons, Locke, Blackett appear to have avoided the worst depredations of excess profits duty. Between 1914 and 1919 the company's total declared net profit was £117,000 but the final net total of excess profits duty paid was said to be £4,514 (after taking credit for £21,445, refunded on account of the year 1918, when a loss was established by writing off capital investment against the profit and loss account, and further refunds of £10,000 received in July 1922 and £14,511 received in December 1923 for excess profits duty overpaid). The chief results brought about by excess profits duty may therefore have been its effect in causing cash flow problems when it had to be paid and in temporarily reducing the surplus available for investment (with a consequent loss of interest). However in delaying the availability of this capital beyond the post-war boom of 1919-20 the duty may have reduced the level of undesirable investment which took place and presented the companies with much needed funds in the depression from 1921 onwards.[2]

For most of the companies rising profits came easily, as a result of the rise in price of pig lead, with little fluctuation, from about £18 per ton in 1914 to £34 in 1918, a rise which was not dissimilar to that in the general level of prices over the period. Consequently the lead manufacturers, with their policy of selling manufactured goods at a mark up on the price of pig lead at the time of sale of the goods, rather than at the time of the purchase of their pig lead content, made easy profits in money terms. To some extent these were offset by a decline in output. Imports of lead, by far the major source of supply, had averaged over 200,000 tons annually in the years before 1914 and actually reached a peak in 1915 of over 250,000 tons, but in 1916 and 1917 they fell to little more than 150,000. This was largely a result of supply interruptions. In the period immediately before 1914 a considerable proportion of the world supply of lead had come into the hands of a convention of lead producers, controlled by German interests, for whom Henry R. Merton & Co. was the British agent (although part of its capital was controlled by the Metallgesellschaft AG).[3] After the first year or so of war, in which economic disturbances were limited, the convention may have directed supplies to Germany and shortages certainly began to develop. Early in 1916 the British and French governments commandeered the output of the Penarroya mines in Spain, which caused several companies to institute an immediate search for alternative supplies. In addition to world disruption in supplies, there came the specific disruption in the UK caused by increasing government control which developed into a monopoly of the importation of lead.[4] From late in 1915 lead works began to be taken by the government as controlled establishments, a process completed in the first half of 1916. From then on raw materials were largely supplied by the Ministry of Munitions and there was some direction of output. In June 1917, for instance, all the lead firms received an instruction from the Ministry that no pig lead was to be used in the manufacture of white lead until further notice (although returns, the non-corroded metal, could be reset in stacks). At FB & W the white lead stacks were closed and at every firm there was a considerable reduction in output. At Cooksons, for instance, stack output fell from 4,000 tons in 1914 to 1,400 in

1918 (largely as a result of which the process cost per ton soared from £4 to £13) and chamber output fell from 7,000 to 4,000 tons (the process cost rising from £2 14s to £6 9s).

Other commodities were also seriously affected. Lead sheet and pipe output, already depressed in 1913 as a result of the downswing in the building cycle, fell from 84,000 tons in 1914 to only 33,000 in 1918. At Elswick, Walkers, Parker's receipts from lead sheet and pipe, despite much higher prices, were lower in 1918 than they had been in 1914. There were, however, compensations mainly in the form of demand for ammunition. The lead companies received large orders from the Ministry of Munitions for both shrapnel bullets and bullet rod and it was on these contracts which much of their profit was made (Walkers, Parker made such a large profit on one contract that it was subsequently renegotiated and the company had to make a refund to the Ministry). At Elswick, where the annual value of the sales of lead shot had averaged a little over £10,000 before the War, the receipts for the shot department (largely from the sale of bullets) reached £½m in 1916 and almost doubled in the following year. In 1917 and again in 1918 the firm also manufactured more than 60,000 tons of bullet rod for the Ministry.

The War brought not only considerably increased profits but many changes and problems. From 1916 lead supply was perhaps the chief of these. The companies spent much time in negotiating new contracts and UK ore, which had declined dramatically in significance since the 1880s, once again was in demand. Those companies, such as Walkers, Parker with its links with north Wales mining companies, which already had a small supply of British ore, were fortunate. Other companies spent much time negotiating for ore and frequently took shares in mining companies in order to guarantee supplies, although this was a dangerous activity since many fairly worthless mines were reopened as the price of pig lead rose. In general, however, the lead companies took interests in established mines. Brimsdown took over the Brough Lead Co. in Derbyshire, while Cox Bros took shares in the Mill Close mines in Derbyshire and Walkers, Parker took shares in the Greenside mines in Cumberland. The supply of home-produced lead was insufficient to do much about the shortfall from abroad, although, until the government took control of all lead supply in September 1917, it did something to alleviate shortages. Subsequently there were numerous complaints from the lead companies about the Ministry's policy of supply and, although lead was made freely available to meet bullet contracts, lead for other purposes was often not delivered by the Ministry to an agreed schedule. This was not entirely due to administrative ineptitude in the civil service, since the wise decision had been taken to appoint A.J. Foster of FB & W as Controller of Lead at the Ministry. It is to be presumed that he understood the problems faced by the manufacturers and that many of the delivery delays lay in supply problems, since visits from several manufacturers to Foster appear to have done nothing to improve the situation.

Ministry control did have some advantages however. In its anxiety to increase output of munitions the Ministry was prepared to assist in the financing of new plant (often to the extent of 50 per cent or more of the cost) while it allowed the firms to write off the remaining capital cost against profits. In addition, in January 1916 the Ministry of Munitions established a welfare section to encourage controlled establishments to erect canteens, etc. The lead manufacturers already had canteen facilities which were required by the Special Rules but they willingly took the opportunity to erect new and better facilities, largely at government expense (for instance FB & W expected to receive £4,000 towards the £5,650 which its canteen was to cost). Locke, Blackett agreed to build new canteens and washing and lavatory facilities both at Gallowgate and St Anthony's in October 1916, FB & W did so in March 1917 as did Cooksons later in that year.

To be offset against the advantages were other problems, many of them minor ones, which went alongside those relating to the supply of pig lead. The most significant was a shortage of labour. Recruitment for the armed forces attracted many men away from the lead works, although the disadvantage of this from the point of view of war production was recognised in June 1915 when lead manufacturing establishments were granted exemption from recruitment activities. At Cooksons 38 of the office staff left to join the forces and were replaced largely by female clerks, whose employment became common at other firms. There was also some loss of process staff to the war effort and complaints of labour shortages by most firms were common. Usually the loss of labour was small-scale and gradual but in May 1915 Walkers, Parker noted the loss of half its Bagillt labour force as a result of the fact that the Ministry of Munitions had taken over a factory at Queensferry for the production of explosives and had offered the higher than usual wage rate of 10d an hour. By September 1915 the firm had 30 women employed at Bagillt on the manufacture of bullets because of the shortage of male labour. As a result of the labour shortage employees were in a better bargaining position and there were renewed attempts by some of the general unions, including the Dockers' Union at Bagillt and the National Amalgamated Union of Labour on Tyneside, to increase their representation at the lead works. As the lead works became controlled establishments, determination of wage levels by the Ministry of Munitions occurred, frequently as a result of the appointment of outside arbitrators to settle wage claims. Judging from the comments made by several of the companies this led to larger and more frequent wage increases than would otherwise have occurred.

Apart from the supply of labour and raw materials the War brought minor problems. Other costs, apart from those of pig lead and labour, also rose rapidly. In 1916 Cooksons paid £78 per ton for acetic acid against only £20 in 1914 and in 1918 wood for stack floors cost the company £1 1s 4½d per ton against only 2s per ton in 1913. At other times such dramatic price increases would have caused serious problems but it is clear that revenue rose more than sufficiently to meet such rises. Problems also arose because of shortage of materials. For example, all spelter was commandeered by the government for use in the manufacture of cartridge cases, but this failed to recognise that it was needed in the process of desilverising lead and supplies had to be negotiated for this purpose. There were fears about the possibility of air raid damage to works but the only recorded instance was of minor damage to Cooksons' Howdon works during a Zeppelin raid in June 1915. Nevertheless black-out precautions had to be taken and this hit Walkers, Parker's lease of its shot towers for electric signs. On three occasions Liptons wrote to apply for a reduction of the rent while restrictions on electricity use were in force, but Walkers, Parker was reluctant to lose revenue and only agreed to a reduction when threatened with legal action. Undoubtedly there were many other minor incidents and quirks produced by the War such as the consent, obtained in October 1917 by the Lambeth manager from the Walkers, Parker board, 'to the payment of 10/- per week, without any definite tenancy, for the use of an archway under the Waterloo Bridge Road capable of sheltering the whole of the Company's Lambeth workpeople on the occasion of air raids'.

Ultimately the War was nothing but beneficial to the lead manufacturers, although the extent of its impact depended largely on opportunism. Some companies such as Cooksons and Walkers, Parker rapidly took advantage of the opportunities to manufacture shrapnel bullets and bullet rod (and Walkers, Parker actually developed new plant for the purpose, the plans for which were sold to the Ministry), while Locke, Blackett was late to offer to supply bullet rod and never made much out of it. FB & W set up plant to manufacture zinc oxide and used some of its spare capacity, after the closure of the white lead department, to manufacture aircraft wings. The war effort required specialist lead products and some firms did well from the production of red lead suitable for submarine batteries and litharge for proofing airship fabrics.

There was obviously most scope for innovation for war products and this was also financially attractive because of the tax advantages, but planning for the future was not entirely forgotten. In November 1917 William Eckford was sent to the USA to investigate the Carter 'quick process' for the manufacture of white lead and an attempt was made to obtain a licence from the National Lead Co. to work that process in the UK. But while National Lead had earlier been very happy to purchase the USA rights to Eckford's red lead process, the company was a much harder bargainer when it came to the sale of its own rights. National Lead would only license the Carter process in return for a 50 per cent shareholding in Walkers, Parker and although Evans McCarty of National Lead attended a Walkers, Parker board meeting to explain his company's desire 'to acquire an interest in some well known and established English Lead Manufacturing Company', nothing came of the negotiations. While there was some interest in preparing for peace little could be done because of war-time restrictions and LL & J's comment on its Burdett Road plant would probably be true for most of the lead works, 'owing to nothing having been done during the War, the whole building is very much out of repair, and extensive repairs are necessary'. As in other parts of the economy, the post-war boom of 1919-20 saw much increased investment. Even a small firm such as Cox Bros spent part of its war-time profits on a new sheet lead mill and a pipe press, costing respectively £14,000 and £3,200, while Cooksons spent £30,000 on new plant between 1919 and 1921. The experience of wartime profitability encouraged the manufacturers to expect continued prosperity and gave them the capital which enabled them to expand capacity to meet post-war demand. The reality, once the boom broke in 1920, was to be rather different from the expectations.

The final impact of the War was its encouragement of merger in an industry in which small to medium size firms were the norm, unlike the United States where National Lead was already in the top 50 firms by capitalisation in 1900.[5] The lead industry had largely missed the merger boom of the 1890s[6] and had controlled competition by means of cartels. The War, bringing with it supply difficulties in raw materials, produced a new reason for co-operation between the companies in order to obtain a strong organisation to bargain for lead supplies. This need became particularly strong when the Ministry of Munitions took control of all lead supplies and in their dealings with the Ministry over this and other matters such as taxation, labour and wage rates, the firms would have found a mutual interest.

The earliest evidence of moves towards a merger comes in September 1916 with letters from F.C. Hill of LL & J to Sloane of Mersey, which show that the White Lead Convention was sponsoring a merger. Accountants were obtaining statistical details on the various companies and, Hill explained:

> The idea in their mind is, — supposing under the scheme we were all absolutely combined; with works like your own, which probably have the capacity of doubling what you are now making, that they would, if your costs are less than smaller works, be developed to their fullest extent and the smaller works closed ... I have an impression that your costs are as low if not lower than anyone elses.[7]

Together with the letters went a cyclostyled sheet which listed the advantages of 'closer arrangements'. They included, 'combination of all best brains in all concerns', 'A general laboratory', 'Elimination of some plants; and concentration of manufacture in larger units having unused capacity', 'Unified raw material purchases, reduction of management and overhead charges by concentration', 'Entire elimination of internal competition and jealousy' and 'Control of selling prices'.

This may well have been too ambitious a scheme and nothing seems to have come of it. The prosperity of war conditions, even though the document listed the problem of post-war competition as a major threat, may well have inhibited a move towards merger. But another development, closely related to war conditions, was occurring almost simultaneously. A Walkers, Parker board meeting of 8 November 1916 referred to the secretary's discussions with H.J. Enthoven about 'conditions under which his Firm, having considerable influence and control of certain sources of supply, would be prepared to assist the Company to the best of their ability and generally to promote more intimate working relations between the Company and themselves'. Enthovens, a major lead refiner and desilveriser, was able to supply silver lead through its subsidiary, Orchardson and Enthoven, of Cartagena, while there is clear evidence that Walkers, Parker was in supply difficulties. Enthovens purchased a shareholding in Walkers, Parker and three new directors were appointed to the Walkers, Parker board, H.J. Enthoven, Andrew Weir[8] and M. le Baron Jules de Catelin, the latter representing Enthovens' interests in French lead smelting. Enthovens rapidly seized on what was apparently a chance enquiry from Walkers, Parker for assistance, to develop a programme of amalgamation within the lead industry. At the beginning of 1917 Dr Heberlein and Richard Tilden Smith,[9] both connected with Enthovens, attended a board meeting of Walkers, Parker and outlined a scheme 'for the acquisition of a controlling interest in a Firm of Lead Manufacturers'. The firm was FB & W, which Enthovens had approached in December 1916 with an offer for takeover. There is no clear evidence as to why this firm was chosen, but its chairman's position as director of lead supply at the Ministry of Munitions may have been considered sufficient reason.

In February 1917 FB & W agreed an offer from Enthovens and in the following month the Walkers, Parker board agreed to purchase the shares. At this stage the latter appears to have become to some extent enthused with the panorama which was being opened up by the ideas and initiatives which were coming from Enthovens. For a long time Walkers, Parker had been the slumbering giant of lead manufacturing — now it was being taught how to use its potential. Apart from increasing the size of its influence in the industry we have seen that it was encouraged to think in terms of up-to-date technology — one result of which was the despatch of Eckford to the United States. Moreover the future was not to be limited to lead since Enthovens had purchased an interest in the Swansea Vale Spelter Co. Ltd and the firms were to develop zinc production.[10]

As a result of these expectations the decision was taken to increase the capital of Walkers, Parker to £400,000, with Enthovens underwriting the increase of £196,250. This was a far cry from the unhappy quarter century before 1914. £36,000 was paid to Enthovens for that company's holding of 20,222 ordinary shares of £1 in FB & W (which suggests that A.J. Foster did well out of selling most of his shares in what had been an unprofitable concern). In addition plant extensions at Elswick, Chester and Bagillt[11] were to cost £46,000 and those at FB & W £20,000 (although in both cases one-half of the cost was to be paid by the Ministry). Further takeovers were also on the way. In July 1917 Walkers, Parker's chairman, David Thomson, together with Heberlein and Reginald Foster,[12] inspected Alexander, Fergusson's at Glasgow, as a result of which Enthovens made an offer for 50 per cent of that company's capital. The offer was accepted and Walkers, Parker paid just over £27,000 to Enthovens in January 1918 for 35,000 £1 'A' ordinary shares in Alexander, Fergusson. This purchase may well have been part of Enthoven's strategy to obtain a broad geographical coverage for the 'allied firms', the title by which they were usually known in internal discussions. The next step, the takeover in April 1918 of James & Co. Ltd, can only have been prompted by the desire for rationalisation and the reduction of competition. James was a small and not particularly profitable nor technologically advanced company which appeared to offer no attractions. The shares were purchased by Enthovens and then sold to FB & W and in August 1919 the Ouseburn works of James was closed and its staff transferred to Hebburn. Although for reasons of customer loyalty the James name was kept separate until 1934, this was the first serious rationalisation in the industry.

In the meantime there had been a further merger development, with the purchase by Enthovens of a majority holding in LL & J, which had given the allied firms a dominating presence in the London area. It is difficult to be sure whether this step, which occurred in July/August 1917, was merely part of the process of drawing the lead firms together or whether it was the cementing of an earlier established relationship. We have seen that Enthovens and LL & J had jointly set up the London Lead Smelting Co. Ltd in 1896 and when, in February 1917, Walkers, Parker began to get silver lead from Enthovens' Spanish subsidiary, it was noted that LL & J had been drawing supplies from this source for 'many years'. At the very least, therefore, the two firms had had close links but in July 1917 LL & J agreed to adopt new articles of association and to change its capital structure to one of 12,500 shares, divided into not more than 6,249 'A' shares and not less than 6,251 'B' shares and H.J. Enthoven was appointed to the board to represent the latter.

It is clear that the allied firms, in the words of H.C. Webster, chairman of Alexander, Fergusson, 'control a considerable and important share of the trade in all lines'. The only really significant outside firms, albeit successful ones, were Cooksons and Mersey. There seemed, therefore, every reason to expect a significant amalgamation of the lead industry, although the allied firms were still only linked by shareholdings and interlocking directorates, while overall policy was directed by what a LL & J minute described as the 'Senior Committee of the Associated Firms'. Walkers, Parker, with its large wartime profits, seems to have been used to finance the operation, while Enthovens acted as the initiators of change and organised the setting up of a technical committee, of which Tilden Smith was appointed chairman. A Walkers, Parker minute of May 1918 noted, 'The desirability of establishing a system of manufacturing costs at the several works of the Allied Firms was also approved, and Mr. Enthoven promised that his Firm, as Managers of the Group, would organise the lines upon which this could be carried out.'

Clearly the overall organisation of the allied firms was on a largely informal basis at the end of the War and this must have been found unsatisfactory since in July 1919 Associated Lead Manufacturers Ltd (ALM) was incorporated,[13] with a capital of £1.5m in one million £1 preference shares and 500,000 £1 ordinary shares – a contemporary account estimating that its output of 125,000 tons p.a. accounted for more than 60 per cent of UK output of lead and lead products. It was a holding company intended to purchase the shares of the allied firms and, although this was not done, the directors of ALM met and determined overall policy. The first directors appointed were A.J. and Reginald Foster, H.J. and Cecil Enthoven and F.C. Hill and E.M. Johnson. The absence of any director appointed to represent Walkers, Parker probably suggests that that company was already experiencing the disillusion with the merger arrangements which it was soon to express. But in October 1919 the merger received a further extension with the addition as directors of C.A. and H.C. Rowe of Rowe Brothers Ltd (who were almost certainly brought in to the merger because of the potential of their quick white lead process).

There were two major problems which faced ALM, rationalisation within the merger and the need to re-establish the British manufacturers' position in foreign markets which they had not been able to supply during the war period. The one positive step taken in rationalisation was the closure of the James works on the Tyne but there was serious discussion of other action which might have been taken but for the break-up of the organisation. Tilden Smith, for instance, wanted concentration of lead production at Avonmouth as a suitable site for the importation of lead; a works organisation committee was set up to rationalise production and accounting periods of the various firms were standardised. ALM also took the initiative in trying to deal with the potential of foreign competition now that the War was over. The international cartels had lapsed and there was no evidence that they would be reorganised. Moreover before 1914 there had been difficulties with Belgian producers. ALM therefore decided to pre-empt problems and in July 1919 was negotiating the purchase of two Belgian works 'in view of the serious menace to the British Sheet and Pipe trade through the possibility of dumping in this country'.

In the immediate aftermath of the War there was a great deal of concern as to how the international trading situation would develop, especially with regard to Empire markets. Tariffs had risen during the War and this meant that the price of British products was raised in foreign markets with consequent attractions to local manufacturers to produce competitive products. Australia, before the War a white lead market dominated by British producers, was the country in which the problem was most marked. The immediate cause was that during 1917 Lewis Berger & Co. Ltd[14] had stolen a march on the other British producers by setting up a white lead works in Australia. In conjunction with Cooksons the allied firms responded in May 1918 by registering an Australian company, British Associated Lead Manufacturers Pty Ltd (BALM).[15] An initial capital of £20,000 was raised by offering shares to the members of the UK White Lead Convention according to their quota and subsequent increases in capital were partly financed in the same way. As a result the company was effectively controlled from London, even though most of the debenture capital raised when it actually moved towards production was Australian. In October 1918 the report of a committee set up to explore the Australian market convinced the British manufacturers that they were right to go ahead. It noted that 'at a time like the present when the cry of "Australia for the Australians" is so much in vogue, the chances of being able to export from the United Kingdom to Australia in the future are so small as to be practically negligible'.

A site was purchased on the Parramatta River at Cabarita not far from Sydney and the building of a stack white lead factory commenced. It may seem that the use of the stack process was a typical example of British conservatism but it was in fact a response to demand. Both Lewis Berger from its new plant and Champion Druce, the current owners of the Islington works (which had the premier British brand on sale in Australia), sold stack white lead and BALM's Australian advisers were clear, 'It is to be sincerely hoped that you will decide upon the "stack" process ...'. Delivery of white lead commenced in August 1921 from the 18 stacks originally erected but by June 1922 there were 26 and annual sales settled at about 2,500 tons during the 1920s. In addition the Australasian United Paint Co. Ltd of Port Adelaide was purchased for £51,000 in 1919, which gave BALM an established outlet for its product. BALM succeeded in obtaining about one-third of the Australian market for white lead but it was considerably less successful than Lewis Berger. This company's Australian activity was largely financed by the Sherwin-Williams Corporation of Montreal, whose marketing techniques were much superior to those adopted by BALM. Nevertheless a major step had been taken as a result of the initiative of the allied firms. The White Lead Convention had been discussing the Australian situation since 1916 but had taken no action and it was the merger companies which led to the creation of BALM, which was not only to pay good dividends on the capital subscribed by the British lead manufacturers but was also to play an important role in future negotiations with ICI.

BALM was, however, to be the only lasting memorial to the early merger attempts. Although the merger had shown signs of being formalised with the formation of ALM in 1919, there were clearly serious difficulties which were to lead to its failure. Symptomatic of this is the fact that the two Rowes who joined the board in October 1919 attended only briefly and resigned in 1920. The major difficulty appears to have been financial. Shareholdings in the allied companies had to some extent been built up by using Walkers, Parker's surplus cash, which was only attractive to that company if it produced a return on its capital investment. However, FB & W, the first company purchased, was in a financially unsatisfactory state. During the War its white lead capacity had been closed and in 1919 it was estimated that the company needed £100,000 for pig lead to refill its stacks but wartime profits had been nowhere near large enough to meet this call. Moreover the wartime diversification into the manufacture of aircraft wings had been converted at the end of the War into the manufacture of wooden toys, under the name 'Bairntoys', and then doors and windows. Neither was successful and both were soon closed with the writing off of considerable losses.

As a result FB & W paid no dividend on its ordinary shares and it seems that Walkers, Parker began to doubt the wisdom of its investment in that company. Consequently, when in October 1919 FB & W decided to issue 100,000 £1 ordinary shares to finance the recommencement of its white lead works, Walkers, Parker refused to take them up. This produced two problems. FB & W desperately needed some of the money because of serious cash-flow problems and this was temporarily solved when Walkers, Parker agreed to guarantee an increase in the company's overdraft to £75,000. More important, however, was the threat to the long-term development of the merger. Walkers, Parker was much the largest company in the industry. If it defected the merger's power would be much reduced and Enthovens would have to find another source to finance its merger activities.

The situation deteriorated further in February 1920 when, at a Walkers, Parker board meeting, Enthoven and A.J. Foster argued for an increase in that company's dividend but this was rejected by the old-established directors under the chairman, Thomson. From then on there was a clear division on the Walkers, Parker board, which only ended in victory for the old directors because the Baron de Catelin was rarely able to attend.[16] Since it was clear that existing arrangements had to be modified an agreement was negotiated by which ALM would purchase Walkers, Parker's shares in FB & W for the price originally paid, £91,000. For ALM the problem was where to raise the money. It had made use of some of the surplus capital of LL & J to provide a loan to meet the immediate needs of FB & W but internal resources were apparently insufficient to buy the shares. Once again Tilden Smith appeared on the scene, towing the Burma Corporation, with its interests in lead mining, as a possible purchaser of the ALM shares in all the lead companies. In January 1921 two representatives of the Burma Corporation were elected directors of ALM, the previous directors resigning, and in February a circular was sent to the shareholders of all the allied firms informing them of the holding in their companies which was being acquired by ALM.[17] At this stage something went wrong and Burma dropped out, although the reason is not clear. There was certainly opposition from Walkers, Parker to the terms of the offer and it may have been felt that without that company's inclusion a merger was worthless.

This meant that there was no money to buy from Walkers, Parker its shareholding in FB & W and the relationship between ALM and Walkers, Parker went from bad to worse. In June 1921 Thomson took the opportunity of having a clear majority at a Walkers, Parker board meeting to elect the Elswick manager, Donald Murray, a director, against the protests of H.J. Enthoven. Now there was a permanent majority of four to three against ALM on the Walkers, Parker board and it was suggested that the ALM appointees should resign. They refused to do so, pointing out that they represented a large shareholding in Walkers, Parker and that they believed that it had not been in the best interests of that company to refuse to negotiate with the Burma Corporation. The majority then passed a resolution to take legal action to force Enthovens to keep to the agreement to buy the FB & W shares.

This brought Andrew Weir, by now Lord Inverforth, back into the reckoning. It would not have done for a firm, with which someone of his prominent position was connected, to be dragged into the courts over such a matter and it may also be that, with the completion of his government work, Inverforth was able to apply his mind to the problem of the lead industry. It may well have been his fertile and dynamic mind which had originally conceived the idea of an amalgamation of the lead manufacturing industry and in the summer of 1921 he came forward to solve the crisis. He proposed personally to take over the ALM interests in the lead manufacturing companies and detailed negotiations took place over Walkers, Parker's holding in FB & W. Compared with the £91,000 which ALM had agreed to pay for the shares, Walkers, Parker asked Inverforth for £75,000 but, having had an investigation of the affairs of FB & W undertaken, he wanted

> a substantial contribution by your Company in cash, say £25,000. In view of the highly critical financial position of that business [FB & W], the large amount of working capital required, its special liabilities under long term contracts and the heavy trading losses now going on, much time and money will be involved in any effort to bring it round, and it seems reasonable that you should assist with cash as well as with any expert advice which may be requested.

This was rather hard on Walkers, Parker which had purchased the shares for the good of a merger with which Inverforth was intimately associated because of his position on the board of Enthovens. The shares were offered to Inverforth at 5s each, but this was refused and the Walkers, Parker directors of FB & W then tried to call an extraordinary general meeting of that firm to propose that it should go into voluntary liquidation.[18] Although they had equal representation on the board with the supporters of ALM, they were defeated by the casting vote of the chairman, A.J. Foster. Hoist with his own petard Thomson now had to rely on negotiation in order to deal with his firm's holding in FB & W. A letter from Sir John Mann,[19] financial adviser to Lord Inverforth, had made it clear that Walkers, Parker's proposals for the sale of the shares 'do not offer sufficient inducement to him [Inverforth] to involve himself in these unfortunate complications. We cannot proceed upon any basis which places a value upon the ordinary shares in Foster, Blackett & Wilson.' Walkers, Parker wanted to extricate itself from an unfortunate situation and therefore had to agree to Inverforth's terms. The shares were transferred to Inverforth free and Thomson and the other Walkers, Parker directors resigned from the board of FB & W. The one concession which Walkers, Parker achieved was the resignation of Enthoven, Foster and de Catelin from its board and their replacement with only a single director, Sir John Mann, to represent the holding in the company which Lord Inverforth had taken over from ALM. It had, however, been a costly exercise since Walkers, Parker had to write off more than £91,000, the costs of the FB & W affair, from its £100,000 reserve fund.

Thus ended the first serious attempt to merge the UK lead manufacturing industry. Born out of the difficulties of the lead supply situation it was fostered by a firm which had its main interests in lead smelting and refining, H.J. Enthoven & Sons Ltd. Whether or not Lord Inverforth was the initiator of this move, he ended up by purchasing the shares in the various lead companies which had been involved in the merger. He, therefore, had a considerable personal interest (estimated at about half a million pounds) in the future development of the industry and it was to be expected that his interest would lie in future attempts at merger and rationalisation. Although Inverforth remained on the board of Enthovens, that company was not involved in subsequent lead merger activities,[20] probably as a result of a change in the factors encouraging merger. In the war years the significant factor was the limitations on the supply of lead, which was of most immediate consequence to a firm such as Enthovens. On the other hand the lead manufacturers were making very good profits, even on reduced output and had no great incentive to look to amalgamation. This position continued into the post-war boom but from the slump of 1921 the factors changed. Lead was once again freely available and remained so until the Second World War and the major reason for Enthovens' interest in amalgamation disappeared. On the other hand for the lead manufacturers profits were falling and competition, both between themselves and from abroad, was rising. Moreover demand for some of their products, including sheet and pipe and white lead, was not expanding and was soon to go into decline. The continuing existence of the UK lead conventions ensured a fairly orderly response to these conditions but cartelisation was only a holding operation and merger was the only way in which individual firms could grow when total market size was stagnant.[21]

After February 1922, when its agreement with Walkers, Parker to buy that company's holding of FB & W shares was rescinded (Inverforth having completed the purchase personally), ALM became dormant. It was not until September 1924 that another entry in the minute book of the company records that, at an extraordinary general meeting, two new directors, A.A. Lough and H.G. Judd, were appointed, the previous directors having resigned.[22] Since Judd was a partner of Sir John Mann it seems likely that Inverforth had something to do with the change. He certainly had a shareholding in LL & J and in January 1924 that company and Cooksons, the two large, successful and forward-looking firms in the industry, had agreed to amalgamate. In October Clive Cookson and H.C. Lancaster were appointed directors to represent their respective firms and, after an increase of capital to £1.85m, a full board of directors was elected (Lough and Judd resigning, having done their work of reconstruction). Sir John Mann was to represent Inverforth's holding, while F.C. Hill came from LL & J and A.J.H. Smith, H.S. Tasker and Frank Reid from Cooksons. In addition C.A. Rowe and E.C. Philp were appointed from Rowe Bros which was to join the merger as from 1 January 1925.

That Cooksons had four out of the nine directors appointed and that Clive Cookson became chairman of ALM point to the dominance of the firm in the merger and it seems clear that Cooksons provided the commercial initiative behind the merger. Not only was Clive Cookson[23] the first chairman, he was followed by his deputy, H.S. Tasker,[24] who was managing director of ALM from 1924 and he, in his turn, was succeeded as chairman by J.L. McConnell,[25] also a Cookson man. Not until 1957 when Roderick Lancaster succeeded McConnell did a non-Cookson man become chairman and he was succeeded in 1962 by Roland Cookson,[26] nephew of the first chairman. For most of the 50 years from 1924 to 1973 the Cookson organisation provided ALM chairmen and for the first part of that period the majority of the firm's senior management. Indeed it is likely that the initiative for the merger came from Cooksons and probably specifically from Clive Cookson. In an account of the development of the merger McConnell claimed that Cooksons had first approached LL & J and mooted the possibility of a merger in 1920, which would have been at the time that the earlier ALM was breaking down.

In order to facilitate the merger the Cookson family sold their lead and antimony manufacturing interests to a new company, the Cookson Lead & Antimony Co. Ltd (usually known as CLACO) while Cookson & Co. Ltd retained the family's other investments in overseas companies, etc. The price paid by CLACO to Cooksons, £849,000, may be taken as the Cookson share of the assets taken over by ALM and it is probable that the price involved some overvaluation of the assets. McConnell certainly thought that both Cooksons and LL & J (for which firm £835,000 was paid) 'were somewhat overvalued', while Kenneth Cookson writing in 1923 to his brother, Clive, about the price to be paid to Cooksons for CLACO, commented, 'You have made a v.g. bargain I think.'

Already before the merger Cooksons were showing the initiative which was to make the merger potentially attractive and to give them such a controlling voice in it. Following its pre-war policy of recruiting graduates, in 1920 the board decided to create 5,000 officers' shares which were to be issued to senior employees as the directors saw fit. Of only sixpence nominal value, they were to rank for dividend alongside the £10 ordinary shares and were clearly intended to buy the allegiance of employees by giving them a stake in the firm's performance. The firm was also continuing to look to overseas investment, both in its traditional area of gaining control of raw material supplies and also with a view to obtaining manufacturing plant on the Continent. Early in 1923 Cooksons had an option to purchase 'the Lead Works belonging to Gamichon Frères, Paris' and although the option was not taken up after examination of the works and the firm's books, it seems probable that it was Cooksons who were responsible for the subsequent ALM purchases of continental works.

LL & J by contrast show less initiative but considerable thoroughness in doing their existing work efficiently and the firm's net profits, never less than £70,000 between 1919 and 1924 (a period in which Cooksons' net profit only once exceeded £70,000), point to a highly successful company. A reorganisation in the firm's management in 1921 shows that hard work was apparently expected even of sons of directors. With effect from the end of December 1921 C.F. Hill resigned as managing director of Burdett Road and Millwall works in order to run the Mersey operation (William Sloane, whose daughter he had married, having largely retired from active participation in that company). This move necessitated some reorganisation. Alan Hill, brother of C.F., was to go to Millwall and he 'was quite prepared to be at the works punctually at 8 o'clock every morning & put his heart and soul into the Millwall work' while Keith Johnson, son of the chairman, R.B. Johnson, was to manage Burdett Road, 'the hours being 9 to about 5.20, Saturdays 9-12'. The firm was technically very efficient, had spent considerable sums on up-to-date equipment and was prepared to innovate. Shortly after the war experimental work was done at Millwall on a 'quick' white lead process commencing with litharge, while research was undertaken on the manufacture of paints and on refining techniques. The most significant of the employees on the research side, W.T. Butcher, was a metallurgist of some significance who published a number of research papers and subsequently became manager of the ALM research laboratory.

Like Cooksons, LL & J had managed to grow large without losing family control and good relationships between employers and employees. In May 1919 the difficulty of retaining employees at the Millwall works was mentioned at a board meeting. By September the company was involved in estimates for the purchase of local land and the erection of houses for its employees. A separate company, Locke's Housing Society Ltd was set up, governed by LL & J directors but with an advisory committee of employees (and it says something for the firm that James

Rugless, a Poplar Borough Councillor, was accepted as a member of the committee). By July 1921 the first four of 36 houses being erected were completed and the committee recommended that they be offered to employees who were then living in Edmonton, Ilford, Chingford and Kings Cross. Rents of the four- and five-roomed houses were set at the same level as charged for local council houses. In 1923 a sports ground was purchased at Millwall and made available to the company's employees free of charge, while in the slump in the early 1920s the firm found 'relief work for our hands in making a road and doing roofing, rather than discharge them'. Paternalistically, therefore, LL & J had similar attitudes to those of the firm with which it was to merge. After the War Cooksons had continued to develop provision for employees. In 1919 the firm commenced a welfare scheme with a club for employees. Initially there were separate schemes for Hayhole and Howdon works but they were amalgamated in 1924. Contributions (6d per week per employee) were provided equally by the men and the firm and both had equal representation on the committee of management. The income provided pensions, death and sickness benefits, subscriptions to local infirmaries and the support of social and sporting activities. Clive Cookson (universally known as 'Mr Clive' by the men) took a great interest in the club and frequently attended its functions. From this time on the 'welfare' administered the annual works trip which continued to prove very popular.

There was, therefore, considerable similarity in background, performance and attitudes between the merging companies. The obvious question which arose was whether they intended to be satisfied with what they had achieved or whether they should extend the merger and the answer was decidedly the latter. While there were tentative discussions, as we shall see, with a number of companies, it may be considered surprising that the first addition to the merger was Rowe Bros & Co. Ltd, less a traditional lead manufacturer than a builders' merchant which was developing manufacturing interests. In the years before the War, however, Rowe had developed an interest in lead chemicals with its investment in the Barton red lead patents but more importantly was involved in experimental work on a quick white lead process. It was undoubtedly for the potential of this latter development that Rowe was one of the first companies to be approached to join the ALM merger. By the early 1920s Rowe was becoming a thorn in the flesh of the other white lead companies, and J.L. McConnell noted that the firm was a 'heavy price cutter', able to sell white lead very cheaply because of its new process (and to some extent forced to do so because the quality was very variable).

The story of Rowe's development in white lead is one of the failure of other companies to see where their best interests lay. In 1901-2 Rowe twice approached Walkers, Parker to corrode white lead for it, presumably at a special low price, but was refused. Since this was at a time when Walkers, Parker was making considerable losses and trade was poor, one would have expected that company to have jumped at the opportunity. It should have been obvious that Rowe would either find an alternative source of supply or begin to manufacture on its own account but, true to its pattern of poor management, Walkers, Parker did not respond and in 1909 Rowe purchased the Runcorn works mentioned earlier. With the works were taken over the patents to a quick white lead process which had not worked satisfactorily and it appears that Rowe had considerable difficulty with it in the years before the War.[27] Even in 1919 Rowe had only a 2½ per cent quota in the UK white lead convention but it had just opened new and larger works at Bootle and shortly thereafter the company left the convention and began to compete vigorously, presumably because it was beginning to get the process working effectively.

The purchase of Rowe, which took place as from 1 January 1925, therefore seemed attractive. The terms, however, were not very favourable. The interests of Rowe Bros & Co. Ltd in the manufacture of lead chemicals were sold to a new company Rowe Bros & Co. (Bootle) Ltd, which was wholly owned by ALM, for 210,000 £1 ordinary shares and 140,000 £1 cumulative preference shares in ALM. Even in paper, which was already beginning to look rather wet, £350,000 was a very large sum to pay for a little-known company with a far from totally satisfactory process. McConnell subsequently wrote that 'the necessity to absorb Rowe Brothers meant a large goodwill payment in their case'. The goodwill between Rowe and its partners lasted less than a year. In November 1925 the three original Rowe directors were replaced on the board of the Bootle company by nominees of ALM. Clive Cookson became chairman and immediately appointed ALM men to key positions to institute standardised procedures in the running of the firm. In particular C.A. Klein and other chemists from Brimsdown (by this time also included in the ALM merger) were brought in to get the manufacturing process right. Reorganisation was clearly needed since the board of directors 'expressed its disappointment' with the results for 1925 and the dividend paid to ALM (the full amount of Rowe's net profit) was less than 2 per cent. By the end of the 1920s reorganisation had lifted the financial fortunes of the Bootle works and, as if in acknowledgement of this, the name of the company was changed in 1929 to Librex[28] Lead Co. Ltd. Bootle's future was secure, with a product which was rapidly expanding its share of ALM's output of white lead and with a good transport network which included the nearby Liverpool docks.

Well before Rowe had been finally brought into the merger, negotiations had begun with other lead manufacturers. The first of these to be taken over was Brimsdown, a company which was certainly not desirable from the point of view of its profitability and, since its white lead process had been functioning for some 25 years, ALM could hardly have thought that it was a potentially serious challenger, unlike Rowe's quick process. Apart from the attraction of eliminating some competition the chief reason for purchasing Brimsdown appears to have been for its technical reputation in research and development and one of the early benefits was the ability to send a team of the firm's chemists to Bootle to get the Rowe quick process working efficiently. Among the specific activities at Brimsdown in which ALM may have been interested was the development of a waterproof sandpaper and research on titanium dioxide as an alternative paint base to white lead, both of which developments came out of C.A. Klein's involvement with the controversy over lead poisoning in the painting trades. The development of a waterproof sandpaper is attributed to W. Hulme, a Brimsdown chemist who had previously researched at Manchester with Chaim Weizmann. It was a response to the problem of the lead dust caused by dry rubbing down of old paintwork, which had been established as the major cause of lead poisoning among house painters. The Brimsdown laboratory also experimented at various times with the production of a number of lead compounds, including nitrate, sulphate and peroxide and there was even an experiment with soap manufacture, which N.J. Reed, one of the chemists, remembers 'came to a sudden and inglorious end when a specimen of shaving soap removed the skin from the Works Manager's face!'.[29] On the basis of their own solid achievements, Cooksons and LL & J may well have thought that they could channel this spring of initiative to more effective financial ends.

Financial reorganisation Brimsdown certainly needed. After one capital reorganisation in 1916 (when £100,000 in debentures, loans and unpaid interest was written off) the company underwent another in 1919 when the long-suffering preference shareholders were enfranchised by the capitalisation of the arrears of preference dividend. From the second half of 1920, under the new scheme, the 7 per cent preference dividend was actually paid and net profit for 1920 showed a return of

5 per cent on capital employed. At the company's AGM in July 1921 the chairman, Stephen Miall, commented, 'it is obvious that in the 21st year of our existence we have now reached years of discretion and we trust that some of the escapades of our early youth will never be repeated'. The cyclical downswing, however, halved the profit in 1921 and brought a loss in 1922 almost as large as the 1920 profit. In these circumstances a take-over bid must have been of interest to Miall. He subsequently informed the shareholders that

It was on March 30, 1922, that I was invited to consider the amalgamation we are now discussing — or, at any rate, an amalgamation which has much in common with it. From that day until this negotiations have been continually pending, sometimes proceeding hopefully, sometimes stagnating hopelessly.[30]

Finally, however, in February 1925, just two months after the Cookson, Rowe and LL & J merger had been formalised, an offer (in ALM paper) was made for Brimsdown of 15s for each preference share and 3s 6d for each 5s ordinary share. Miall told the shareholders that while it was not 'a handsome price for such good assets as are represented by our ordinary shares . . . You will receive in exchange shares which are also represented by valuable assets: there is no water in the capital of the Associated Lead Manufacturers, Limited.' Unfortunately, as we have already seen, Miall's estimate was incorrect and the existing over-capitalisation of ALM was only made worse by the purchase of an unprofitable company whose only merit was its technical know-how. One positive result of the Brimsdown takeover was that since it was the only one of the four merged companies which was a public limited liability company a stock market quotation for ALM had to be arranged as part of the agreement and shares became freely negotiable. This provided the opportunity for the partners in the old family firms to sell to the general public some of the shares which they had received in ALM, and their involvement as directors of ALM gave them intimate knowledge of its financial performance. Judging from the rate at which they reduced their holdings, many of them were highly pessimistic about the company's future. For instance M.M. Johnson reduced his holding of ordinary shares in ALM from 38,000 in 1926 to 250 in 1930 and H.C. Lancaster reduced his from 21,000 to 500 by 1929. Other holders were less pessimistic; F.C. Hill reduced his holding from 11,000 to 4,000 by 1930; the Rowes reduced theirs from 210,000 to 127,000 by 1930 and the Cooksons only began to reduce theirs in 1930 to

372,000 ordinary and 260,000 preference shares (compared with 494,000 and 334,000 respectively in 1929).[31]

The next step in the development of the merger came in May 1925 when Lord Inverforth's holdings in the lead companies were offered to ALM. By this time Inverforth may well have lost interest in his earlier desire to merge the industry and have recognised that this was being achieved by a new rival. In addition he would have found that his purchases were not financially fruitful. After he had taken over FB & W, Inverforth had set Sir John Mann's firm of accountants to investigate the company's financial position and the report made unattractive reading. As a result of its problems FB & W paid no dividend on its shares in the years in which Inverforth owned them and his desire to sell his interests to ALM is therefore understandable. From the point of view of ALM Inverforth's holding offered a simple way of gaining control over a considerable proportion of the remainder of the industry.

In the early summer of 1925 ALM offered £519,000 (40 per cent in preference and 60 per cent in ordinary shares) and then raised the offer to £580,000 when Inverforth showed no interest in the original figure. By August, however, Inverforth was in negotiation with the National Lead Co. of America which was presumably following up its earlier interests in 'some well known and established English Lead Manufacturing Company'. Nothing came of this and finally in June 1926 agreement was reached with Inverforth for the sale to ALM of part of his interests (excluding those in Enthoven and Alexander, Fergusson). ALM received 26,000 £1 preference and 89,000 £1 ordinary shares in FB & W and 2,466 £1 ordinary and 62,186 £1 ordinary (15s paid) shares in Walkers, Parker, in exchange for £404,126 in its own capital made up of 161,650 £1 preference and 242,476 £1 ordinary shares. McConnell was to comment that 'Lord Inverforth exacted a slightly excessive price for his interests'. As a result the overcapitalisation of ALM was increased and, from 15 December 1926 when the exchange of shares was formalised, Inverforth became the largest single shareholder in the company. Apart from the fact that ALM paid £115,000 to Inverforth for FB & W, whose shares he had regarded as valueless when they belonged to Walkers, Parker, it also took over from him that company's debts in the form of loans and interest payments due of £141,373. It seems clear that ALM was building up some potentially difficult problems.

Subsequently ALM purchased Inverforth's holding in Alexander, Fergusson, with which went voting control of the firm and in December 1929 the remainder of the ordinary share capital of that most significant of the Scottish lead manufacturers was purchased. This was a satisfactory acquisition since Alexander, Fergusson was a profitable company which was to offer ALM a comfortable return on the purchase price. Although ALM had also purchased from Inverforth a minority holding in Walkers, Parker, that company did not respond favourably to overtures to the effect that it might join the merger. There had been several attempts to persuade Walkers, Parker and in February 1925, before the purchase of Walkers, Parker shares from Inverforth, ALM made another approach to the company, pointing out 'the extent to which the competition of White Lead manufactured by the Rowe process would surely be felt by them in the future'. Apparently only Sir John Mann, who appears not to have found his position on both boards of directors anomalous,[32] favoured joining the merger and the Walkers, Parker chairman, G.S. Pawle (who had been appointed after the resignation of David Thomson), seems to have been strongly opposed to amalgamation. Clive Cookson wrote to A.J. Foster on 7 September 1926:

> I am very upset at the way that my advances to W.P. have been received. They seem to distrust everything that I do, and they almost make me feel that they think I am a villain. My intentions are absolutely honest and straightforward, and I realise that, at any rate for the time being, anything in the nature of amalgamation with them is out of the question, and all that I want is a good trade agreement, but they seem to think that I have some other object and always misinterpret my intentions and my approaches. My last turn down was so rude and so flat that I cannot again approach them . . . I feel very sore and very badly treated.

During the 1920s Walkers, Parker had capitalised on its wartime prosperity and had made a remarkable recovery from its dreary performances from 1889 to 1914. Profits were of the order of £30,000-40,000 p.a., sufficient to pay dividends of 10 or 11¼ per cent on the ordinary share capital. To a considerable extent this performance may be attributed to the replacement of amateur directors with men having experience of the lead industry. Water Burr, having lived through all the vicissitudes d management changes since 1889 was appointed to the board in 1910, while the managers of the Elswick, Chester and Lambeth works (respectively Donald Murray, William Eckford and A.G. Simkins) were appointed between 1921 and 1925. These men were able to take quick decisions with regard to matters of lead supply, capital expenditure and output levels which they knew from their works' experience to be necessary – a totally different situation from that of the 1890s and 1900s. As a result Walkers, Parker had commenced a new lease of life. Capital expenditure, for instance new blast furnaces at Elswick and Chester, a new rolling mill and expansion of the red lead capacity, had helped to bring the firm up-to-date in technology and reduce manufacturing costs. It is not surprising that the board of directors did not welcome the ALM proposals when the prospects, given continued independence, appeared good.

Walkers, Parker still remained essentially a conservative company.[33] During the 1920s it built up a large reserve fund of over £100,000 which might have been more productively deployed than using it to purchase government stock. Certainly this was the view of some of the shareholders in the company who, on several occasions, endeavoured to obtain an increase in dividend, while in 1928 one shareholder suggested a capital repayment, pointing out that the net assets of the company were worth about £3 per share while the share price was only 30s. The company was also conservative in its approach to rationalisation, the obvious problem being what to do about the Bagillt works. Despite the difficulty of obtaining sufficient ore to work the furnaces to capacity and despite the knowledge that the works was losing money, it was kept open, since 'the closing down of Bagillt would result in serious depreciation of the plant'. An attachment to book values was therefore allowed to depreciate current profitability. Late in 1926, however, the Walkers, Parker board recognised the stupidity of such a position. New plant, to replace the Bagillt capacity, was ordered for Chester and Bagillt was finally closed in the following year. Closure led to the writing off of £30,000 from the value of plant at Bagillt (a relatively small loss considering that the works had a net trading loss of about £8,500 in 1925) and attempts to sell the site. Since the upper part of the works had been unused for years and industry in Flintshire was in serious decline it must have been obvious that this would be difficult and, although some land was sold to Courtaulds, the site was not finally disposed of until the 1950s.[34]

Having failed in direct approaches ALM began in 1928 to purchase Walkers, Parker shares on the market. At an ALM board meeting in June Sir John Mann was authorised to purchase Walkers, Parker shares from a financial agent with whom he was in contact, since 'The [ALM] Board felt it was wise to pursue the policy of ultimately acquiring effective control of Walkers, Parker & Co. Ltd.' How Sir John, as a director of Walkers, Parker, salved his conscience is unknown. In October 1928, however, at his instance, the two boards were brought into discussion and a pooling agreement was provisionally approved, with agreed market shares in various products and fines and compensation for surpluses and shortfalls. It would seem that negotiations on this matter were not very satisfactory and early in 1929 ALM was once again in the market, buying Walkers, Parker shares at prices between 30s and 35s, and by this time it controlled about 40 per cent of the total voting capital. At the end of February 1929 Pawle, the Walkers, Parker chairman, suggested to Sir John Mann an amalgamation, perhaps because he felt that this would lead to better terms than if control came by the creeping method of gradual share purchase.

ALM's first offer was, however, unacceptable and relations became strained but it was finally agreed in April 1929 that auditors should be appointed (Deloittes on behalf of ALM and Kemp, Chatteris for Walkers, Parker) to agree merger terms on the basis of net asset valuations. In July it was reported that the respective accountants' ideas 'were somewhat far apart' but negotiations were continuing and in October an offer was approved by the Walkers, Parker board with the exception of Pawle who, wisely as it turned out, was still holding out for cash.[35] The offer placed a value of about £2 1s 0d per share on the Walkers, Parker capital, less than Inverforth had received for his holding and below the net asset value of £2 15s 7d (of which £1 2s 2d was in cash and other liquid assets).

Nevertheless the financial press regarded the offer as a favourable one and there was never any doubt that the shareholders would accept it. In a leading article on 8 November the *Financial Times* referred to Sir Mark Jenkinson's recent comment that 'no industry in this country, apart from the chemical trade, has been rationalised' and noted that this was only true of the basic industries, while 'The lead industry is an outstanding example' of one in which progress was being made. The takeover of Walkers, Parker would be 'the coping-stone' and, in a phrase which the paper had well-oiled at this time, 'the technical advantages of the merger [are] so clearly established'. Finally the paper continued, the benefit to Walkers, Parker shareholders 'is so substantial that we have no hesitation in advising them to close immediately with an extremely fair and reasonable offer'. The shareholders accepted and Walkers, Parker was officially controlled by ALM from 1 January 1930. Although ALM directors were appointed to the Walkers, Parker board it was recognised that the existing board and management were efficient and no changes occurred other than the non-re-election of Pawle, while Burr became chairman and he and Eckford joined the ALM board.

ALM had succeeded in obtaining control of its most significant rival

and the greatest threat to its plans to control competition in lead manufacture. Market shares are difficult to estimate but by 1930 ALM probably accounted for more than three-quarters of the UK output of white
lead, perhaps two-thirds of red lead and around one-third of sheet and
pipe. Although the acquisition of Walkers, Parker had been the most
significant development, in the late 1920s ALM had also made several
other takeovers. In April 1926 Locke, Blackett had been approached
with an offer which entailed that company's ceasing production although
without ALM's buying the land and buildings but nothing came of this.
In November 1928, concerned at the growing power of ALM and his
own company's poor performance,[36] T.H. Leathart offered to sell to
ALM. According to J.L. McConnell the terms (£1 7s 6d for each Locke,
Blackett ordinary share and par for the preference shares, to be paid in
ALM paper) were much more favourable to ALM than in the earlier
mergers and the ALM board was informed that, if Locke, Blackett were
liquidated, substantially the whole of the purchase price would be
recovered. It is clear that the takeover, which was effective from 1
January 1929, was undertaken purely to eliminate competition, since
Locke, Blackett offered no other benefits.

ALM also took over the London Lead Oxide Co. Ltd (LLO), in
1927, although with some financial contribution from other members
of the Lead Oxide Convention. LLO had been set up at Milton Wharf in
Gravesend in 1919 by E.J. Clark to manufacture red lead. Output had
been increased to about 4,000 tons p.a. by 1927 and LLO, which was
undercutting the Convention price for red lead, was a decided nuisance
to the other lead manufacturers. It was also a very profitable company,
making a net profit of nearly £7,000 in the unfavourable year of 1926,
on a capital of £35,000 and paying a dividend of 10 per cent. Fortunately for ALM Clark was in ill health (he died soon afterwards),
wanted to settle his affairs and accepted an offer of £1 13s 10d per £1
share. It appears that the Lead Oxide Convention was involved in the
negotiations because it was intended that the company's output should
be reduced after it was taken over by ALM and to allow for the benefit
of reduced competition the members of the Convention contributed to
the purchase price. LLO was to prove a profitable acquisition; an
efficient, single-product company, it produced annual profits for ALM
of £6,000-10,000 during the 1930s and was only closed during the
rationalisations of the 1970s. Also taken over were two manufacturers
of solder, the Oidas Metals Co. Ltd (in 1928) and A.T. Becks & Co. Ltd
(in 1933). Their works were closed and they were amalgamated with
LL & J, the merger's chief producer of solders. In addition in 1929

ALM, in conjunction with Alexander, Fergusson, purchased the ordinary share capital of T.B. Campbell & Sons Ltd, a lead sheet and pipe manufacturer with works at Kinning Park, Glasgow.

It might be thought that at the end of the 1920s the rationalisation of the group of companies which had come under its control would have been sufficient concern for ALM but a further amalgamation was about to take place, bringing with it a change of name to Goodlass Wall & Lead Industries Ltd (GWLI). J.L. McConnell expressed the position thus:

> The paint industry was an important consumer of white and red lead and as prices of these latter products were artificially high, there was always the danger that one or more of the powerful paint companies (e.g. Pinchin Johnson) would commence making their own lead products. As a defensive measure, it was decided to enter the paint trade and on 1 December 1930 the old established business of Goodlass Wall & Co. Ltd., Liverpool, was acquired under a 'Scheme of Arrangement'.

Goodlass, Wall was a conservative, old-fashioned and not particularly well-run concern. In 1928 Leeds Fireclay Co. Ltd had purchased a large holding in the company and its chairman, C.F. Spencer, had been put on the board of Goodlass, Wall in order to rationalise it.[37] Spencer either intended initially to sell at a higher price as soon as possible or discovered that the reorganisation problems were too severe and decided to get out. He probably achieved good terms. As he explained to the Goodlass, Wall shareholders, the ALM capital was to be written down from £2.9m to £2m[38] (by writing each £1 ordinary share down to 11s 3d, which did something to reduce the overcapitalisation which had been obvious since 1924 but also justified G.S. Pawle's reluctance in 1929 to accept ALM paper). On the other hand the Goodlass, Wall capital was written up by £500,000 to £1m. Spencer pointed out that average ALM profits were about £200,000,[39] although the full benefits of rationalisation in lead manufacture were yet to emerge, while those of Goodlass, Wall were £68,000. Goodlass, Wall, Spencer continued, was an insufficiently large company to deal effectively with increasing competition and rising tariff barriers: 'The export trade is decreasing, and the decrease will become progressive as tariff barriers have been erected in so many countries. The scales of tariffs seem elastic, and the menace of increasing tariffs is disconcerting.' Fusion with ALM with its wide overseas experience would, Spencer argued, enable Goodlass, Wall to set up factories abroad. Despite these arguments there was a number of hostile questions from shareholders about the terms of the merger and the low level of dividend which ALM had been paying, although the merger was finally agreed.

From the point of view of ALM the merger offered not only a defensive base against possible entrance by paint companies into the manufacture of lead pigments but also a major outlet for its own production. Goodlass, Wall was a paint manufacturer of significant size which owned a number of retail paint outlets and at a time when the overall market for lead-based paints was contracting, direct control over final market outlets was obviously going to be beneficial to ALM. Moreover under Spencer's influence Goodlass, Wall in 1929 had purchased from the Valentine Varnish and Lacquer Co. Ltd the UK and Australian rights to that company's lacquer and varnish paints (including 'Valspar'), which offered an attractive area for diversification. Amalgamation also offered Clive Cookson the opportunity to tighten up on the overall control of the lead organisation. The ALM board had reached a total of 14 members in 1929 and although they were all reappointed to the newly incorporated ALM, there was no power there and only six of them were put on the GWLI board.[40] Like D'Artagnan and the three musketeers, Cookson ran the new organisation through Tasker (managing director of GWLI), McConnell (secretary of GWLI) and Reynolds (who was sent from Cooksons to take control of Goodlass, Wall at Liverpool). Once again the *Financial Times* produced its standard phrase in blessing the merger, 'the technical advantages of the projected merger [are] so clearly established', although *The Times* noted that 'To date ALM has not had a very successful career.'[41]

The immediate future was to be little more satisfactory than the immediate past, partly as a result of the cyclical depression of the early 1930s but also as a result of the over-optimistic expectations from the merger. ALM shareholders had been informed that the reduction in the capital value of their holdings 'will render unnecessary in future the special provisions which have been made out of profits in the past, and thus make available for distribution a greater proportion of the profits of your Company than hitherto'. This was not forthcoming and there is some evidence that the accountants' report had been seriously misleading. At the first meeting of the board of GWLI Cookson had to produce a reconciliation of the difference between the figure of £200,000, which the shareholders had been informed was ALM's average profit for 1928 and 1929, and the actual figure, which was said to be £100,000.[42] Moreover the trading results for Goodlass, Wall to 30 August 1930 were described as showing 'a considerable shrinkage' and the directors of that company were asked for 'a full report'. Most importantly perhaps, the write-down of ALM capital and the write-up of Goodlass, Wall capital had occurred, according to Clive Cookson, because the accountants, Deloittes, 'brought it very strongly to our notice that . . . it would not be possible for the Company [ALM] to pull its full weight in the earnings of a new undertaking unless some of our capital values were written down'. Subsequent profits did not show the accountants' values to have been justly determined and the possibility of suing Deloittes was considered. As the first lesson which he subsequently stated should be learned from the GWLI merger, J.L. McConnell wrote:

> It is dangerous to rely solely upon the recommendations of outside professional accountants as to merger terms. In our case, we had no opportunity of looking into the affairs of G.W. Liverpool in 1930 before amalgamation. The accountants were misled and possibly inefficient and in consequence the price paid for G.W. was rather excessive. A number of useless assets had to be written off and results for the first few years were catastrophic.

In theory forward integration into paint manufacture was a sensible move for ALM[43] but the problems it brought, on top of the need to rationalise its own lead holdings during the most severe world depression that had ever occurred, were to tax the powers of Clive Cookson and his cohorts.

Notes

1. The complexity of the negotiations over excess profits duty and the difficulty of interpreting from the accounts what sums were actually paid, rather than merely set aside for payment of the duty, make these figures rather insecure estimates. The duty was introduced in 1915 with effect from the commencement of the War on the basis of 50 per cent of any excess profit over the average of the best two of the three immediately post-war years. It was raised to 60 per cent in 1916 and to 80 per cent in 1917, while a munitions levy was also added.

2. Nevertheless it has been estimated that excess profits duty raised 25 per cent of all government tax revenue during the War. See B.E.V. Sabine, *A History of Income Tax* (1966), p. 152. Walkers, Parker claimed that in the ten years 1914-23 it had paid £568,000 to the government and only £174,800 to its shareholders, *The Times*, 13 March 1924.

3. See A. Plummer, *International Combines in Modern Industry* (3rd edn 1951), p. 41, E.J. Cocks & B. Walters, *A History of the Zinc Smelting Industry in Britain* (1968), pp. 16-17, *Financial Times*, 22 May 1909 and 10 March 1910,

The Globe, 6 and 24 Nov. 1914, and *The Times*, 13 and 17 Dec. 1917. The Convention controlled most of the output from Spain, Australia and Mexico but probably less than 50 per cent of world output. For subsequent developments see W.Y. Elliott, *et al.*, *International Control in the Non-ferrous Metals* (New York, 1937), pp. 591-662.

4. *London Gazette*, 6 April 1917 and *The Times*, 10 April 1917. From 1 Sept. 1917 the Ministry of Munitions also took control of all UK lead ore and secondaries. For wartime control see *History of the Ministry of Munitions*, vol. VII *The Control of Materials*, part III (1922), pp. 89-99.

5. See A.D. Chandler (jun.), 'The Beginnings of "Big Business" in American Industry', *Business History Review*, XXXIII, 1 (1959), p. 30. For the pre-1914 development of large companies in the UK see P.L. Payne, 'The Emergence of the Large-scale Company in Great Britain', Hannah, 'Mergers in British Manufacturing' and A.D. Chandler, 'The Growth of the Transnational Industrial Firm in the United States and the United Kingdom: a Comparative Analysis', *Economic History Review*, XXXIII, 3 (1980).

6. The White Lead Conference of 1893 had, however, given serious consideration to the subject but collapsed before anything positive was done. In July 1893 E.M. Johnson wrote to Col. Thompson, President of the recently formed National Lead Co. of America, asking a series of questions about merging the white lead industry in the UK and received a reply based on the American experience. A committee of the Conference then drew up a report stressing the advantages of amalgamation: 'The command of a large capital & all the power implied thereby – i.e. the power to undersell rivals – to tire them out – to buy them up if advisable – to secure complete control of all valuable Patents & Trade Marks . . .' There would be scale economies with the closure of some plants and the running of others at high capacity, while there would also be the advantage of 'The increased control over labour. The necessity of combination against the Trade Unions is now merely a truism in England.'

7. Although nothing came of it, this is an early example of merger proposals aimed at rationalisation rather than just the restriction of competition. See Chandler, 'The Growth of the Transnational', pp. 402-5, where he argues that mergers in Britain 'remained federations of autonomous family enterprises' which did not bring administrative centralisation.

8. Andrew Weir (1865-1955), 1st Baron Inverforth, 1919. Senior partner in Andrew Weir & Co., shipowners of London and Glasgow; director of H.J. Enthoven & Sons Ltd, Surveyor-general of Supply at the War Office and Member of the Army Council 1917-19, Minister of Munitions 1919-21. See *Who Was Who 1951-60* and DNB.

9. Smith is a shadowy figure who was much involved in the early negotiations for a merger of the lead manufacturers. He was a financier with an office at 70 Lombard Street, was involved with the Ministry of Munitions and was a director of and large shareholder in Burma Mines Ltd (from 1920 Burma Corporation Ltd), which was also to be involved in negotiations over the amalgamation of the lead manufacturing companies. Smith appears to have known many civil servants and he was frequently left by ALM to make negotiations with government ministries. At the same time that he was involved with negotiations for the merger of the lead industry he was responsible for the establishment and early development of the Avonmouth zinc smelting works. See Cocks and Walters, *History of Zinc Smelting*, particularly Ch. 4.

10. The links were obvious. The Parkes desilverising process required zinc while zinc oxide based paints had gained in popularity as a result of the investigations of the Home Office paint committees and imports had been reduced because of the War. For the air of euphoria see David Thomson's report

to the Walkers, Parker AGM, *Financial Times*, 13 March 1919.

11. Heberlein had made a report on Walkers, Parker's works 'with a view to laying down a policy for future extensions and development'. He was paid the huge sum of 1,000 guineas for producing the report – a further significant comment on the way in which wartime conditions eased earlier financial pressures.

12. Reginald was the son of A.J. Foster. The presence of one from each of the three companies perhaps implies a lack of mutual confidence.

13. *The Times*, 15 July 1919. Company no. 156,850.

14. For Berger see T.B. Berger, *A Century and a Half of the House of Berger 1760 to 1910* (1910). The firm was a paint manufacturer which had taken over a white lead works in Sheffield.

15. Details on BALM here and subsequently are taken largely from H.J. Barncastle, British Australian Lead Manufacturers Pty. History of the Company. 3 vols. (Ts., 1955 in ICI Library, Millbank, London). Barncastle was managing director of BALM from 1931 to 1951 and in 1918, when he was working for M.H. Lauchlan & Co., Australian representatives for LL & J, he produced a report to the British manufacturers on which the decision to go ahead in Australia was made. I am grateful to the ICI Librarian for permission to quote from the typescript.

16. The death of one of the old Walkers, Parker directors caused a crisis in March 1920. G.S. Pawle was put forward as a replacement by Thomson and his election forced through on Thomson's casting vote as chairman, despite a letter from de Catelin requesting that the decision be delayed until he could attend (had such a delay occurred the ALM element would have put in its own nominee and the merger might not have broken up).

17. *The Times*, 29 Jan. 1921. It seems clear that severe overcapitalisation would have occurred. The circular gave the proposed capital as 1.5m £1 ordinary shares and 950,000 8% debentures of £1 each, but that the average profit of the companies during the very profitable war years was only £200,000. Walkers, Parker shareholders were advised, on the basis of an accountant's report based on the figures, not to accept the offer for their shares. They eventually got much better terms when their company was finally sold to ALM in 1929.

18. If FB & W had been closed at the same time as the James' works in 1919 the ALM merger might have gone through successfully but instead FB & W proved to be the rock on which the merger foundered. Closure of the firm in 1919 was justified both by its history of failure and by the need to reduce capacity in the lead industry. However it is certain that A.J. Foster's importance as chairman of the various lead conventions as well as of ALM ensured that his family firm was held sacrosanct.

19. Senior partner in Mann, Judd, Gordon & Co., London accountants, he had also been involved in the Ministry of Munitions and was Controller of Munitions Contracts, 1917-19. See *Who Was Who 1951-60* and *History of the Ministry of Munitions, vol. III Finance and Contracts* (1922).

20. Enthovens, now established at Darley Dale in Derbyshire, following a move of its smelting operations from Rotherhithe, is a subsidiary of Billiton (UK) Ltd., itself owned by the Shell Transport & Trading Co. Ltd.

21. For a discussion of the development of the merger movement in the 1920s and the factors affecting merger see Hannah, *Rise of the Corporate Economy* and L. Hannah & J.A. Kay, *Concentration in Modern Industry* (1977).

22. The situation is highly confusing but it is clear that the old ALM (dominated by Enthovens) was carried on by a new ALM (dominated by Cooksons), certainly without any formal change of title or company number, using the same minute book and even taking over the earlier company's book

debts for preliminary expenses.

23. See *Who's Who* 1970.

24. Hubert Sanderson Tasker (1885-1959), joined Cooksons as a sales trainee in 1909 but transferred to management and was made general manager of the company in 1919. Appointed managing director of ALM in 1924, he acted as Clive Cookson's executive arm in the negotiations leading to the establishment of a new series of international lead manufacturing conventions in the period 1925-7, and in the takeovers of Walkers, Parker in 1929 and Goodlass, Wall & Co. Ltd in 1930. Appointed managing director of Goodlass, Wall & Lead Industries Ltd in 1930, vice-chairman in 1940, he became chairman 1947-51. He was also vice-chairman and then chairman of British Titan Products Ltd, member of the Council of the Federation of British Industry, vice-chairman of the British Non-ferrous Metals Research Association, fellow and one time president of the Institute of Metals. See the obituary notice in *Bulletin of the Institute of Metals*, 4,22 (1959), p. 169.

25. John Lawson McConnell (1892-1980), chartered accountant, served in the First World War winning the MC and Croix de Guerre. Joined Cooksons after the War and appointed company secretary 1920. In 1924 he was appointed company secretary of ALM and was the technical adviser to Cookson and Tasker on merger negotiations. Subsequently became secretary to GWLI in 1930, director 1935, joint managing director 1940 and chairman 1952-7. He had a talent for recruiting able men and considerable enthusiasm and time for encouraging them.

26. See *Who's Who*.

27. Even as late as 1918 the UK Lead Oxide Convention stated, 'The process has been proved for Orange and Red Lead, but not for White Lead, and it is doubtful whether it can be successfully employed for White Lead Manufacture.'

28. Librex was taken from the initial and second letters of the towns in which Rowe Bros & Co. Ltd had its builders' merchants and lead works Li(verpool), Br(istol) and Ex(eter). Rowe in fact also had works at Birmingham.

29. ALM, *Southern Area News Bulletin*, April 1964.

30. *Stock Exchange Gazette*, 26 March 1925.

31. These figures are drawn from the share registers which are available in Companies House, London.

32. Links such as this, not deliberately created, were quite common in the small world of lead manufacturing. C.F. Hill, for instance, after 1921 was on the boards of both Mersey and LL & J and after the 1924 merger he was on the boards of ALM and its competitor, Mersey.

33. *The Times*, commenting on the Company's annual report for 1926, noted for instance that, 'The secret of this Company's good record lies in the conservatism of its management.' Capital, debentures and loans were all smaller in 1926 than in 1913, a reserve of £100,000 had been created 'in cash or high-class securities', fixed assets had been written down from £309,000 to £252,000, while stock (which was valued at £146,000) had been taken at £10 per ton against a current price of £28, which, the newspaper commented, was 'a valuable hidden reserve'. *The Times*, 1 March 1927.

34. Although eventually sold the site has not been developed and, although partly obliterated by the reconstruction of the Flint-Holywell road, still has clear remains of the old lead flue systems on the south side of the road.

35. Pawle was, however, out on a limb. On 14 Oct. 1929 he wrote to Burr, 'Fighting against a powerful combination single handed is a hard task for a man of 74.' He went on, however, to note that his predecessor had been less fortunate, having made the indiscretion of supporting the earlier ALM and eventually sustaining as a result a loss to Walkers, Parker of almost £100,000, on which Thomson's health had broken down. Pawle then noted that in 1922 it had been

proposed that Walkers, Parker should convert its shares into £1 6s in shares of ALM but 'Since then we have been paying regular dividends averaging $11\frac{1}{8}\%$ while the A.L.M. paid 2½% last year for the first time in its history.' At the subsequent AGM of Walkers, Parker Pawle made a pathetic figure. When it came to the re-election of directors he said 'I do not suppose anyone is going to propose me. I will just put my name forward, and they can if they like, but if they do not like they needn't.' No proposal was made and Pawle then 'stated that he had a very important appointment to keep and then left the meeting'.

36. There had been losses of £7,500 and £8,500 for the years 1926 and 1927. With a little over 50 per cent of the ordinary shares of Locke, Blackett, the Leathart family had voting control of the company.

37. *The Times*, 7 Dec. 1928.

38. The actual share capital of ALM sold to GWLI was £3m, made up of the purchase prices of the various companies less considerable write-offs already made by 1930. The figures for the purchase prices were: Alexander, Fergusson £125,000, Brimsdown £163,000, T.B. Campbell £58,000, Champion Druce £94,000, Cooksons £845,000, FB & W £295,000, Harburger £46,000, Librex £350,000, Locke, Blackett £112,000, LL & J £835,000, London Lead Oxide £48,000 and Walkers, Parker £587,000: totalling £3,558,000, to which has to be added a little over £3,000 for the capitals of ALM (Engineering), British Titanium, British Waterproof Abrasives, etc.

39. In fact net profits in 1928 and 1929 were only £100,000 p.a. Goodlass, Wall shareholders were misled into believing that ALM was considerably more profitable than was in fact the case.

40. They were Cookson, Inverforth, Tasker, Foster, Mann and Rowe. *The Times* (14 Aug. 1930) noted, 'An incidental feature of the plan sufficiently rare to be worthy of notice is that no compensation is being paid to any director for loss of office.'

41. *Financial Times* and *The Times*, 14 Aug. 1930.

42. See Appendix VII for details of ALM financial statistics. The figures quoted there are drawn from published annual reports and, where these are unavailable, from company ledgers. Despite the published figure for 1929 certified net profits by Thomson McClintock were 1927 £41,118, 1928 £100,573 and 1929 £106,413. These were said to have been struck after the writing down of goodwill and base stocks and after deduction of management expenses. One obvious distinction from the published net profit figures is that the latter were struck after deduction for depreciation.

43. From 1931 GWLI became the holding company for both lead and paint sides of the business. A new ALM was incorporated as a wholly-owned subsidiary with a share capital of only £1,000, and for a few years its board met and took decisions with regard to the lead side (which were then taken to the GWLI board for approval). From 1934, however, ALM became moribund and decisions were taken by the lead committee of GWLI and forwarded to that company's board. In order to avoid confusion between the holding company and the lead side, ALM will be used to describe the lead side of the merger up to 1949.

10 THE LEAD MANUFACTURING INDUSTRY IN THE INTER-WAR YEARS

This is the Hour of Lead —
Remembered, if outlived,
As Freezing persons, recollect the Snow —
First — Chill — then Stupor — then the letting go —
      Emily Dickinson (1830-86), Poem 341,
      'After great pain'

The lead manufacturing industry in the inter-war years; price fluctuations, lead supply, output of particular products, overseas markets, the re-establishment of both home and international conventions in white, red and blue lead products, and the problems faced in working them. The development of publicity organisations to promote the sales of lead products. The development of ALM within the broad experience of the lead manufacturing industry up to 1949: centralisation of policy making, economy measures, technical organisation, research and development; rationalisation with works closures and reallocation of manufacture of particular products between plants; commercial organisation remains largely in the hands of each company and is not rationalised until 1949; diversification and the development of abrasives and titanium products; profitability, welfare policy, wages and salaries and pensions; the impact of the Second World War, rising profitability and post-war expansion.

The economic background to the period in which the merger took place was a complex one and there were special considerations relating to the position of several of the main lead products. The aim of this chapter is to explore these factors and assess the way in which the merger developed in its first quarter century of existence.

The lead manufacturing industry was seriously affected by the general economic pattern of the period. A short, sharp boom, 1918-20, in which demand rose rapidly as attempts were made to overcome war-time shortages, was followed by a sharp depression in the early 1920s and by another steep depression in the early 1930s after a period of relatively slow growth in the later 1920s. From 1933 the economy began to recover and the housebuilding boom of the mid-1930s was especially beneficial to the lead manufacturers, while the gradual build-up towards and the Second World War itself offered (as in 1914-18) continued profitable opportunities.

Among specific factors affecting lead manufacturing were fluctuations in the price of pig lead, which, as in the past, continued to have a major impact on profitability as a result of the principle of selling manufactured goods at a mark-up on the price of raw materials; although the significance of this problem was reduced by the growing practice of setting aside a profit reserve in good years to cope with the effects of serious price falls. Having been controlled by the Ministry of Munitions at £30 per ton for most of the war period, pig lead fluctuated violently in price when it was decontrolled — from £40 at the beginning of January 1919 to £22 10s at the end of June and £53 in February 1920. In such circumstances the timing of purchases was crucial and it is not surprising that there was considerable variation in the companies' profit performances in this period. With the depression of the early 1920s the price fell steeply to a little over £20, recovered to £30 by 1924 and then began a prolonged and almost unbroken decline to £11 in 1934 (little above the low levels of the 1890s). The upturn of the mid-1930s took the price above £20 by 1937 and for the whole of the War it was once again controlled, on this occasion by the Ministry of Supply, at £25 per ton, while after the War the Ministry allowed the price to rise by a series of steps to equate with a free market price which exceeded £100 in the late 1940s.

There was no ideal solution to the problem of price fluctuation, given the delay between purchase of raw material and sale as a manufactured product (even to sell at a mark-up on the price of pig lead at the time of the sale of the manufactured product would only have shifted the relative positions of profits and losses along the time schedule of price fluctuations). ALM recognised this and in 1927, after discussing the impact of recent price falls in pig lead the board resolved on greater care to match purchases to sales in order to avoid stock fluctuations and the facilities offered by the London Metal Exchange were used to that end. Secondly it was decided 'that every possible effort should be made to ensure changes of prices of lead manufactures at more frequent intervals on a rising Lead Market'.

After an hiatus during the First World War, the supply of lead was quite free during the inter-war period, although there were once again shortages in the 1939-45 War. The German cartel was of only limited significance after the War and ALM, which had links with the Australian lead producers through BALM, had no difficulty in obtaining its supplies of raw material. World supply of lead was expanding, fortunately since widespread industrialisation was increasing the level of demand, while new technology, such as the motor vehicle, was creating additional demands for the metal. In addition to the use of newly mined metal, this demand was increasingly being met by the use of secondary lead, by the remelting and refining of scrap lead. With a shortage of primary metal during the War, major attempts were made to find and reuse scrap metal and these were continued after the War, fostered by the growing supply from used vehicle batteries and the demolition of buildings. There are no UK figures available for the use of secondary lead and the consumption figures given in Table 10.1 are based on primary metal. They do, however, show that over the whole period the consumption of manufactured lead increased considerably, from less than 200,000 tons p.a. before 1914 to well over 300,000 tons in the later 1930s, although with output slumps in the early 1920s and early 1930s.

Table 10.1: UK Consumption of Primary Lead, 1920-38 (metric tons)[1]

1920	162,000	1925	257,300	1930	260,000	1935	332,400
1921	131,300	1926	254,600	1931	260,000	1936	351,400
1922	156,800	1927	279,200	1932	240,000	1937	340,200
1923	193,200	1928	245,000	1933	274,000	1938	347,000
1924	224,100	1929	274,300	1934	333,800		

Home supplies of primary lead continued their pre-war decline, accounting for only about 10,000 tons of pig lead p.a. in the early 1920s but recovering to a peak of over 50,000 tons in 1934 in the heyday of the Mill Close Mine. Australia was the dominant overseas supplier, while the USA, Canada and India became more significant than the traditional source, Spain.

While total output expanded, the proportions contributed by individual lead products changed. In particular white lead, which had always made up a considerable proportion of total consumption of pig lead,[2] went into stagnation and then decline. The Mersey annual report for 1932 noted, 'The total tonnage of white lead consumed continued to fall and for the year 1932 was 27,421 tons compared with the basis year [1927] 43,940 tons.' Although there was a recovery to 38,000 tons in 1935 during the housebuilding boom, consumption fell to 34,000 tons in 1938, collapsed during the war years and only averaged about 20,000 tons p.a. in the late 1940s. To some extent bad publicity over lead poisoning had turned consumers towards alternative paint bases such as zinc oxide but of much greater long-term significance was the development of titanium dioxide as a paint base and its entry into the market in branded paints such as Dulux. These had great advantages in ease of spreading and a higher gloss finish than did white lead based paints. Similarly consumption of tea lead declined, being replaced by the cheaper aluminium foil and although the housebuilding boom of the mid-1930s in particular and construction in general continued to place demand for lead sheet and pipe, there were the beginnings of competition from other materials and especially from copper for piping. Serious impact was not, however, to occur in this field until after 1945 and in the mid-1930s the output of sheet and pipe was almost back to its all-time high levels of the early 1900s.

Table 10.2: UK Lead Manufacturers Association — Annual Deliveries of Lead Sheet and Pipe, 1919-48 (tons)

	Sheet	Pipe	Total		Sheet	Pipe	Total
1919			54,920	1934	59,964	79,358	139,322
1920			72,029	1935	65,973	81,507	147,480
1921			66,257	1936	68,927	81,271	150,198
1922	35,375	33,425	68,800	1937	64,473	73,233	137,706
1923	41,205	38,096	79,301	1938	62,939	72,311	135,250
1924	40,480	43,328	83,808	1939	63,534	65,434	128,968
1925	41,295	47,525	88,820	1940	33,953	34,883	68,836
1926	41,286	46,726	88,012	1941	30,729	31,469	62,198
1927	43,818	48,185	92,003	1942	15,566	20,243	35,809
1928	44,774	43,687	88,461	1943	9,437	11,993	21,430
1929	45,218	49,202	94,420	1944	11,975	15,801	27,776
(1929)[a]	(49,008)	(57,289)	(106,297)	1945	30,716	34,077	64,793
1930	54,337	56,126	110,463	1946	33,042	48,265	81,307
1931	47,239	58,524	105,763	1947	35,028	47,060	82,088
1932	44,919	58,485	103,404	1948	35,887	42,004	77,891
1933	55,403	73,647	129,050				

Note: a. In 1929 the Association recruited a number of firms which had previously stood outside the trade association. The figures in brackets are for all firms (old and new) for 1929 and they show that the old association accounted for almost 90 per cent of UK output (only a very few small firms remained outside the association after 1929).

The influence of the building cycle is obvious. Sheet and pipe deliveries held fairly firm during both depressions in the early 1920s and early 1930s and then rose rapidly to the mid-1930s. The stability of these blue lead products in depression years, when many capital goods industries experienced severe declines in demand, was a major factor in helping the lead manufacturing firms to weather these periods with relative ease.

While several traditional markets for lead were under threat others, chiefly related to electricity, began to open up. Firstly came the sheathing of electric cables, which accounted for 76,000 tons of lead in 1945 (25 per cent of total UK consumption of lead) and 120,000 tons in 1948 (more than a third of UK consumption). Although there are no figures available for consumption of lead in the output of cables before 1945, there can be no doubt that the significant increase came in the 1930s with the development of the National Grid and it is likely that consumption in the late 1930s was at a similar level to that of 1945. The technology of cable manufacture was, however, electrical, an area in which the lead manufacturers had no expertise and, although they did a big trade in supplying lead alloys to the cable manufacturers, they did not produce cables themselves. A second development was the growing importance of batteries which required both blue lead for the grid and lead oxide for paste. Again there are no consumption figures for such purposes before the Second World War but in 1945 batteries took 50,000 tons of lead for grids alone. While the grids were manufactured by the battery manufacturers themselves from bought in pig lead, they tended to purchase their supplies of lead oxides from the lead manufacturers, at least in the early days of the industry. Growth of battery manufacture therefore provided a useful boost to demand for lead. In the early 1920s the London Lead Oxide Co. Ltd, for instance, had a weekly order for 20-30 tons of red lead to be sent to the Lucas Battery Co. at Birmingham.[3]

The international climate was unfavourable to an expansion of export sales in the inter-war period and the chief problem was the protection of those overseas markets which existed. Some of the measures taken to achieve this end inevitably meant a decline in the level of UK exports. The setting up of BALM to produce and market white lead in Australia was one factor in this decline and the forces which produced BALM, the rise of tariff barriers in overseas markets, led to other similar steps. Rising Indian tariffs led ALM into negotiations with an Indian firm, D. Waldie & Co., in 1928, although it was not until 1934 that ALM and Waldie agreed to form a joint company, after a visit to India by H.S. Tasker, which 'had convinced him that the Company's trade in White Lead and Lead Oxides in that country would eventually disappear were arrangements not made to manufacture these products on the spot'.[4] A factory to manufacture oxides was erected at Konnagar on the Hooghley River, about ten miles from Calcutta, and ALM had a majority holding in the new firm, known as D. Waldie & Co. (Lead Oxides) Ltd. ALM also took a majority holding in the Bangalore White Lead Syndicate Ltd, closed the old works of the company and erected new plant at Konnagar, where lead pipe and zinc oxide production were added shortly afterwards.

There were also steps into Europe, undertaken less to secure an existing market than to provide a negotiating counter towards the formation of a new international cartel. In the mid-1920s ALM officials were in negotiation both with Harburger Chemische Werke Schoen & Co. and Bleifarbwerke Co. of Wilhelmsburg, majority control of the former company being finally obtained in 1927. The purchase of the Harburg company was to prove of considerable value to ALM. It became quite profitable and in 1929 plant extension was undertaken in order to enable Harburg to take over the output of another German firm, Renninger. Surplus funds continued to be invested in German industrial companies and in 1936 ALM added white lead to its existing production of lead oxides in Germany with the purchase of Westdeutsche Bleifarbenwerke Dr Kalkow AG. This company had been in existence since 1908 with works at Offenbach, near Frankfurt, to manufacture white lead by the Kalkow process. The firm was offered to ALM by its owners, two German Jews, F. and S. Cahn, who wished to leave Germany because of the rise of Hitler and who subsequently came to play a significant role in the management of ALM.[5]

In addition to the establishment of overseas production, the network of foreign agencies was extended and did something to protect sales but total exports declined. This was especially true for white lead, whose exports halved from 12,700 tons in 1927 to only 6,000 tons in 1935, although there were also smaller declines in exports of lead sheet and oxides. By the mid-1930s total lead exports were about 25,000 tons p.a. compared with 45,000-50,000 tons in the years before 1914 and after 1945 exports became insignificant. While the UK performance in overseas markets was deteriorating, it is important to note that German exports were expanding rapidly. The German surplus of exports over imports (the latter hardly ever exceeded 5,000 tons p.a.) of manufactured lead goods had risen from 25,000-30,000 tons p.a. in the 1890s to 45,000-50,000 tons in the years before 1914. The War brought a collapse but by 1922 the German export surplus had reached 57,000 tons and although the figure fell as a result of the internal disturbances of the next two years, a surplus of 70,000 tons was exceeded in 1926 and 80,000 tons in 1930. Thereafter the depression and rising tariffs reduced the figure considerably, although by 1936 the German export surplus once again exceeded 50,000 tons. Chiefly as a result of this evidence the British manufacturers made renewed attempts to reduce competition by the resurrection of the international cartels which had collapsed with the onset of the First World War.

That the international conventions had not been re-established immediately after the War was probably a result of the facts that in the post-war boom demand was rising, the full extent of German competition was not felt until after 1924 and the major British manufacturers were heavily involved with internal reorganisation. In addition, with the exception of 1920 when imports exceeded exports by 2,500 tons, the position for white lead, the major lead commodity in international trade, was not unfavourable to British manufacturers. In most years in the early 1920s exports of the commodity exceeded imports by several thousand tons as they had done before 1914 but in 1925 and subsequent years there were deficits rising from 3,000 to 6,000 tons p.a. and this change sparked off a determination among the British manufacturers to re-establish order in international markets. Inevitably it was to an agreement with the German manufacturers that the British firms looked as the key to an international agreement, although American exports of white lead to the UK actually exceeded the German level (but since they were largely controlled by one firm, National Lead, negotiations with the Americans were less difficult).

The precise initiative for renewed international conventions may well have come from the reorganisation of the UK conventions. These had continued to function during and after the War but by the mid-1920s they had become beset with a number of problems, the chief of which was the rise of new producers. In white lead the development of the quick process by Rowe Bros and its takeover in 1924 by ALM meant difficulties for the White Lead Convention, while the Lead Oxide Convention was troubled by the rise of the London Lead Oxide Company's output of red lead from around 15 tons a week to a claimed peak of 150 after the introduction of a continuous furnace. At the end of September 1924 Mersey withdrew from the White Lead Convention over the question of an increase in quota for Rowe Bros and, although prices were maintained mutually by the firms, the Convention virtually ceased to function. An interim committee was set up, dominated by ALM members, to draft a new convention agreement but its report was unacceptable to Mersey, which, with white lead as its only product, must have felt that it was being squeezed between the competing giants, ALM and Walkers, Parker. Finally in March 1926 a new scheme was drawn up by Sir William McLintock, partner in Thomson, McLintock, accountants and secretaries to the conventions, which in principle satisfied Mersey, although the company was to have a slightly smaller quota of the UK market. Ultimately it seems that Mersey recognised that it did not have the power to fight against the convention proposals, but it did receive a special inducement.

Mersey had previously been approached in 1924 to join the ALM merger but had refused for two reasons which, with the advantage of hindsight, seem remarkable prescient, '(1) Sentiment, and a doubt of the Merger especially in view of the extraordinary price to be paid to Rowe. (2) Merger paper not acceptable.'[6] The new difficulty in getting Mersey to assent to terms for the White Lead Convention therefore led ALM to decide that it had to come to some accommodation with the firm. Mersey was offered and accepted a pooling agreement, based on the sharing of profits according to a fixed division of sales between the two companies,[7] which was to be kept secret from the remaining members of the White Lead Convention. In 1936 the White Lead Marketing Co. Ltd was set up to take over the pooling agreement, which was extended for a further term. It adopted a more positive role than its predecessor and rationalised sales areas. Mersey's sales to the London and Hull areas were supplied by ALM firms from, respectively, Millwall and Hayhole, while Walkers, Parker's Chester stacks were closed and ALM's trade to the Potteries was supplied by Mersey. Under the new agreement GWLI and Mersey exchanged a director, McConnell joining the Mersey board and C.F. Hill that of GWLI and relations between the two companies became close. Mersey was already paying 12½ per cent of the cost of ALM's research activities in white lead and from 1936 onwards McConnell provided Mersey with advice on such matters as lead supplies, the appointment of salesmen and pension schemes.

While the offer of the pooling agreement was sufficient encouragement to Mersey to join the White Lead Convention, there were other problems to be overcome before that body could be formed. Of particular significance was the relationship between ALM and Walkers, Parker, which was strained because of ALM's takeover ambitions. Walkers, Parker, like Mersey was unhappy about the claim for increased quota for Rowe Bros, suspecting that ALM was transferring trade from its other companies to Rowe because of the latter's lower production costs. While ALM inevitably claimed that this was untrue there is no doubt that the lower costs[8] at Bootle would have made such a rationalisation tempting. Walkers, Parker finally agreed to co-operate with ALM over the formation of the convention on the ground that it was necessary for the protection of both companies but suspicion abounded and an ALM minute on one occasion commented on the unhelpful attitude of Walkers, Parker, 'culminating in an outbreak by Mr. Eckford [which] made further negotiations between the Company and Walkers, Parker & Co. Ltd's present representatives impracticable'. The relationship between the two firms on convention matters must have made it even more attractive to ALM to get Walkers, Parker into the merger, since the latter was too large a firm either to ignore or ride over roughshod.

From January 1927 the UK White Lead Convention was officially reformed and attention now turned, with a secure home base, to the formation of an international cartel. This had become necessary because of the rising level of imports attracted by the fact that UK prices were being held above world levels.[9] After protracted negotiations[10] the eventual agreement, made in December 1927 and to run for six years from 1 January 1928, gave the German section 8.95 per cent of the UK and British colonial markets, the American section 10.46 per cent and the British 80.59 per cent. As a result of the agreement import prices of white lead to the UK were raised and imports, which had risen from 6,000 tons in 1922 were stabilised at between 11,000 and 12,000 tons between 1927 and 1931. Giving both foreign and home producers the opportunity to stabilise their sales at higher prices the Convention appears to have been mutually satisfactory and it was to be renewed in 1933 for a further term.

The formation of UK and international conventions was clearly beneficial to a more orderly trade but there were still problems caused by non-signatories. When the Walkers, Parker board was informed in December 1927 of the finalisation of the international convention it was told 'that Champion, Druce & Coy . . . would still have to be dealt with specially'. Like a number of other small firms, Champion, Druce, which had the Islington white lead works originally begun by Walkers, Parker, was affiliated to the Convention, not having a quota but maintaining prices and paying a share of expenses. The firm had, however, built up a significant market share in Australia and was cutting prices in order to compete in that country with BALM and Lewis Berger. Moreover, Champion, Druce was even considering the erection of manufacturing plant in Australia. Such activities threatened the newly-established international convention and had to be contained. Clive Cookson took up negotiations with C.D. Druce and after considerable haggling the latter agreed in June 1930 to sell his firm for £303,000, an extortionate price including a huge element of goodwill. Druce sold at the perfect time just as the depression began. The company made a net loss of £1,500 in 1930, only exceeded a net profit of £10,000 from 1934 onwards and no dividend was paid throughout the 1930s.

The sale of Druce's partnership was made to a limited company of the same name, incorporated on 5 July 1930, in which the members of the UK White Lead Convention took shares, largely according to their quotas. As a result ALM was the largest shareholder (with 47,232 of the 78,628 shares issued) and tended to dominate Champion, Druce's policy, chiefly through Tasker, who was appointed chairman. The firm's prices, both at home and abroad, were brought into line with those laid down by the Convention and output was to be reduced to the 'lowest economic' level. Negotiations were entered into in order to sustain viable exports of Champion, Druce white lead to Australia. The Australian producers (anomalously in the case of BALM which was ultimately controlled by the UK manufacturers) were offered the choice of buying the goodwill of Champion, Druce or setting up a pooling agreement, backed up by the threat of the establishment of manufacturing plant in Australia. That it was threat rather than promise is surely suggested by the fact that ALM was the largest single shareholder in both BALM and Champion, Druce and it is inconceivable that Clive Cookson would have allowed such a ridiculous position to occur. In any event the possibility was never tested since the Australian producers agreed both to manufacture white lead for sale under the Champion, Druce brand (in order to overcome the problem of import duty) and to the establishment of a pooling agreement. In these circumstances it is surprising that the Islington white lead works was not closed and its sales made up by taking supplies from Convention members who certainly had surplus capacity. As an alternative a scheme of capital investment was instituted in order to reduce production costs and by the end of 1935 the Islington white lead works was possibly the most modern stack works in the country with every stack serviced by overhead electric conveyor.[11]

Apart from having to deal with Champion, Druce the Convention also faced the problem of an American renegade, the Eagle, Picher Co. Late in 1930 this company was selling white lead through its UK subsidiary, Harrison & Clark, although it did not have a Convention quota. Despite its considerable power, the National Lead Co. of America showed little interest in putting pressure on Eagle, Picher to cease this activity and its general manager, Evans McCarty, reacted angrily to a suggestion that the British section of the Convention be allowed to export to America in order to attack Eagle, Picher's trade there. The problem must have been solved, presumably negatively as a result of the isolationism which the depression brought to the USA, since it disappeared from discussion by the British firms early in 1931. The Conventions, both UK and international, experienced numerous other problems but they were all solved and eventual harmony achieved. Quotas were adjusted to satisfy some companies, negotiations brought some outsiders in while others, such as Terrell's Paint and Varnish Co. Ltd, were bought out in order to eliminate their competition. Through all this the role of the English members of the management committee of the International Convention (and especially Foster, its chairman, Cookson, Tasker and Reynolds) was considerable. They were frequently used to settle disputes between continental manufacturers and their position was clearly a powerful and respected one.

If the introduction of the International Convention controlled imports of white lead to the UK, then the imposition of a 20 per cent tariff on white lead in March 1932 helped to ensure that the level of imports, already declining because of the depression, was maintained at a low level.[12] After 1932 the peak level of white lead imports was 5,700 tons in both 1935 and 1936 compared with 11,000-12,000 tons p.a. between 1927 and 1931. Indeed C.F. Hill was so confident of the new position when the International Convention came up for renewal late in 1933, that he proposed that the new convention should not include a quota in the UK market for foreign producers. The ALM board considered this proposal before its representatives attended the convention negotiations but decided to reject it, 'In view of the fact that the break-up of the International White Lead Convention might involve a diminution in profit of £100,000 p.a.'. Price competition to ensure that imports were eliminated would, it was felt, invite foreign manufacturers to establish UK plants and this was too great a risk for the potential gain of an extra sale of 5,000 tons. The International Convention was therefore renewed with a lower UK quota for foreign producers in order to reflect changed circumstances.

At home the convention strengthened its position by a number of steps, chief among which was the introduction in March 1932 of the Covenanted Distributors' Agreement. It was described as 'a scheme whereby paint grinders and merchants receive a rebate on their purchases of White Lead, providing they agree to maintain Convention prices on resale and buy only Convention White Lead'. While it had some success in restricting outside competition to the convention, a number of independent white lead companies was formed in an attempt to benefit from the higher prices brought by the tariff and the convention. In most instances these firms had difficulty in establishing a market and were brought under the control of the convention. The only problem of significance was Novadel Ltd, a company set up in 1938 with works at Gillingham by the Dutch white lead manufacturers,

Van der Lande, in order to manufacture within the British tariff barriers. Its output was never large, always less than that of Mersey, but its price competition did lead some firms to resign from the Covenanted Distributors' Agreement.

In both home and international spheres it was the White Lead Conventions which were of greatest significance. White lead was historically the most significant lead product in international trade and it was a mainstay for all the large lead manufacturing companies. Negotiations for cartels therefore came first in white lead and once these had been provisionally agreed the participants (together with a few specialists in the product) turned their attention to red lead, which was primarily the product about which the Lead Oxide Conventions were concerned. From its inception in 1914 it may be inferred from the lack of information on the subject that the UK Lead Oxide Convention worked smoothly once the problems relating to the Barton and Eckford patents were solved. Chiefly this may be ascribed to the fact that there were no outsiders of any significance and one of the few pieces of evidence of the Convention's existence comes in a reference in November 1925 to the fact that one outsider, Henry Wiggin & Co. Ltd, was giving up its red lead production in return for compensation from the Convention. While this may have been cause for minor satisfaction, the growth of output from the London Lead Oxide Co. Ltd (LLO) brought serious problems both at home and abroad, since that firm was endeavouring to sell part of its increasing output on the Continent. In October 1926 a Walkers, Parker minute noted that 'it had been agreed [between that company and ALM, who between them dominated the Convention] that, in view of the serious disturbance of the trade caused by the severe competition of the London Lead Oxide Company, it was imperative that an active selling campaign be at once instituted, in order if possible to bring that Company into line'. With low cost production from its new continuous furnace LLO was not easily to be beaten by price cutting and by the spring of 1927 the Convention had decided that LLO would have to be purchased and then 'sold by the Convention to the Associated Lead Manufacturers Ltd., to be carried on by them on the basis of a quota to yield them 10 per cent or 15 per cent on their investment, the balance to be borne by the Convention as a whole'.

With the competition of LLO removed it was possible to redevelop an international convention, to which thoughts in the UK were already turning, since imports of red lead were rising and the white lead negotiations were under way internationally. It took surprisingly long to reach an international agreement, which was not signed until May 1929, chiefly as a result of the inability of continental producers to reach agreement among themselves. With the home market largely tied up between two companies the British negotiated from a position of strength. Late in 1928 they were able to sell at what were described euphemistically as 'keen competitive' ('non-remunerative' having been crossed out) prices in the Belgian market, in order to aid the potential French members of the proposed convention whose home market had been attacked by the Belgians. Even after a convention agreement was eventually reached it was continually threatened by disagreements among continental producers, although the British market was unaffected.[13] The UK convention continued to operate smoothly. In 1931, with the closure of its Morledge works, Cox Bros of Derby gave up its red lead production and in future drew supplies from other manufacturers in the convention – a piece of rationalisation which offered the opportunity for fuller utilisation of plant elsewhere. As with white lead, however, higher prices and profit margins attracted outsiders. In the case of red lead the chief of these was the International Paint & Compositions Co. Ltd which had begun to manufacture (through a subsidiary, Paint & Pottery Chemicals Ltd) in order to supply its own raw material requirements but which had clearly then decided to sell red lead to other firms. In February 1933 an ALM minute noted that it had become necessary to purchase the red lead interests of International Paint for a sum of £24,000 since 'the loss of profit involved by open competition with this firm would reach this sum within 3 or 4 months under International Convention conditions'.

The third area in which cartelisation occurred was sheet and pipe, although, in view of the limited international trade in these products before the early 1920s and the wide diversity of producers, this was largely confined to the UK for much of the period. In October 1922 a Walkers, Parker minute noted that in 1919

in order to dispose of the Belgian competition the U.K.L.M.A. had invested approximately £40,000 of its Defence Fund in the purchase of [D'Autricourt's] works at Bruges, the intention being that by cutting tactics the other Belgian Manufacturers would be forced to respect United Kingdom prices by the formation of an Association to work with the English makers, but although some three years had elapsed this had not been achieved and there was great dissatisfaction in consequence.

However by 1925 a Belgian association had been formed, including the UKLMA works and Belgian sales to the UK controlled. But a new problem had appeared with increasing sales in the UK by the French Penarroya company and late in 1925 UKLMA made a second overseas purchase, taking a controlling interest in Pelgrims & Bambeeck at Arras for £13,600, with the specific aim of using its output to attack the French market. This led to an agreement with the Belgian and major French producers to limit their exports to the UK market and increase prices. They were given a joint quota of 8,000 tons (about 10 per cent of the total UK market), although this figure was never achieved as a result of a price increase agreed by the convention and compensation of £2 per ton was paid for the shortfall. Overall the continental activities of UKLMA appear to have paid off. In direct financial terms the Bruges works made profits which were more or less cancelled out in the late 1920s by the losses made by the Arras works, whose sales were largely made in France where price competition was severe. More generally, however, UKLMA's activities in Belgium and France had controlled exports from those countries to the UK, although in the later 1920s competition from Holland and Germany became severe.

With both international white and red lead conventions formed it became possible in 1929 for the major British firms to press for an extension of the sheet and pipe agreement with Belgium and France to cover other countries. While it is probable that their colleagues in the white and red lead conventions sympathised with this position it seems that the problem lay in the multiplicity of small firms on the Continent which were not members of their national conventions. There were particular difficulties with the German producers and in February 1930 ALM decided that it should 'be pointed out clearly to the German representatives when in London next week that it was the serious intention of ALM to commence the manufacture of Lead Pipe in Germany failing a satisfactory and prompt outcome of quota negotiations'. This appears to have had no effect, for in June 1930 ALM shipped a spare pipe press to its Harburg red lead works and it was noted shortly afterwards that the Germans had considerably reduced their quota demands. Nevertheless it was not until November 1931, after ALM had eventually erected the pipe press at Harburg in order to speed the otherwise tardy negotiations, that an international convention was signed.

That the International Sheet and Pipe Convention was almost entirely concerned with control of the British market, with no discussion of continental markets (where considerable competition continued) is a reflection of the internal control which the UKLMA possessed, backed up by the bargaining power of ALM's size and its continental production. The quota for continental exports to the UK was agreed at 11,753 tons, against British imports which had been 14,000 tons in 1930 and were to total more than 20,000 tons in 1931. A reduction of this order clearly justified the considerable British effort which had gone into the organisation of an international convention, while the immediate return was an increase in the margin at which sheet and pipe were sold over pig lead from £8 to £9 10s per ton, since part of the agreement was that foreign producers should not undercut British prices. Although the price rise must have made it difficult for foreign producers to sell on the British market, it is difficult to believe that the decline in imports, to only 5,500 tons in 1932 and 3,400 in 1934, was due largely to the introduction of the International Convention and it seems more likely to have resulted from the introduction of the British import tariff. Indeed when the International Convention came up for renewal in June 1934 the British group resigned and submitted new terms in the light of the changed trading conditions. To these terms the British representatives stuck, despite an initial refusal by the Belgians to accept them, which, at a time when UK demand for sheet and pipe was rising rapidly, suggests that the British were confident that their home market was now largely immune from overseas pressures. The import figures for the later 1930s, which show that sheet and pipe imports had fallen below 3,000 tons p.a. by 1938, show this confidence to have been justified.

Ultimate control over imports had only come, however, as a result of the UKLMA's internal position. There had always been a number of outsiders, especially in pipe production, small firms with low overheads which undercut the prices set by the UKLMA. There was continual concern in the Association about the activities of such firms as R.E. Roberts & Son Ltd of Bolton, which was described in 1922 as 'a very troublesome outsider'. During the 1920s, however, the UKLMA began to make efforts to reduce outside competition by buying up and closing down firms which came on to the market. There were also attempts to persuade outsiders to join the Association, although this often caused difficulty with the rearrangement of quotas. In 1922, for instance, Baxendale & Co. Ltd of Manchester and the Leicester Lead Works Ltd joined the Association but the major change came in 1929 when the UKLMA was renewed after long negotiation and some price-cutting aimed at influencing outside firms. At least ten new firms, some of which, such as Roberts, had been serious problems for some time, were brought into the Association in 1929 (increasing the total UKLMA turnover by about 12½ per cent) and several more were added to its membership in the following year. To some extent this campaign was a result of the contemporaneous moves towards an international agreement, since it was obviously recognised that the elimination of internal competition would increase bargaining power over imports. The two sets of negotiations were, however, reinforcing. Many small British pipe producers obtained their supplies of lead sheet cheaply from the Continent but the international agreement of 1931 included a clause preventing foreign manufacturers from supplying non-members of the UKLMA. With their raw material supplies cut off, a further nine firms joined the UKLMA in December 1931, by which time the Association could note confidently that 'There were still a few firms outside, but negotiations were on foot with them.'

It seemed that the almost impossible, the cartelisation of an industry composed of many small units, might be achieved but the reduction of competition attracted new firms, such as British Lead Mills Ltd, and some of the existing firms continued outside the Association. Of these perhaps the most significant was H. Polan & Co. Ltd[14] of Sheffield, which responded to the loss of its supply of sheet from Belgium, initially by buying sheet lead through the agency of a friend (to avoid recognition by the Association) and then in 1933 by erecting its own sheet mill, from which it no doubt supplied other pipe manufacturers which were outside the Association. Nevertheless the UKLMA did manage to restrain competition and increase profit margins and its success owed much to the large manufacturers (especially to ALM but also to those outside the merger such as Sheldon, Bush and Rowe Bros) which were chiefly responsible for the running of the regional associations.

Apart from some success in achieving their regularly reiterated aims of obtaining profitable price levels and distributing the available trade among member firms according to their quotas, a by-product of the Conventions was growing co-operation generally among the lead manufacturers. Initially, perhaps, this was forced by concern over foreign competition and evidence of stagnating markets but generally it came to be seen that co-operation was mutually beneficial. This is not to say that relations between Convention members were always smooth — there were many occasions when accusations of price-cutting led to dissension and in the late 1930s one firm was expelled from UKLMA membership — but there was increasing willingness to pull together for the sake of the industry. The Conventions themselves, with their fighting, defence and publicity funds, inevitably enforced co-operation on their members but the willingness to co-operate was seen in the setting up of separate organisations chiefly for publicity purposes. In September 1926 the UKLMA formed a company, the British Lead Manufacturers Association Ltd (BLMA), to develop and handle the licensing of trade marks and to adopt schemes for promoting lead products. Its membership included all the firms which were in the UKLMA and their subscriptions covered the company's expenditure. The BLMA was responsible for an advertising campaign (chiefly in journals of the building trade) which cost about £4,000 in 1928, while it also helped to ensure the continuation of training in lead craft skills. In October 1928 Hugh Davies, Staff Inspector of Technical Schools, told A.J. Foster that 'the plumbing classes had been marked down for suppression by the Geddes Committee and had only been saved by the fact that the U.K.L.M.A. were giving the free use of the lead'. By the mid-1930s, however, the large companies felt that insufficient was being done in the way of lead promotion and the GWLI board resolved 'that trade co-operation, with the object of improving the quality of Sheet Lead and Lead Pipe was in principle desirable, particularly bearing in mind the increased competition from Copper Pipes'. While this was a view broadly echoed by the UKLMA it was less attractive to some of its smaller members who by the end of the 1930s were making a very comfortable living from the cheap production of pipe from remelted secondary lead as against the policy of the large firms of refining the lead in order to improve its quality.

Nevertheless the view of the large firms prevailed, with the formation at the end of 1935 of the Lead Industries Development Council, significantly with Tasker as chairman. The Council was intended to merge 'the various existing propaganda and development sections of the lead trade', although it ventured into new areas with the setting up of a Technical Information Bureau on lead which was available to builders, architects and technical education institutes as well as member firms. To a considerable extent the Council was devoted to the white lead section of the trade, which, being oligopolistic, was easy to organise and had a long history (through such organisations as the White Lead Corroders' section of the London Chamber of Commerce) of defending and advertising itself. The calculation of subscriptions to the Council shows the importance of white lead. In 1935 Tasker estimated that sheet and pipe accounted for 70 per cent of lead consumption covered by the three conventions, white lead 18 per cent and oxides 12 per cent, while total yearly propaganda expenditure was 16 per cent on sheet and pipe, 80 per cent on white lead and only 4 per cent on oxides.

The two sets of figures were then blended (giving a weight of only 50 per cent to expenditure) in order to produce the following figures for contributions to the Council's expenditure: sheet and pipe 52 per cent, white lead 39 per cent and oxides 9 per cent. White lead was in effect being subsidised by the blue lead manufacturers (or, in other words, ALM was being subsidised by the small pipe makers). The white lead section was the only one which had non-technical publicity, with the issue of posters, blotters, etc. Poster sites were taken on railway stations and road hoardings, newspaper advertisements were placed, calendars were given to firms of decorators, a film advertising white lead was made and lectures were given. In particular the new, hard-gloss, white-lead-based paints were pushed in an attempt to stem the advance of titanium-based paints. It is not possible to give figures for the Council's expenditure but it certainly ran into tens of thousands of pounds annually, since with the onset of war and a reduction in the supply of lead it was decided that the Council's budget for 1940 should be £21,000 less than in 1939.

It can be seen that the broad picture of the lead manufacturing industry in the inter-war period was one of increasing competition and the ways in which that competition was met and reduced. In this respect the developments of amalgamation through ALM and cartelisation through the various conventions went hand in hand. By the early 1930s to all intents and purposes ALM was synonymous with two of the conventions, those for oxides and white lead.[15] In these areas of few firms the reduction of competition through merger was possible. In the third area, the production of blue lead, it is noticeable that ALM made no attempt to rationalise the industry apart from the influence which its representatives had on UKLMA and the publicity bodies. Here, perhaps, lay the distinction between ALM and the conventions. The latter were largely negative organisations controlling competition by limiting the output of existing firms, while the former had the opportunity to go further and adopt positive schemes of rationalisation. Because of the significance of ALM within the lead industry as a whole it is difficult to distinguish their respective roles and we have already seen that ALM played a directive part in overall development. But having looked at the general pattern of change in the industry as a whole it is well to turn to an investigation of the way in which ALM dealt with the specific problems and opportunities brought by the merger.

From the first it is clear that there was to be no major attempt to rationalise into a single organisation the firms which were taken over.

ALM was a holding company which would direct overall policy and receive the net profits of its constituents. Initially its board comprised representatives from each of the constituents, although from the reorganisation of 1930 the GWLI board was smaller and more high-powered. In most instances the constituent companies retained their own boards, although with the addition of a member (or members) of the main board, frequently from the Cookson group. Moreover it was made clear that company boards would implement the broad policy decisions of the main board and when this was not done retribution followed swiftly, as with the removal of the Rowe family directors from control over the Bootle works in 1925. Gradually the powers of the individual boards declined and by the mid-1930s were insignificant. At Walkers, Parker, for instance, the board met only annually and in between meetings decisions were taken by McConnell (who was not a member of the board) and the company secretary.

From the initiation of the merger in 1924, considerable control over individual firms was exercised from the top. In most instances Thomson, McLintock replaced the existing auditors to the individual companies, as well as being appointed auditors to ALM and this enabled standardisation of accounting procedures which facilitated financial comparisons of the various companies' performances. A management committee of ALM (dominated by the Cookson interests) was set up in 1924 to control all capital expenditure, application for which had to be made on standardised forms with a justification for the expenditure and this came down to small items costing only a few hundred pounds. With finance continually stretched by existing overcapitalisation and new takeovers, careful control of expenditure was inevitable and economy measures were enforced. An early example was the issue in March 1925 of a memorandum headed 'Expenditure on Travelling, Entertainment and other ordinary expenses of the business', which gave detailed guidance on the level and type of expenditure allowed to various categories of staff.

While economy was certainly a watchword it was not taken to the point of strangling development and expenditure was undertaken especially in areas where it reduced production costs.[16] In 1925-6, for example, a new mechanical red lead furnace was erected at Cooksons' Hayhole works, which, together with associated buildings, cost £19,000. It was expected to reduce production costs by 25s a ton compared with the existing furnaces and subsequent cost figures show that this figure was in fact exceeded in 1926 with a process cost differential of 28s 2d in favour of the new furnace. Thereafter, however, with more efficient use of labour the costs of both processes fell and the differential narrowed to such an extent that it no longer justified the capital expenditure on further mechanical plant and as late as 1939 output at Cooksons of red lead from older furnaces exceeded that from the newer mechanical furnace. In examples such as this ALM was economically rational in adopting new technology where it reduced output costs but otherwise sticking to traditional processes. Such a policy was not always carefully worked out however. With numerous companies brought in to the merger and, as we shall see, several works closures, a policy of moving spare plant from one works to another which required it was adopted. As a result superannuated plant was sometimes kept going when new plant would have offered higher productivity which would soon have repaid the additional capital cost. This was particularly the case in such areas as the manufacture of pipe where the labour cost was the same whether a new or old machine were used but where the rate of extrusion of the former was considerably higher. Old pipe presses were retained, moved between works and occasionally taken over via the UKLMA as small firms closed, with the result that in the late 1940s all the Elswick presses were of nineteenth-century vintage and two new presses could have done the work with a considerable saving in labour cost.

Broadly it might be said that the merger companies remained backward in engineering technology throughout the inter-war period, with limited exceptions such as the development of continuous-process red lead plant. ALM clearly recognised the necessity to obtain overall control of engineering throughout the group and in 1928 set up a subsidiary, the ALM Engineering Co. Ltd, based at Bootle. It was partly intended to develop technological ideas from within the group (its technical director, H. Waring, for instance, developed a dust and fume collector) but also to act as a purchasing and organising body with regard to engineering matters. It would seem likely that the individual lead companies did not react with enthusiasm and the Engineering Company appears to have had little success in rationalising the group's technology. By 1932 its board of directors was meeting only annually in order to declare a dividend to be paid to the holding company (of a few hundred pounds each year). In 1938, however, ALM once again felt the necessity for reorganisation on the engineering front and ICI was approached with the result that L. Close joined the merger as chief engineer.[17] Having been with ICI, a very dynamic company, he was distinctly unimpressed by what he saw in his first inspection of the lead factories to advise on engineering and plant rationalisation. To Close the company appeared to regard engineers as fitters and repairers rather than designers of efficient equipment (an attitude which was, of course, embraced by much of British industry at the time). Close put forward a number of modernisation proposals which would have been costly and in this respect the company was probably grateful for the advent of the Second World War. Close had been a TA officer and his being called up to full-time service meant that the problem he presented to the company was shelved.

In the autumn of 1935 another young appointee, Anthony Makower,[18] a metallurgist who was subsequently to become a managing director of GWLI, reported on the company's smelting plant and the techniques for recovery of secondary lead. Like Close on the engineering front, he argued that existing practice was inefficient and he claimed that the merger had no idea as to whether or not metallurgical operations were undertaken at a profit. In particular he criticised the differing methods of treatment of secondary lead at different works and only practice at Millwall found favour. Some modernisation took place as a result of Makower's report but it was not until the end of 1939 that serious expenditure took place with the demolition of the old blast furnaces at Chester and Elswick, which had not been in blast since the early 1930s, and the installation of additional Newnam hearths to smelt increasing supplies of home-produced ores.

It is noticeable that critical reports, like those of Makower and Close, were made by young men who saw things in blacks and whites and did not have to concern themselves with many of the problems which faced the decision takers. Among these undoubtedly were the question of cost and the problem of imposing standard practices and new technology on a diverse range of companies with historically differing methods of production. To the historian and the young 'meritocrat' the opportunities for rationalisation may have seemed great but to those responsible for the smooth running of the merger firms those same opportunities probably seemed hedged around with uncomfortable possibilities.

While modernisation and rationalisation presented some difficulties, research and development were centralised. We have seen that Brimsdown had been taken over largely because of its research reputation and when its manufacturing capacity was closed in 1927 the Brimsdown laboratory was carried on as a research capacity for the whole group and in 1933 a new research laboratory was built at Perivale in Middlesex. It took over the Brimsdown establishment and also metallurgical research which had previously been concentrated at

Millwall and the cost of the research department more than doubled between 1930 and 1937. At Perivale C.A. Klein, as technical director, continued the build-up of a graduate-based department concentrating on project research which might offer production possibilities. Among the developments was tellurium lead, an alloy which increased creep resistance as compared with ordinary lead, offered increased resistance to chemical attack and was less prone to fracture under freezing conditions (offering householders greater protection against burst pipes). Group output of tellurium lead was about 1,500 tons p.a. in the later 1930s and National Lead took out a licence to produce it in the USA. Nevertheless, although the research effort was productive it was not always aimed at the practical problems the group was facing in meeting demand for lead products. In September 1933 ALM was approached by Oldham & Sons Ltd which was 'being forced to take up the use of Grey Oxide[19] on account of the financial saving involved as against the use of Red Lead and Litharge' for battery paste. Two months earlier ALM had been approached by Lucas to produce grey oxide but nothing had then been done despite the knowledge that an outside company, Base Metal Products Ltd, had commenced its production. Eventually in 1934 an expenditure of £4,500 was made on the erection of plant at Bootle to manufacture grey oxide under licence by the Japanese Shimadzu process and by 1937 some 600 tons were being produced annually. ALM was, however, reluctant to develop the new product any further than demand made necessary, partly because profit margins were low for grey oxide, and a 1940 sales manual advised that, 'The sale of this material cuts across the sale of lead oxides and it should never be introduced unless asked for.'

While the output of new products, as with grey oxide, could be concentrated on a single efficient plant, most of the traditional lead products were produced at several plant locations in the group and, although this continued to a considerable extent, there was some attempt at rationalisation in order to reap economies and increase profitability. This was most obviously needed in the production of white lead where capacity had expanded, as a result of developments such as the Bootle quick process. The basic policy of ALM was to close surplus stack capacity, continue the cheaper production of chamber white lead at Hayhole and expand the output of the cheapest white lead from the Bootle plant. Consideration still, however, had to be given to the conservatism of many painters and white lead grinders who continued to demand stack white lead because of its 'fuller body'.

The most obvious candidate for closure was the unprofitable plant of Brimsdown and it seems likely that its closure was intended when that company was taken over by ALM in 1925. In August of that year the Brimsdown board received instructions from ALM to make plans to close down both the Brough Lead Co. Ltd (a Brimsdown subsidiary) and Brimsdown's own white lead production, and output ceased in 1927. The next step was the closure of the white lead stacks at FB & W at Hebburn. Again this must have been a decision taken as soon as that company was taken over in January 1927 since production ceased in July of that year and all the employees were laid off. In addition further reorganisation was undertaken at Hebburn with the decision, taken as early as February 1927, that there should be a broad division of interest among the Tyneside companies with Cooksons and Locke, Blackett undertaking the merchants' trade while FB & W developed the retail side of the business. As a result FB & W began to develop a number of paint and decorating outlets including the Leith Paint Co. Ltd, Donald Leslie & Co. and John Truelock & Sons Ltd (of Manchester) and, even before the merger with Goodlass, Wall in 1930 (which swung it further in that direction), FB & W was clearly developing in the paint field. This was a sensible development for a firm which had long been uncompetitive in lead manufacture and which had a debit balance of £72,000 on its profit and loss account, to which a further £8,000 loss was added as a result of trading in 1926. Red lead manufacture ceased in 1930 with customers to be supplied in future by Cooksons and in the next few years the remaining lead activities of FB & W were transferred to other ALM works. Finally in 1934 FB & W and James & Co. Ltd were formally amalgamated as Foster, Blackett & James Ltd to produce paint alone. The new company was linked more closely to Goodlass, Wall and by the end of the 1930s was making net profits of £15,000-20,000 p.a.; its translation into the paint trade had clearly made it more profitable than it had ever been as a lead manufacturer.

The next stage in the rationalisation process came with the closure of the Locke, Blackett works, again a decision which had already been taken by ALM when the takeover was undertaken in 1928. Locke, Blackett suffered from the fact that its Gallowgate works was poorly sited from a transport point of view, the company was unprofitable and ALM already had excess capacity on Tyneside. Two months after the formal date of takeover on 1 January 1929, ALM's instructions to cease filling the white lead stacks at Gallowgate took effect and the stacks were closed later in the year. This was followed by the gradual cessation of production of other lead products both at the Gallowgate and St Anthony's works and by 1932 all production had ceased and most of the process workers were laid off, although a few were transferred to other Tyneside works. Plant was either sold or transferred to other works, the St Anthony's lease surrendered and the Gallowgate site put up for sale. The sales operation of Locke, Blackett was continued for some years after its production units were closed, in order to retain customer goodwill, but from 1 January 1938 Locke, Blackett was formally taken over by Cooksons and ceased trading. When this decision was taken in the previous year it had been stated 'that it was the ultimate policy of G.W.L.I. to merge the Tyne Companies into one body, the object of doing so being the concentration of work, greater efficiency and economy', towards which the takeover of Locke, Blackett by Cooksons was seen as a first step.

There can be no doubt that in the first decade of ALM's existence major rationalisation of its Tyneside capacity had taken place. This was not limited to the complete closure of Locke, Blackett's productive capacity and the cessation of lead manufacture at Hebburn, for Cooksons was not to emerge unscathed from rationalisation. In 1926 ALM instituted an enquiry into the manufacture of blue lead products on Tyneside and as a result of its report the lead smelting and refining works at Howdon was closed in June 1927 and the site remained untouched until after the Second World War, although antimony production continued on the adjacent site. Further rationalisation came as a result of the recommendations of the Tyneside Production Committee set up by ALM. It produced a number of reports during 1930, as a result of which the white lead stacks at Elswick were closed and all Tyneside production of white lead was concentrated at Cooksons' Hayhole works by 1931. By the late 1930s Cooksons' output of white lead was 8,000-9,000 tons p.a. (compared with more than 10,000 tons pre-1914) and there was only a single Tyneside producer as against five in 1918. There was to be further rationalisation after the Second World War with the final closure of Cooksons' stacks in 1947, although chamber output of white lead continued into the early 1960s. In October 1930 manufacture of litharge ceased at Elswick and in future customers were to be supplied from Millwall, while Cooksons was to draw its sales requirements of sheet and pipe from Elswick. A further major step in the rationalisation of the Tyneside firms was taken in 1934 with the completion, at a cost of more than £9,000, of an extension to the offices at Elswick which subsequently contained not only the staff of Walkers, Parker but those of Locke, Blackett and of Cooksons (who gave up their offices in Milburn House in the centre of Newcastle). This enabled considerable economies and facilitated working arrangements between the companies but still meant that the Tyneside works were run as individual units apart from control at GWLI board level of overall policy. To some extent this was not surprising since until the end of the 1920s the companies had been in competition, in some cases for more than a century. It took a long time to remove old rivalries and at least initially rivalries may actually have intensified as a result of the rationalisation brought about by the merger. It was clearly recognised by the other companies that it was Cooksons which was dominating the rationalisation and directing the closures and even the objectivity shown by the closure of one of Cooksons' own works did nothing to reduce the resentment. It merely confirmed that Cooksons was a hard-headed organisation − one of the standard comments of the period among those employed in the lead industry on Tyneside was that 'Walkers, Parker was engraved on your heart but Cooksons was engraved on your backside'. It may be that the choice of the Walkers, Parker works at Elswick as a site for the combined Tyneside offices was a deliberate attempt to reduce the animus against Cooksons.

Away from Tyneside there was less need for rationalisation, particularly in white lead production, since ALM had fewer contiguous plants but the Chester stacks were closed in 1938 and in future Chester drew its supplies from the quick process plant at Bootle (which by that time was also supplying many west coast customers of the Tyneside and London companies in the merger) and in 1939 the Westferry Road, Millwall stacks were closed and London production concentrated on Burdett Road. In all by 1939 ALM had dramatically altered its productive capacity in white lead and possessed only three plants, at Bootle, Hayhole and the old stack plant of LL & J at Burdett Road, which was itself closed in 1947 ending more than a century and a half's production of white lead by the stack process by constituent companies of the group.[20] By the end of the War quick process white lead from Bootle accounted for about two-thirds of the group's total white lead output.

Further rationalisation of smelting and desilverising works, as a result of the decline in imports of ore and bullion lead and their replacement by pig leads refined abroad, occurred in London. To some extent this was forced on ALM, which was only the tenant of the Bridge Road, Millwall smelting works of LL & J. In 1927 the freehold of the site had been sold and LL & J was given notice that its lease would not be renewed when it fell due in 1931. This was probably welcomed by ALM which arranged for the transfer of Bridge Road's production to its Westferry Road site on the Isle of Dogs. Conveniently placed with a frontage to the Thames this works offered considerable scope for expansion as the major London centre for the merger. It was much further downriver than the other Thames-side site, Lambeth, with consequently better facilities for water movement of goods and raw materials, was close to the major London docks and had better road provision than Lambeth. In addition the Lambeth works of Walkers, Parker was threatened by a proposal to erect a new Charing Cross Bridge, which would have required the compulsory purchase of the Lambeth site. Although this eventually came to nothing it had had the effect of making the Lambeth works something of a backwater. Both Walkers, Parker and subsequently ALM neglected capital expenditure at the works and although it made a considerable contribution to the firm's wartime output, after 1945 it had nothing to offer that was unique to the merger in London (apart from its shot production for which demand had fallen) and its closure in 1948, with subsequent use of the site for the Festival of Britain, came as no surprise. For some time previously Lambeth had been something of a satellite of Millwall and it was at the latter works that development occurred. In the late 1930s plans were laid for modernisation of the Millwall works but these were shelved with the advent of war. During the War, however, the opportunity occurred to extend the site. Previously this had been unnecessary because it had been thought by ALM that it could always extend the works on to the football field on the opposite side of Westferry Road. Wisely, however, Roderick Lancaster pointed out that this land was likely to be scheduled for building purposes after the War and pressed for the purchase of the neighbouring site belonging to the Manganese Bronze & Brass Co. Ltd, which had been severely bombed in 1941. The purchase was made in 1944 at a cost of £27,500 (which was effectively reduced because a firm of contractors was actually prepared to pay to demolish the bombed buildings on the site — so great was the shortage of building materials during the War). This provided scope for expansion and by the end of the 1940s, with a newly erected blast furnace and a new rolling mill, Millwall was the most modern works in the merger.

Finally there was some rationalisation at Glasgow, where GWLI controlled Alexander, Fergusson and T.B. Campbell, both of which were sheet lead rollers. In 1932 the companies agreed that Campbells should build a new factory at Kinning Park and should supply Alexander, Fergusson with sheet lead. By 1935 the ancient rolling mill at Alexander, Fergusson's McAlpine Street works had been dismantled and the premises leased and the firm was drawing all its supplies of sheet lead from Kinning Park. This caused some problems and in 1938 Alexander,

Fergusson formally took over the Kinning Park works.

It can be seen that GWLI (and ALM before 1930) were not just holding companies but that they took vigorous steps to rationalise the output of their constituents and made some attempts to integrate activities where possible. Up to 1930 policy decisions were taken by the ALM board but after the merger with Goodlass, Wall rationalisation was undertaken through the appointment of a series of committees, composed chiefly of board members but also including relevant senior personnel. The most significant committees were those for lead and paint, the former comprising Tasker, Reynolds, McConnell, Klein, Eckford, Roderick Lancaster[21] and J.H. Stewart.[22] This brought together the three powerful Cooksons' men[23] together with a representative from each of Brimsdown, Walkers, Parker, LL & J and FB & W. But it is doubtful whether the last four were there as representatives so much as to complement the organisational and financial abilities of the Cookson men. Klein was a technical expert, Eckford and Lancaster were works managers of proven technical ability, while Stewart was sales manager for the lead side of GWLI. This was a committee wielding considerable power which was responsible for overall decisions on rationalisation and development. There were also other committees concerned with standardisation of practice on such matters as salaries and patents.

Together with the closure and rationalisation of many of the lead works and the centralisation of the Tyneside offices at Elswick, there came other developments. A minor one was the centralisation of all insurance requirements within the group, since 'Grouping the insurance of the merger as a whole in this way a very substantial saving is possible' and the same was done for bank accounts. Of greater significance, the London offices, previously in a number of places for the different companies, were centralised; initially in 1924 on the Cooksons' offices at Dashwood House, 69 Old Broad Street, and later, when these became too small with the advent of new companies to the merger in the later 1920s, at London House, 3 New London Street from 1928. In 1938 the offices were transferred to Ibex House, Minories, although war led to evacuation to Limpsfield Court, Oxted, Surrey (purchased in 1939 for £9,500), but Ibex House was reused after the War together with offices at 14 Finsbury Circus for the board and the smelting and alloy division. Finally in 1958 the London offices were once again united in Clements House, 14 Gresham Street. Office centralisation offered some scope for staff economies but also facilitated integration between the companies, an area in which rationalisation was limited.

As president of ICI, Sir Harry McGowan told the directors of its four constituent companies, 'I want us all ... to forget our old Companies.'[24] No such directive seems to have been made at ALM and considerable rivalry, albeit on a friendly basis, continued to exist between its constituents, while customers were kept in the dark with regard to the merger and as late as 1948 the Elswick switchboard had separate lines which had to be answered with the name of the relevant company (Walkers, Parker, Cooksons, etc.). Unlike ICI, which was conceived with the intention of producing a single, powerful, rationalised unit in the chemical industry, ALM, conceived only a couple of years earlier (perhaps without the same grandiose intentions, although by 1930 with sufficient domination to have achieved them), did not produce a single unit until 1949. It is difficult to produce any conclusive answer as to why it did not do so but there were clearly several factors of significance. The merger companies had for many years been in active competition one with another and remained fiercely possessive of their independence within the merger. It remained necessary to negotiate any changes very carefully and Tasker considered that his job was largely that of a diplomat. Undoubtedly the piecemeal nature of the takeovers of the later 1920s made it more difficult to consider a single structure than with the ICI merger in 1926. ALM appears to have begun with agreement between two strong companies (Cooksons and LL & J), without any determined intention of buying up the rest of the industry and thereafter it grew rather like 'Topsy' (in particular the takeover of Locke, Blackett appears to have been totally unpremeditated). From such beginnings it would have been very difficult to adjust expectations to the contemplation of unitary structure.

The merger in 1930 with Goodlass, Wall may have made complete integration even more difficult by adding further problems to an already complex picture. It was soon discovered that the Goodlass, Wall staff was unnecessarily large and the office accommodation so ramshackle that it needed replacement. More worrying was the low level of profitability and lack of efficiency and, despite the appointment of lead members to the paint committee and the Goodlass, Wall board, little seems to have been achieved and there is the impression that the lead men lost interest in the paint side in the later 1930s and left it to get on on its own so long as it produced reasonable profits. A.P. Bevan, chairman and managing director of Goodlass, Wall, appears not to have satisfied the lead men and was under frequent pressure 'to run the business on a more economic and profit-earning basis' as a 1933 resolution of the GWLI board phrased it, after Clive Cookson had commented 'that the Liverpool results from the date of the amalgamation up to the end of 1932 were extremely disappointing'. Gradually Goodlass, Wall men were eased out of prominent positions. In 1935 Bevan became president of the Paint Federation, which it was expected would take up much of his time and the opportunity was taken to appoint Reynolds as an area director for Liverpool and Glasgow. A Cooksons' man had now been insinuated into the paint side and was to replace Bevan as managing director in 1937. Similarly, Goodlass, Wall directors who had clearly not proved themselves of great use, were pruned from the GWLI board, only Bevan of the original four remaining by 1936. When he retired he was not replaced and the original Goodlass, Wall organisation had failed to throw up anyone of board level ability. That company was left to be run by Reynolds and up to 1949 the lead men made up the GWLI board.

While much was certainly achieved in terms of works' closures and office reorganisation of the lead companies by the end of the 1930s, their structure remained largely unchanged. The limited exceptions to this statement lay in export markets and in some marginal rationalisation of home sales. In December 1924 the ALM Export Company Ltd was set up, in effect a subsidiary of Cooksons, which continued to own rather more than 50 per cent of its shares as late as 1939. It was a successful and profitable venture, although in 1936 C.N. Ribbeck noted that

> The export trade of the merger is nominally in the hands of this firm, but in practice Walkers, Parker, Locke, Lancasters, Foster, Blackett & James and Locke, Blackett still handle their own export trade to a great extent. Almeco deals with export orders for Cookson's, Librex and London Lead Oxide.

Here again the difficulties in obtaining co-operation from previously competitive firms, which were entrenched in their own practices, meant that the economies offered by a specialised sales organisation, which built up agencies throughout the world, were not utilised to the full. With regard to home sales some semblance of order was maintained by the existence of reserved lists of customers. Travellers and sales staff were given lists of the customers who traditionally belonged to particular firms within the merger and other companies were not allowed to approach them for business. Inevitably this did not always work on the ground, whatever the theory, although the competition was usually polite. Stories are told of salesmen from different companies in the merger meeting outside the offices of customers and tossing a coin to decide which should go in first.[25]

There can be little doubt that the system was less than efficient since each firm had its own travellers, many of whom were competing over the same areas (and even if they respected the reserve lists of customers it still meant that more than one merger traveller was covering the same ground). At the end of the 1920s Cooksons undoubtedly possessed the most advanced sales organisation, with the development of the 'Flying Column', a group of travellers who initially covered the whole of the country visiting painters and builders' merchants in order to expand the sales of Cooksons' products. Early in the 1930s these men were given permanent territories and L.V. Cooper,[26] for instance, was given the south-east to open up, where he came into competition not only with T. & W. Farmiloe, whose 'Nine Elms' was the brand leader in white lead paint in the area, but also with the products of LL & J which was strongly entrenched in the area because of its London base. The men on the ground, such as Cooper and J.R. Deller, clearly imply that considerable economies could have been made in the sales organisation, while senior management in the group was more concerned with maintaining customer goodwill and not losing the confidence of the staff in the individual companies. Moreover senior management was probably less aware than were the travellers of the amount of duplication which went on. There is no doubt that there was sound argument for maintaining the existing system. In the trading conditions of the inter-war period with excess capacity and concern about competition in many products, it would have been a highly risky venture to have unified the sales structure. Many firms genuinely believed, for instance, that one company's white lead was dramatically superior to that of another[27] and it would have endangered customer goodwill to have phased out individual manufacturers' names and broken the relationship built up by travellers with their customers.

Eventually, however, changes came. That customers accepted the takeover of Locke, Blackett by Cooksons in 1938, without any adverse reaction, appears to have urged ALM to consider more significant change. In the same year Urwick, Orr & Partners, management consultants, were retained to report on sales reorganisation in the group's twelve constituent companies.[28] Nine of the companies operated sales forces in the Home Counties, eight in Scotland, Lancashire and Cheshire and the West Midlands, seven in the East Midlands and Ireland and six or fewer in other regions. As Urwick, Orr's report commented, 'To all intents and purposes the twelve Associated Companies with which we are now concerned have been operating as independent sales entities.'[29]

Attention was chiefly drawn to the inefficiency of the reserved customers scheme which meant that a Cookson traveller might not sell white lead to an LL & J customer but might call on him to sell antimony products. It was proposed that a unified sales force be set up based on three geographical areas with offices in London, Chester and Newcastle, which would involve a reduction in the number of travellers from 66 to 47 (with an expected saving of £4,000 p.a.), who would be placed under three area sales managers instead of the existing situation with '14 executives responsible for Sales Promotion and Sales Control'.

This was no doubt an admirably rational scheme but it had to overcome the doubts of those who were concerned about the loss of company individuality and the potential impact on customers, since it was felt that clumsy handling of commercial reorganisation would risk a loss of goodwill. During 1939 sales unification was discussed by the board of GWLI and eventually postponed because of the advent of the War and it was not until 1 January 1949 that trading by the individual companies ceased. Had unified control come two decades earlier it would undoubtedly have reduced production costs and selling prices. Some of the opportunities for savings in selling costs have been outlined above but in addition there were administrative costs of maintaining separate commercial organisations. Extra office staff were required for separate paper work and filing systems, different headed paper had to be used for different customers, while different coloured orders had to be used to ensure customers' brand loyalty. On the production side there was surplus capacity in the group as a whole and production was not concentrated on the lowest cost works. Even where one group company took its output of a specific product from another works, different stocks of labels had to be kept in order to ensure that the tin or keg was labelled with the name of the selling company, irrespective of where it was produced and a mass of paperwork was generated by inter-company sales. Had greater economies been made in the early 1930s the group would have been in a better position to withstand the competition from products such as titanium dioxide and copper pipe. Ultimately, however, as one employee of the group put it, lead was a dinosaur, inevitably to be replaced by more efficient materials and unified organisation of the group from the early 1930s might have been of value in enabling its senior members to employ their undoubted talents in pursuing policies of diversification rather earlier than was in fact done.

Nevertheless some diversification, if rather hesitant, took place, most of it being based on work done by the Brimsdown chemists. The first significant step was the development of 'Beekay'[30] waterproof sandpaper.

This was of considerable interest to all the white lead manufacturers because of the outcry over the number of cases of lead poisoning among house painters and the threat of a ban on the use of lead in paints. A company, British Waterproof Abrasives Ltd, was set up as a subsidiary of ALM to produce waterproof sandpaper but R.W. Greef & Co. Ltd, which held the English patents to 'Wetordry', a waterproof sandpaper developed in the USA by the Minnesota Mining & Manufacturing Co. (3M), challenged the ALM patents as an infringement of its own. In January 1928 the disputants agreed to form a new company, British Wetordry Ltd, in which ALM, Greef and 3M would have equal shares, with works at Hebburn taken over from FB & W. Other American manufacturers, however, threatened to commence manufacture of waterproof sandpapers in the UK and it was eventually agreed in 1929 that all participants should join together to form Durex Abrasives Ltd in the UK (in which company the three original participants were each to have an 8 per cent share in exchange for their shares in British Wetordry Ltd). A plant was acquired in Birmingham (thus do international combines affect the prospects for employment in depressed areas) and production commenced in 1930. As a result of an American anti-trust suit in 1949 3M(USA) bought from the other American participants their shareholdings in Durex Abrasives Ltd and changed the name to 3M (UK) Ltd. From 1953 GWLI held a 6.67 per cent share of the capital of 3M (UK), although this holding was reduced after 1969 and finally eliminated in 1973.

It is difficult, on the limited evidence available, to pass valid judgement but it does appear that ALM possessed a viable product with reasonably sound patent rights but allowed the opportunity to develop abrasives in Britain to slip into the hands of the powerful American producers. A similar comment may be passed on the eventual position which occurred with regard to the second significant diversification by ALM, that into titanium. By the early 1920s the possibility of using titanium dioxide as a paint pigment was well established and it was clearly a possible future alternative to white lead. Once again the fertile chemists at Brimsdown were working on the subject and a patent (no. 243,081) for the manufacture of titanium dioxide was taken out by Brown and Klein in 1924. Another subsidiary of ALM, British Titanium Co. Ltd, was set up in 1926 to produce titanium at the Brimsdown works on a pilot plant basis. By January 1927, however, when the ALM board came to further discussion of the subject, fears were expressed that, if serious production of titanium were undertaken, the National Lead Co. of America would attack ALM's patents. There followed some months of discussions and negotiations. A barrister's opinion was sought (from a certain Mr Stafford Cripps), on the patent position and, reassuringly, this was regarded as satisfactory. There were, however, other problems. The UK market for titanium dioxide was chiefly supplied by the Americans and it was felt unwise to disturb their control at a time when negotiations were going ahead with National Lead, among others, to form the International White Lead Convention. Moreover ALM was now uncertain whether it should proceed under the Brown/Klein patent, while it looked at the Blumenfeld process, which was being worked in Italy but, significantly, had been rejected by ICI. There was also concern over the level of competition likely to be encountered from imports and in April 1927 Lord Inverforth approached the Board of Trade 'as to the possibility of bringing the manufacture of Titanium Oxide within the scope of the Safeguarding of Industries Act'[31] but he received no encouragement.

In July 1927 ALM finally approached National Lead with the offer of joint exploitation of the Brown/Klein process but National Lead made it clear that it was not interested and regarded its own process as superior. ALM then decided to drop further development of titanium dioxide, which might have been the end of another potentially promising British development. In April 1932, however, under the impact of the recently introduced British import tariff, National Lead decided to manufacture titanium in the UK and invited ICI, the Imperial Smelting Corporation Ltd and GWLI to join in financing the new company. As a result British Titan Products Ltd (in which National Lead had a 49 per cent share, part of which was in the name of R.W. Greef & Co. Ltd, and each of the other three partners 17 per cent) was formed with a plant at Billingham, County Durham.[32] Initial output of titanium dioxide was about 1,000 tons p.a. but capacity had been raised to 7,000 tons by 1939, and after the War was dramatically increased to almost 30,000 tons at Billingham by 1971 with even larger plants at Grimsby and Greatham, near Hartlepool, while wholly- or jointly-owned plants existed in a number of countries abroad. As a result of the 1932 tariff and its earlier interest in titanium, GWLI had had the good fortune to be taken under the wing of a powerful international combine which gave it a share in a profitable new field (in 1939 British Titan made a net profit of £139,000). After the War an anti-trust decision in America forced the withdrawal of National Lead and by 1978 Lead Industries Group (LIG) and ICI had become equal joint owners of Tioxide Group (to which British Titan had changed its name). While GWLI and its successor, LIG, have found their investment in titanium to be very profitable, it has had its side effect, since it encouraged ICI to develop hard gloss paints which ate steadily into the market previously supplied by white lead paints.

The influence of the National Lead Co. of America has been seen not only in the case of titanium but in a number of places in the development of the UK lead industry and this is a suitable point at which to outline its considerable influence on ALM. The British and American companies had had something of a mutual understanding since the beginning of the First World War. In 1922 when Cooksons were negotiating a licence from National Lead to sell orange lead in the USA, Clive Cookson replied to a letter from Evans McCarty, which had pointed out the futility of price competition, 'I note all that you write about competition and am in absolute agreement with it, and of course the very last thing I want to do is needlessly and uselessly to in any way upset your own home arrangements.' In February and March 1928 ALM sent representatives to visit the American factories of National Lead,[33] the main aim being to exchange information on red lead practice as had been earlier set out when the Barton patents for the USA were sold to National Lead. It turned out, however, that National Lead was prepared to exchange information on all matters of mutual interest to the two companies and an agreement was signed in 1929 at the same time that National Lead acquired a shareholding in ALM.[34] As a result National Lead was given the opportunity to appoint a director to the ALM board and the first was Evans McCarty, National Lead's general manager. Although unable to attend with any regularity, National Lead's director undoubtedly had some influence in encouraging ALM to move towards rationalisation and increased concentration in production, activities which had been developing in America, especially in the area of white lead production, for 40 years. The importance of advertising was stressed (National's advertising budget was equivalent to £50,000 p.a.) and this may have been a spur to the white lead publicity campaign followed by ALM in the early 1930s. Details were provided of National's management structure and sales organisation and ALM borrowed the pattern (to quote a report on National's sales organisation) that 'where customers express a preference for a particular brand, it is usually supplied by the Company working in the buyer's territory, but the containers are marked as though they came from the favoured seller'. In production the extent of mechanical handling and the saving in labour costs it produced in American plants was stressed and this was an additional factor in encouraging the modernisation of many ALM works in the 1930s.[35] The most lasting impact of the relationship with

National Lead, however, came through the exchange of information agreement. Although National Lead may at one time have been backward in red lead technology there can be no doubt that the flow of information was chiefly to the UK and the most significant development was the use of white lead to stabilise PVC,[36] which after the Second World War was to provide an important market to check the decline in white lead sales.

To return to the topic of ALM's diversification, there were other developments before the War. In 1926 one of the Brimsdown chemists put forward proposals for the manufacture of gums and pastes and these were supported by the setting up of British Gels Ltd, a company which proved unsuccessful and was sold in 1928. More satisfactory was the diversification into paint which came increasingly in the 1930s. Many of the white lead corroders had sold what was called 'paint', as far back as the eighteenth century, although this was merely white lead ground in linseed oil. Most of the production of paint had been previously undertaken by the white lead grinders (who produced paint to be sold both retail and wholesale) and by painters themselves, mixing their own paints from dry white lead. As alternative paint bases,[37] including zinc oxide, lithopone and barytes became more generally used, however, the white lead manufacturers began to develop ready mixed (including tinted) white lead paints in order to maintain the market. This began just before the First World War but most of the manufacturers developed them in the 1920s (Cooksons had 'Crescent' brand and LL & J 'J' brand). This was much to the disgust of the traditional paint grinders and led to a good deal of ill-feeling (indeed one of the reasons mentioned for the takeover of Goodlass, Wall was that it was a defence against the possibility of one of the big paint grinders moving into white lead production).

From 1930 ALM began a major campaign[38] to expand paint sales direct to merchants and painters and by the end of the 1930s sales by the lead firms (excluding the Goodlass, Wall operation) exceeded 200,000 gallons p.a. Although the group's sales of genuine white lead paint continued to rise there was growing pressure from the new range of titanium-based paints but even this was held in check by the introduction in 1936 of a hard-gloss white lead paint, sold under the trade name 'Magnet', developed by one of the Brimsdown chemists, H.E. Tapley. Again this caused trouble with the paint manufacturers and one of them, Walpamur, responded by importing its white lead requirements from Canada. After a couple of years, in which period ALM had had the opportunity to establish its new product, the disagreement was solved by giving the paint grinders the formula for the manufacture of hard-gloss lead paints in return for a price agreement on their sale. For two decades until about 1950 the expansion in sales of lead-based paint was to prove a useful profit earner for ALM but thereafter the competition from titanium-based paints was to become severe.[39]

Broadly it is clear that although there was some diversification in the later 1920s and in the 1930s, it had little overall impact on the product strength of the merger. Output remained heavily concentrated on the traditional lead products, some of which were under the threat, but fortunately as yet not the fulfilment, of the competition of new products. Rationalisation of existing output plus the benefit of the building boom, therefore, helped GWLI to a fairly comfortable financial existence in the 1930s,[40] although this could hardly have been forecast in the merger's early years. Compared with the 1924 profits of the pre-merger companies of £167,000, net profits did not exceed £100,000 until 1928 (and then only did so by a whisker) largely as a result of the necessity to write down the considerable preliminary expenses, the overvaluation of lead stocks and excessive goodwill.[41] There was no ordinary dividend in the years 1925-7 and comment on the company in the financial press in the late 1920s was decidedly unfavourable. At this stage it was difficult to discern the benefits of the merger to the Cookson, Johnson and Lancaster family shareholders but gradually matters were sorted out. With a slight dip in 1931 the ordinary dividend rose to 7 per cent in 1936 and was maintained at that level for the rest of the 1930s.

Since the merger was largely run by the partners in the two largest of its constituents, it was inevitable that the atmosphere with regard to personnel was dominated by them and continued in something of a paternalistic direction. The major firms, through the Lead Employers' Council, of which Clive Cookson was chairman, had supported the setting up of a Joint Industrial Committee for the lead industry in 1919 and its proposals for minimum wage rates and a 48-hour standard week. One week's holiday with pay for all employees (process as well as office workers), which had been standard practice at Cooksons, was adopted by all merger firms, and in 1947 this was increased to two weeks, the GWLI chairman noting that the lead industry had been in the van with the development of holidays with pay and was once again to the fore in their extension. Salaries and wages were not outstanding, the company appearing to pay around and sometimes deliberately just a little over the norm for the area in which each works existed, in order to attract labour, but salaries were obviously good enough to attract and hold some good management.[42] In 1929 an enquiry was instituted throughout the merger to increase efficiency by pensioning off older men and pensions, usually of between 10s and 15s a week, were given to a number of men over 65 (including one of 80). A committee was then set up to consider the whole question of pensions and proposed a scheme for a pension fund for all employees (based on that run by ICI) but this was rejected by the board because of the estimated cost to the company of £4,000 p.a. Again in 1938 pension proposals were put forward but it was not until 1945, having discussed the subject on several occasions (and incidentally having built up a pension reserve fund of more than £100,000) that GWLI introduced schemes for both process and office staff.[43]

While some of the older workers were pensioned off in the economy drive of 1929 it was inevitable in the circumstances of the early 1930s that there would be redundancies. There is only one account of total works' employment in the merger (and none at all for office staff). This was produced in 1929 for the review of employment then made and gave a total works' employment of 1,192. Since this came after the closures of Brimsdown, the Gallowgate works of Locke, Blackett and Cooksons' Howdon works, it is clear that redundancies of perhaps 200 had already occurred. The figure of 1,192 also excludes employment at Walkers, Parker, which, when it was taken over later in 1929 would have pushed total employment to about 2,000. If the figures for Elswick[44] (the one works for which annual average employment figures are available) are anything to go by, then this figure might have been reduced by about 25 per cent as a result of the depression of the early 1930s and been maintained at this level during the remainder of the decade as tighter manning levels were observed. Redundancies were deliberately confined to process workers and to those who had been taken on most recently. For those who escaped 'the sack' the evidence of the wages books still available suggests that they obtained regular employment even during the worst of the depression. Even unskilled labourers appear regularly to have been paid for a 47-hour week and although there was a 10 per cent wage reduction in 1931, this was less than the fall in the cost of living and real wages rose. In making the wage cut ALM was following the lead of the government which had reduced the salaries of all state employees during the 1931 crisis, as an economy measure. It is, however, worth noting that this was an economy which ALM did not need to make since its profit figures were well maintained during the depression. This was soon realised and from 1933 it was decided to pay a bonus to all employees of two weeks' wages or salary as a partial restoration of the cut (since the dividend on the ordinary shares for that year had reached 5 per cent — twice the level of 1929 — this was hardly a generous step). The 10 per cent wage and salary reduction was retained, in case of a setback in demand, but in 1934 a four weeks' bonus was paid[45] and in 1935 this was increased to five. In 1936 a 5 per cent restoration in wage and salary levels was made together with the continued payment of a bonus supposedly the equivalent of the other 5 per cent and from 1 January 1937 the full pre-1931 wage and salary levels were restored. The cuts had certainly done something to reduce ALM's production costs but in view of the profit figures for the period they appear to have been unnecessary and certainly caused considerable employee disgruntlement (more than 40 years later the cuts were the thing most often remembered of the 1930s by ex-employees).

Nevertheless wage cuts were a common feature of employment in the early 1930s and blame was not necessarily placed by employees at ALM's door. Indeed the long service of many workers and the frequently expressed view that it was a good company for which to work would suggest a reasonable level of employee satisfaction. Inevitably this may have been stronger among salaried employees, who, although they suffered the wage cut of 1931 were deliberately protected from redundancy by a policy of moving them to jobs with different companies within the group whenever possible (and, of course, the lack of rationalisation of company structure meant that little reduction in office jobs occurred). Office staff also received the benefit of a reduction in hours. In 1935 the GWLI board considered a proposal for the introduction of a five-day week but this was too radical a step for the time and it was decided to give all staff one Saturday in two off.

When war came in 1939 its major impact was to lead to the postponement of the implementation of the rationalisation of the merger's sales and company organisation. The second impact was a loss of labour as men were called up for the forces, although the seriousness of this development should not be exaggerated. Between 1939 and 1941 less than 20 per cent of the Elswick male labour force at the beginning of the War was actually called up (41 out of 227). As in the First World War, it was expected that less labour would be needed since the supply of and demand for lead would be reduced. This proved correct and it was noted at the end of 1941 that merger output had fallen by 7,000 tons compared with 1940 — half the fall coming in the output of sheet and pipe because of the reduction in building activity. Nevertheless there were wartime demands for lead products, which to some extent offset the loss of peacetime demand and at the end of 1941 ALM sent around its merger works a circular providing the basic facts which could be argued before local tribunals for establishing reserved occupation status for lead workers. Much of the work was heavy and men were therefore required, while interestingly the health risk of working in lead processes was invoked to show that the merger could not just take on any labour which might be unsuitable for armed service. Most significantly, however, the government's demand for many products was stressed. Lead was added to steel to improve its machineability (the Ledloy process) and the Walkers, Parker shot towers were said to produce the only dust lead shot suitable for this process; lead was necessary for batteries for various war purposes; sheet lead, pipe, traps and bends were important for army installations, etc., since there was a serious shortage of copper; solder was widely used; lead paint was required in literally millions of gallons for camouflage purposes;[46] and there was a large government demand for shrapnel and bullet rod. From January 1942 the labour situation was eased since most of the merger works became controlled establishments classified as munitions works.

As in the First World War all lead supply was controlled by the government and directed to specific firms. As a result of its dominant size in the industry and integrated product structure ALM benefited from government control. It was some of the small lead manufacturers, concentrating on sheet and pipe production for which demand fell most, which were closed during the War as a result of restrictions on the availability of lead, while ALM works received allowances of lead even for sheet and pipe production. Lead supply, little affected by its early stages, became more difficult after the entry of Japan into the War, which restricted supplies of pig lead from the Far East and Australia (which had been much the largest single supplier pre-war). However, it was noted that a fortunate side effect of bomb damage was an increase in the supply of secondary lead. A disadvantage of bombing was damage to merger plants. In May 1941 the Lambeth works, which had been given over to munitions production with sheet and pipe transferred to Millwall, was severely damaged, while late in 1944 the Burdett Road white lead plant was damaged. More serious was the almost complete destruction of Champion, Druce's white lead works at Islington as a result of a V2 rocket hit in January 1945. While the latter event limited white lead supply just at the time when demand was building up again it did give ALM the opportunity further to concentrate output on the cheaper quick process white lead from Bootle, whose output was considerably expanded during the War.

While there were inevitably restrictions on new capital development during the War, some did take place, usually related to the war effort. There was the obvious introduction of plant to produce munitions, such as bullet rod, while additional Newnam hearths[47] were introduced to smelt the increased output of lead ore from the Greenside mines in Cumberland (in which GWLI had a large share stake). Such expansion was easily financed out of the increased wartime profits and, although an increased tax bill made a considerable hole, it proved possible to reduce the overcapitalisation, which had continued to exist after the 1930 merger, by writing £150,000 off goodwill. Compared with net profits of between £250,000 and £275,000 in the late 1930s, GWLI's profit reached £½m in 1944 and 1945 and, under the influence of a rapidly rising lead price, exceeded £1m in both 1946 and 1947.

Full of cash, the company was in a position at the end of the War both to modernise its existing production and consider new ventures. Between 1946 and 1949 capital expenditure totalled £900,000, of which £250,000 went on offices, laboratories, wash rooms and canteens, and a minute of the managing director's committee in 1950 could note that as a result 'the factories will have been largely modernised in plant and equipment and brought up to modern standards with regard to the housing of Technical and Commercial Staff'. There was a continuation of the pre-war policy of the opening of plant in major overseas markets, the most significant development being the opening in 1950 of a works to manufacture nearly the whole range of lead products at Jacobs, near Durban in South Africa.

Both forms of expenditure so far discussed, plant modernisation at home and expansion overseas, were along traditional lines within the lead industry where GWLI had experience. The third outlet for surplus capital differed, since it involved diversification into a largely new field.[48] In 1943 the company negotiated the purchase of Fry's Metal Foundries Ltd, which only overlapped GWLI's production in solder and otherwise consisted of the output of printing metals, bronze, aluminium alloys, etc. Why this particular company was chosen is not clear (although the willingness to pay rather more than £½m, valuing Frys at ten times its current wartime earnings, suggests a considerable determination to buy the company). Nevertheless the general reasoning was clear. Demand for many of GWLI's traditional lead products was at best stable and there was no point in expanding in such areas. Indeed during and immediately after the War GWLI was offered several small lead companies and, in T. & W. Farmiloe & Co. Ltd, one moderately-sized firm and rejected them all. Having rejected expansion in its own field the company wisely decided to avoid conglomerate type growth and concentrated on an allied area of non-ferrous metal production – a policy which was subsequently laid down in a memorandum presented to the board by Clive Cookson in 1946 and confirmed as follows:

> It was agreed that it was not desirable to expand for expansion's sake nor to expand by acquiring any additional business mainly engaged in the Company's existing activities; expansion if carried out should take place if possible in some allied field of industry, and preferably one in which use could be made of the Company's existing organisation.

A principle had been established which was to be the chief guide for development in the next 30 years. In the meantime the major need was the rationalisation of company structure, much delayed by the War.

Notes

1. Metallgesellschaft, *Metal Statistics*.

2. As late as 1927 the International Labour Office (*White Lead*, p. 276) estimated the figure as 20 per cent of total world lead consumption.

3. The order provides an interesting sidelight on transport history. Finished red lead was initially, in 1919, taken by horse and cart to Gravesend station and then loaded on a London train. This was soon replaced by a lorry service from the works to Gravesend station, undertaken, the ex-works manager remembers, at suicidally competitive rates by owner-drivers of ex-army lorries. Soon an initiative among the lorry-owners led to direct transport from the works to London railway termini and this was followed, in the mid-1920s, by a venturesome spirit who collected red lead from the Gravesend works on Friday afternoon and delivered to Lucas on Saturday morning. Interview with A.F. Kerr.

4. In the case of India it was (unlike the case in Australia) less local production than imports from other countries which were causing decline in UK exports. In 1928 it was estimated that the existing trade was shared between Waldie (India) 600 tons, the UK 700 tons, Japan 600 tons and Italy 350 tons, with imports from the last two countries increasing.

5. S. Cahn, CBE, was a director of ALM from 1949 to 1966 and was appointed a managing director of GWLI in 1952 and sole managing director in 1962 until his retirement in 1966.

6. These words were written by C.F. Hill (the son of F.C. Hill), managing director of Mersey, probably in the late 1940s or early 1950s.

7. The pool ratio was ALM 81.45 per cent and Mersey 18.55 per cent but by the 1930s Mersey was consistently below its share because the price of its stack white lead had been fixed at a premium of 10s per ton.

8. No figures are available for 1926, the date of this dispute, but for 1932 ALM estimated the works cost per ton of producing dry white lead at £8 10s for the stack process, £7 10s for chamber and £6 10s for the Librex process at Bootle.

9. See P. Fitzgerald, *Industrial Combination in England* (1927), pp. 118-19. This brought some interesting activities. L.V. Cooper remembers that an employee in Cooksons' export department took advantage of the price differential by getting a friend in another concern to order a large tonnage of white lead for export which was then sold on the home market. Unfortunately for him he was soon found out and given the sack.

10. At much the same time ICI was in negotiation with I.G. Farbenindustrie AG on heavy chemicals and fertilisers (see Reader, *ICI*, vol. II (1975), especially Chap. 3). Boat trains and aircraft to the Continent must have done well from cartel negotiators and it would be fascinating to know whether or not the different industries gained anything from each other's experience of negotiations.

11. See *Official Architect*, July 1939, pp. 731-3, which contains a description and illustrations of the works.

12. Some foreign manufacturers overcame the tariff by buying white lead from English manufacturers, although this was a net gain to the latter. The Mersey report for 1932 noted that in March of that year the firm had 'agreed to manufacture for three Foreign Firms (Rhodens, Bergmann and F.A.G.) through their English Agents, James Beadel & Co. Ltd., Liverpool'. There is considerable debate among economic historians over the impact of the introduction of tariffs in 1932. For a recent contribution to this debate and an outline of the problems involved see J.S. Foreman-Peck, 'The British Tariff and Industrial Protection in the 1930s: an Alternative Model' and F.H. Capie, 'Tariffs, Elasticities and Prices in Britain in the 1930s', *Economic History Review*, XXIV, 1 (1981), pp. 132-42.

13. In 1938, the last full year in which the International Convention functioned, quotas were as follows: UK and British Colonies, British quota 12,600 tons (of which ALM had 12,000 tons), while imports were 2,800 tons; the German market of 12,000 tons was open only to German producers; but ALM had quotas of 1,450 tons in the total of 3,400 for Holland and the Dutch Colonies, and 2,500 in the total of 3,950 for Scandinavia. UK exports therefore considerably exceeded imports.

14. Polans provides an interesting example of the development of the small sheet and pipe manufacturer. Harry Polan came to England from Lithuania in 1894 at the age of about 15. In the depression of 1921-2 he sold the scrap-metal business he had built up because there was little work and decided on housebuilding or pipe manufacture. His wife tossed the coin and the latter was chosen. In 1923 he obtained premises at 162 Rockingham Street, Sheffield and named them the Wentworth Lead Works and business went well under the stimulus of expansion in housebuilding. In addition to the 1933 installation of a sheet mill, a second pipe press was added in 1934 and annual output was about 6,000 tons (60 per cent of it sheet) in the late 1930s. The works was closed during the War as an inessential user but its stock of pig lead (of 6,000-7,000 tons) somehow escaped the Ministry of Supply and had quadrupled in value by the time the firm commenced again in 1945. After the War the trade was built up by the founder's sons who established a reputation for meeting customers' requirements as soon as possible. ALM continued to regard the firm as a price-cutting nuisance and eventually took it over in 1965, to discover that most of its trade had been built up on quality of service and not low price. (ALM, *News Bulletin*, August/September 1979 and interview with Hiram Polan).

15. In 1934 output of lead products by ALM was sheet and pipe 47,981 tons, white lead 26,409 (some of which would have been used within the group for the manufacture of paint), red and orange lead 13,465, pig lead 8,712, litharge 6,435, alloys 5,846 and frit 976 tons. In addition 480 tons of antimony metal, 203 of antimony sulphide and 2,455 of antimony oxide were manufactured at Howdon. It is unlikely that the conventions would have been formed, or at the

very least that they would have worked as effectively as they did, but for the power of the merger. In August 1930 C.F. Spencer told Goodlass, Wall shareholders, 'So strong is this organisation [ALM] in the world's trade, that they have brought about far-reaching international agreements for the regulation of the trade', and J.L. McConnell subsequently wrote, 'I question whether such "Conventions" could have been created at all had A.L.M. not acquired most of the main U.K. makers first of all.'

16. There were many of them but it would be of no interest to itemise them in detail. A few examples may suffice: in 1931 the introduction of a new trommel at Hayhole reduced process costs by 50 per cent and that of a new machine to cast lead straps reduced costs from 2s 4d to 1s 3d per ton, while in 1937 new dust collecting plant and alterations to the Barton oxide plant at a cost of £2,500 enabled a 50 per cent reduction in labour cost on the process.

17. The late 1930s was a period in which ALM began to recruit graduates for management purposes and these began to have a beneficial effect on the company in the post-war period. C.N. Ribbeck, a graduate chemist recruited in 1935, became sales manager of the North West Area in 1950 and subsequently a director of ALM (and, incidentally, followed in the footsteps of a previous Chester partner, Sir E.S. Walker, by becoming mayor of the city in 1972), and A. Hellicar, an Oxford chemistry graduate recruited in 1938, also became a director of ALM.

18. A Cambridge natural sciences graduate, Makower then went to the School of Mines where he met Roderick Lancaster who persuaded him to join LL & J. He became Chester works manager, where he remained during the Second World War. He was particularly responsible for the modernisation and development of smelting and refining and also for the development of ALM's alloys business. He became a technical director of GWLI and was responsible for bringing together the ceramics supply companies purchased in the 1960s and for the building of a new plant at Meir, Stoke-on-Trent. He was a managing director of LIG from 1966-72.

19. Grey oxide is a mixture of litharge and metallic lead (PbO and Pb), with as much as 30-40 per cent of the latter. It is produced by melting pig lead, casting it into strips which are then chopped up and fed into a ball mill where it is oxidised by friction not by roasting. It produced a suitable paste for battery plates which was much cheaper to produce than other lead oxides, and it is still widely used.

20. Champion, Druce at Islington, in which GWLI had a holding, was the last firm connected with the group to cease stack production, in 1950. The last UK firm to produce white lead by the stack process was probably Mersey, whose stacks were closed in 1963.

21. R.L.H. Lancaster, son of H.C. Lancaster, was born in 1900 and studied at the Royal School of Mines. He joined the family firm of LL & J in 1921 and became works manager. He was subsequently appointed to the GWLI board, visited the USA to observe metallurgical techniques, became technical director, a managing director and finally, from 1955 until his death in 1962, chairman of GWLI.

22. Born on 21 August 1884 J.H. Stewart joined FB & W as a tea-boy, but eventually became commercial director of GWLI.

23. It is noticeable that during the Second World War it was these three, Tasker, Reynolds and McConnell, together with Clive Cookson, who were appointed an executive committee of the GWLI board empowered to take all decisions between the infrequent board meetings.

24. Reader, *ICI*, vol. II, p. 25. During the 1930s at ALM, however, individuals, below board and senior management level, regarded themselves as employees of a

particular firm not of the group. This attitude continued after the War and in the late 1970s, when I questioned retired employees, I was often told, 'Oh! I would not know about ——, I was a Cooksons' [Walkers, Parker's, etc] man.'

25. It is really quite remarkable that after an initial three-day sales conference of all the group companies, held at the Charing Cross Hotel in 1930, the experiment of bringing all the group's sales personnel together was not repeated during the rest of the 1930s.

26. Cooper's 'Reminiscences', serialised in ALM, *Southern Area News Bulletin* 1973-4, give a useful and amusing insight into the life of a traveller in the 1930s.

27. There were of course differences, especially in the cases of such products as red and white lead where physical consistency could vary within a considerable range. On the other hand a major element in customer preference was sheer conservatism. As late as 1958 an ALM salesman was complimented on the LL & J 'J' brand white lead paint used on the forward part of the superstructure of the *Queen Mary*. It was contrasted with the Cooksons' paint, which had been tested earlier in the year, which 'was absolute rubbish'. The two paints had been on the same formulation for some time and had for years been made in the same factory, although tinned with the different firms' labels.

28. The twelve were Alexander, Fergusson, Thos B. Campbell & Sons Ltd, Cooksons, Foster, Blackett & James, Locke, Blackett, Walkers, Parker (taken as three companies because of the existence of separate sales organisations at Elswick, Chester and Lambeth), Librex, Brimsdown, London Lead Oxide and LL & J. Their total sales were just over 100,000 tons, including sheet and pipe 45,000, red lead 13,000 and white lead 22,000. 46.1 per cent of sales was made in the Home Counties (28 per cent of total population), 12.7 per cent in the West Midlands (10.1 per cent), 6 per cent in Northumberland and Durham (4.5 per cent), 10 per cent in Scotland (9.8 per cent), 8 per cent in Yorkshire, Cumberland and Westmorland (9.6 per cent), 9.7 per cent in Lancashire and Cheshire (12.4 per cent) and 1.6 per cent in Ireland (8.9 per cent).

29. This was not an uncommon situation at the time. See generally, Chandler, 'Growth of the Transnational Industrial Firm', pp. 404-6. Until it began rationalisation in the early 1930s, Unilever's 49 manufacturing companies had had 48 separate sales organisations, C. Wilson, *The History of Unilever*, vol. II (1954), pp. 345-6. For the position at the Imperial Tobacco Co. Ltd, see Alford, *Development of the UK Tobacco Industry*, pp. 309-12, 330-3 and 364-70.

30. Named after the initial letters of the surnames of R.S. Brown, managing director of Brimsdown and C.A. Klein, technical director.

31. Under this Act provision was made for the imposition of import duties on certain products.

32. For British Titan Products Ltd see Cocks and Walters, *History of Zinc Smelting*, pp. 118-22 and G.A. North, *Teesside's Industrial Heritage* (Middlesbrough, 1975), pp. 108-9.

33. The report contains illustrations and details of plant in a number of American lead works. There were many points which impressed the visitors such as American improvements to Barton pots, where 'As compared with the present Bootle practice, the American method involves 7 less operations', and 'Unlike those in the Merger Factories, the packing machine was provided with automatic weighing apparatus which saved a considerable amount of time and was an excellent arrangement from a hygienic point of view *and might, with advantage, be copied in the Merger works*.' The overall impression was of the considerable extent to which mechanical rather than labour-intensive methods had been adopted in the USA, although there were areas, such as the continuous red lead furnace, in which ALM practice was considered superior. In general the differences between British and American practice were put down to the high cost

of labour in the latter (about twice the British level). This was, no doubt, a factor causing one criticism of American practice, which was of the existence of a two-shift (against the British three 8-hour shifts) system from 7.30 a.m. to 5.00 p.m. and 6.00 p.m. to 5.30 a.m., the gap between shifts being known as 'God's Shift', when the plant was looked after by casual labour. There was also criticism of dusty atmospheres and lack of care over hygiene. It is probable that the ALM representatives who went to the USA and compiled the report were C.A. Klein and H. Waring.

34. On 22 Oct. 1929 Cookson informed the ALM board that National Lead was to buy for cash at par from Cooksons, Rowe Bros and Lord Inverforth 150,000 Preference and 300,000 Ordinary Shares in ALM, with a 12 months' option on a further 50,000 Preference and 100,000 Ordinary. Intriguingly, National Lead had already purchased a further 100,000 Ordinary Shares on the open market. There is no evidence that the option was taken up but the other shares gave National Lead a holding of not far short of 20 per cent of ALM, although its holding in GWLI, a year later, was reduced proportionately as a result of the merger with Goodlass, Wall and National Lead suffered a paper loss as a result of the effective write-down of the ALM capital during the merger. There is no evidence as to why National bought the shares and neither its records nor those of ALM show any evidence that the shareholding was used to influence policy decisions – indeed with only a single, rarely-attending director it would have been difficult to do so. Apart from reinforcing the links which had already been developed between the two companies, National Lead may have been interested in obtaining an interest in ALM because of its earlier activities in titanium dioxide. It is interesting to note that although National Lead had a 49 per cent holding in British Titan Products when it was set up in 1930, its holding in GWLI (which had 17 per cent of BTP) would have given it voting control.

35. The American visit of 1928 had shown the Cooksons' representatives how backward British practice was and the final conclusion of their report was that *'It is clear that most drastic alterations to plant and methods of operation are essential in order to place the Merger in the same state of efficiency as obtains in the American Works and it may well be that but few of the existing Merger Works are capable of being so transformed.'*

36. As a result of exposure to heat and light unstabilised PVC, polyvinyl chloride, one of the new 'plastics' of the 1930s, breaks down, with the creation of hydrogen chloride and resultant discolouration. It was discovered that it could be stabilised by a number of metallic salts (including those of barium, cadmium, calcium and zinc) but basic lead carbonate was the most suitable because of its low cost as well as its good heat and light resistance. It was, however, of little use (being basically an opacifier) when high optical clarity of PVC was required.

37. Among these was antimony oxide, production of which had been commenced by Cooksons in 1919, and the paint based on it was marketed as 'Timonox'. Since the Second World War, with the domination of ordinary paints by the cheaper titanium dioxide, 'Timonox' has been sold specifically on its flame retarding properties and for the same reason antimony oxide is used in the treatment of many plastics and textiles.

38. Advertising was based on the slogan 'White lead paint lasts', which was capped by a competitor with 'Our paint lasts longer'! There was at this time considerable competition and the introduction of many 'new' paints. It can hardly be coincidental that the infamous 'Riddoppo', the 'new novelty super-paint' was introduced by H.B. Cresswell in *The Honeywood File* and *The Honeywood Settlement*, first published respectively in 1929 and 1930.

39. Although profit margins in the 1930s in paints were good there was still considerable competition. L.V. Cooper tells the story of how he persuaded

Brighton Corporation to replace Farmiloes' with Cooksons' paints for the redecoration of the Royal Pavilion but 'Soon after the war and during rationing, Champion, Druce stole a march on us by cutting a flake of stucco off a wall and having it photographed, edge on, immensely magnified. They gave the corporation a framed copy, each repaint was identified approximately with the year and the monarch reigning at the time. Brighton Corporation was greatly impressed and I had difficulty in retaining a share of the business, although by various means we got the lot the following year and have kept it ever since.' ALM, *News Bulletin*, 67 (March/April 1980). It is an interesting reflection on the link between lead and many of the major old buildings that the same issue of the *News Bulletin* carried an account of the reroofing with sheet lead of another Nash building, the Coutts block in the Strand.

40. For details of the group's financial performance see Appendix VII.

41. £35,000 was written off in 1927 and £28,000 in each of the 1928 and 1929 accounts.

42. Methods of payment were, however, archaic. One graduate, recruited from ICI in the 1930s and used to having his salary paid into his bank account, was startled at the end of his first week with ALM to receive a telephone call to the effect that his money awaited him at the gate cabin.

43. That for process workers gave 6d per week (9d from 1947) for each year of service after 20 years. Clive Cookson had 'strongly suggested that Trade Unionists should be ineligible' but was overruled. This should be taken as an expression of his personal views rather than of the company's experience, since labour in the lead industry appears to have been largely non-unionised and there is no evidence that the company experienced any serious trouble from union activity.

44. Elswick male employment during the 1920s averaged about 250 but fell from 253 in 1930 to only 179 in 1931, recovered only slightly to around 200 in the mid-1930s and rose to 227 in 1939.

45. This was represented to the board and no doubt to the employees as a restoration of 8 per cent (of the 10 per cent reduction) — as indeed the previous year's bonus of two weeks had been represented as a 4 per cent restoration. A little simple arithmetic will show that this calculation suffers from a problem of bases — a four weeks' bonus was certainly about an 8 per cent increase in wage or salary but it only made about a 7 per cent restoration of the 10 per cent cut.

46. In 1942 Champion, Druce alone received an order from the Air Ministry for one million gallons. Relations with government ministries offered much scope for subtle dealings. In March 1941, having discovered that the Admiralty was using zinc oxide/white lead paint mixed in the proportion 40/60, GWLI approached the Ministry of Supply 'to urge the Admiralty to use a higher proportion of White Lead with a view to economising the use of Spelter, which was most important from a national point of view'.

47. The Newnam hearth was the modern version of the open hearth ore smelter. Although mechanised so far as its operation was concerned it still involved considerable manual labour, since its two operatives had to shovel a total charge of perhaps six tons on to the hearth (in the face of considerable temperatures) during their eight-hour shift. The last Newnam hearths ceased production at Elswick in 1957.

48. There was also a big increase in the company's investment in British Titan Products Ltd, partly to finance expansion but also to buy out National Lead's holding.

11 THE INDUSTRY AND THE DEVELOPMENT OF ASSOCIATED LEAD MANUFACTURERS LTD SINCE THE SECOND WORLD WAR

Much study had made him very lean,
And pale and leaden-eyed.
>   Thomas Hood (1799-1845),
>   'The Dream of Eugene Aram'

*The development of ALM from 1949 to 1980. Administrative
reorganisation and abolition of the old company names; the
background to the lead industry − changing consumption of particular
products and rationalisation of supply to adapt to changes in demand
− the development of new products − control of competition via
conventions for lead products and secondly by takeovers.
Diversification away from lead and the development of new products
by the holding company in materials such as zircon and fields such as
ceramics which were linked to ALM's traditional expertise.
Administrative reorganisation in the 1970s to isolate profit centres,
which showed the small contribution made by lead to group profit and
subsequent reduction in lead capacity.*

On 1 January 1949 the major lead companies in the group were amalgamated to form Associated Lead Manufacturers Ltd[1] and the old company names were dropped. The step, discussed and planned for implementation before the War, had at last been taken. The fear of losing customer loyalty, however, still existed and it may be seen in a booklet, *The Story of Associated Lead*, issued in 1948. Despite its light-hearted introduction the booklet continually expresses the fear, stressing that the customer would experience little change, other than in company name, and would get a better service:

> When people (or companies) change their names, they invariably do
> so for the very best of reasons; and we are no exception to so general
> a rule. 'The best of reasons', however, may carry many and different
> meanings − anything from matrimonial custom or the nice requirements of an alias to the acceptance of a peerage or to a shrinking
> from the past. While we lay no claim to originality, not one of these
> considerations has influenced us − and decidedly not the last.

The late 1940s was, however, an ideal time at which to undertake the change envisaged. There was little foreign competition in lead products in the years after the War and ALM had come out of the War in a strong position *vis-à-vis* its smaller home competitors (some of which had been closed and otherwise disorganised during the War).[2] In addition there was a shortage of lead in the face of rapidly rising demand for lead products (particularly linked to an expanding building programme). In such circumstances there was little chance of losing customers merely because of a change of name. Moreover the time was propitious from an internal viewpoint. Sales reorganisation (which was the major area influenced by the 1949 restructuring), had it taken place in the 1930s, must have led to a staff surplus but at the end of the 1940s, with the sales structure run down during the War and with rising demand, it was possible to avoid redundancies and reallocate sales areas on a positive, expansionary basis which encouraged morale. ALM was therefore able to undertake the reorganisation it desired without conflict with its principle of avoiding staff redundancies and without losing customer goodwill. In practice the delay in reorganising the sales structure had proved justified.

In a concluding chapter it is not possible to do more than outline the most significant changes which have affected ALM over the last 30 years. In particular it must be made clear that the diversification of ALM's holding company, away from lead, is only briefly outlined.

A starting point for any analysis of the period since 1949 must be the demand and supply conditions within the industry which ALM faced. After the War consumption of lead recovered rapidly with renewed demand from building, cables and automobiles and these have continued to sustain the demand for lead, with cables declining in recent years (as a result of the growing challenge of synthetic materials) and the motor car becoming dominant. Figures for the outputs of specific lead products are given in Table 11.1. It will be seen that total consumption of lead reached a peak in the mid-1960s and then fell back to the 1945 level by the mid-1970s, at which level consumption appears to have stabilised. It is likely that UK consumption will fall further in the 1980s (particularly as a result of declining use of tetraethyl lead) although world consumption is forecast to rise.[4] While there has been an increase in the world output of virgin pig lead, which has gone some way towards meeting the increase in consumption since 1945, there has been a considerable increase in the use of secondary materials, chiefly the result of resmelting of old battery plates. By 1980 almost two-thirds of UK lead consumption came from secondary

Table 11.1: UK Consumption of Lead by Chief Products, 1945-80 (tons)[3]

Year	Cables	Batteries	Oxides	Tetraethyl	White lead	Sheet & pipe	Shot	Total
1945	76,300	50,800	30,600		20,100	73,700	9,300	298,800
1946	87,500	51,900	29,500		23,000	84,900	4,500	322,300
1947	96,800	55,600	29,800		21,600	87,300	2,200	322,300
1948	117,500	56,700	31,300		20,800	84,900	3,800	346,100
1949	123,300	57,300	31,000		16,600	70,800	4,000	333,800
1950	87,500	52,000	37,100		16,600	81,400	5,400	333,400
1951	93,100	62,700	39,500		19,500	73,600	5,700	346,600
1952	110,400	48,500	25,600		21,300	60,900	4,300	298,000
1953	92,600	52,600	24,800	7,400	7,700	71,700	4,500	308,600
1954	87,200	59,100	27,800	15,500	11,200	79,900	5,000	341,300
1955	112,300	61,700	28,700	22,100	12,000	80,100	4,700	378,200
1956	115,500	53,800	26,000	21,500	11,600	75,200	4,400	363,400
1957	116,200	54,400	25,000	21,700	10,300	68,800	4,300	354,800
1958	101,700	56,400	26,200	20,500	9,800	67,300	4,500	341,500
1959	97,700	60,400	27,800	24,100	9,100	69,900	4,200	351,500
1960	97,000	76,200	27,600	27,100	8,400	74,900	5,700	385,500
1961	98,400	69,100	23,400	25,700	8,300	73,900	5,500	374,500
1962	109,600	70,600	25,800	27,300	8,600	68,600	5,600	385,300
1963	110,300	77,300	24,100	30,900	8,400	67,900	5,300	393,300
1964	122,700	87,500	26,800	35,200	7,800	72,400	5,000	427,400
1965	134,700	86,200	27,200	36,700	7,000	70,200	5,800	435,100
1966	115,700	89,000	29,400	35,100	4,300	67,200	5,400	413,900
1967	102,500	87,000	28,400	36,900	4,400	67,300	5,100	393,300
1968	80,700	96,100	28,900	40,300	4,000	66,200	6,200	384,300
1969	66,500	99,300	32,200	39,500	4,500	58,500	6,200	367,100

Table 11.1: Contd

Year	Cables	Batteries	Oxides	Tetraethyl	White lead	Sheet & pipe	Shot	Total
1970	60,400	94,700	31,600	40,400	3,000	53,100	6,300	350,200
1971	51,900	101,300	30,600	44,000	3,200	54,000	6,500	346,400
1972	48,600	100,400	30,500	49,800	3,100	55,100	7,100	355,000
1973	45,800	106,500	37,100	54,400	1,700	55,300	7,400	364,100
1974	44,500	80,100	35,500	56,100	1,100	47,500	7,300	325,200
1975	36,500	66,200	26,000	58,200	1,000	50,300	8,600	293,100
1976	33,500	89,700	32,000	56,800	800	48,200	10,800	320,700
1977	31,100	97,300	30,900	55,400	1,400	47,200	8,400	320,100
1978	31,000	107,700	35,500	60,800	900	47,900	8,100	339,000
1979	26,700	113,900	32,900	58,900	1,200	47,100	7,800	335,900
1980	21,700	98,600	18,600	63,300	900	50,900	6,500	305,500

sources while Australia, together with Canada, continued to be the chief sources for newly-mined lead. The price of pig lead has followed the general inflationary pattern with the overall trend affected by cyclical fluctuations caused by variations in the level of activity in the economy and with exogenous fluctuations caused by events such as the Korean War. We have seen that the price rose steeply at the end of the Second World War and after falling back to £90 per ton in 1950 it rose to £180 during 1951. These and subsequent fluctuations (for instance from a little over £340 on average in 1978 to almost £570 in 1979) have made the running of ALM no easier than it was during the price fluctuations of the inter-war years. From 1 October 1952, however, the London Metal Exchange recommenced dealings in lead and the metal price, having been controlled by the government since the beginning of the War, was once again left to market forces. ALM then resumed its role as a ring dealing firm but in 1980 gave up its ring-dealing seat, although it continues to use the facilities of the market for 'hedging' purposes through brokers.

Although price fluctuations have continued to cause problems the situation has, in another respect, been much easier for home producers since the War. Compared with a level of imports of lead products of around 12,000 tons per annum from 1932 to 1938, UK imports averaged only a few hundred tons in the 1940s and 1950s. In the late 1960s and the 1970s there was an increase in imports (chiefly of sheet lead) from 1,000 to about 2,000 tons per annum but this was significantly less than the level of UK exports which ran at between 8,000 and 10,000 tons p.a. at that time.

In the immediate aftermath of war the level of home demand for lead products, especially for the building industry, was high and there was little concern about competitive products. By the early 1950s, however, this position was changing and the pre-war trend of decline in sheet, pipe and white lead had been re-established. In these years much work was done by the old Lead Industries Development Council to promote lead products, including the publication of a reference handbook for plumbers. As compared with the late 1930s a growing proportion of the Council's financial support was coming from the Australian and Canadian lead producers and in 1953 the Council was replaced by the Lead Development Association (LDA) in order to promote all aspects of lead use. Since then the LDA has done much publicity work in order to increase the consumption of lead and has held international conferences in conjunction with other Lead Development Associations which exist in many countries throughout the world. The LDA has an environmental study group which brings the manufacturers together to discuss problems of industrial health, it provides a technical information service for members, a library and produces publications on aspects of lead use.

Broadly speaking the period since 1945, at least until the early 1970s, was a fairly favourable one in which ALM might develop, with satisfactory demand conditions although with declining demand for certain traditional products. It was concern about the declining demand for white lead and lead pipe which had caused GWLI to diversify with the purchase of Fry's but the immediate post-war conditions of unsatisfied demand caused capacity expansion in traditional lead fields. In most instances, where this involved the installation of up-to-date rolling mills and blast furnaces to replace antiquated equipment, the investment was beneficial and was to earn ALM good returns. In at least one area, however, capital investment as a response to the post-war demand conditions proved unsatisfactory. Despite the long pattern of declining demand for white lead as a pigment, it was decided to respond to the post-war demand, created by the shortage of other pigments, by erecting a quick process plant at Millwall. This was intended to supplement the original Bootle plant (which had itself been considerably expanded during and after the War) and replace stack capacity which was being closed. Even had there been a long-term future for white lead production it would have seemed sensible to concentrate output on Bootle where expertise in the process was available but the decision to erect a plant at Millwall clearly reflected group thinking that the ability to supply from regional production points was advantageous. Quick process plant was erected at Millwall during 1948 at a cost said to have been about £70,000 but before it was completed it was recognised that demand was falling,[5] the plant was never commissioned and parts of it were transferred to Bootle soon afterwards. With all stack plant by then closed, the only alternative to Bootle quick process white lead was that made at the Hayhole chambers. The latter were gradually run down, following the introduction of the quick process at Hayhole in 1958 to provide for the manufacture of stabilisers there and the last use of the chambers was in 1961.

In the cases of other traditional lead products for which demand was declining ALM did not make mistakes comparable to that in white lead. Shot production was concentrated on Chester with the closure and demolition of the Elswick and Lambeth towers. As demand for lead pipe fell, from more than 40,000 tons p.a. in the late 1940s to less than 20,000 in the late 1970s, the company's pipe manufacturing capacity was reduced. It was, however, always much in excess of total demand, partly because of the fact that there are definite advantages to regional production of pipe to meet the demands of local customers. In support of this one can point to the fact that for most of the post-war period a number of small firms managed to remain in business producing lead sheet and pipe. In the case of lead sheet, while the same argument is to some extent true, it has on several occasions been pointed out that ALM had far too many rolling mills and that the company's output could have been produced on two of the half dozen or so mills it at one time possessed. While it is again clear that many of the smaller companies, such as Polans, had thrived on the ability to supply the immediate needs of local customers, it is doubtful that a company the size of ALM had the flexibility to make the best use of regionally-based capacity. Rolling mills are much more capital intensive than pipe presses and to have them working below capacity, as a cost resulting from a policy of regional production, suggests an unwise use of capital. Current thinking at ALM has accepted this position and a new rolling mill, being installed at Elswick at the time of writing, will supply a large proportion of group output following the closure of the Millwall and Chester mills, with distribution depots in several parts of the country.

By contrast the production of lead oxide was an area which benefited from the growing output of batteries and production of grey oxide, chiefly at Bootle, has expanded at the expense of red lead, which was the chief battery paste until the 1930s. Red lead was also being replaced by other preservatives in the coating of iron and steel (and demand from this area was also being reduced by the inroads into iron and steel consumption which were being made by plastics). As a result red lead capacity has been reduced and production at the Gravesend works, originally belonging to the London Lead Oxide Co. Ltd, ceased in 1976 and the site was sold. Nevertheless the electro-chemical characteristics of red lead make it particularly suitable for use in traction batteries, an area in which there has been some increase in demand. Red lead has, therefore, adapted to changing circumstances, as has smelting capacity. Declining supplies of UK lead ores led to the closure of the Newnam hearth furnaces by the end of the 1950s and the cessation of primary smelting by ALM. Meanwhile secondary smelting capacity, from blast and rotary furnaces, had been built up at Chester, Millwall and Elswick to cope with expanded supplies of secondary lead, especially used batteries which became more widely available with the increase in car ownership. This expansion also required the extension of refining capacity as it became necessary to purify secondary lead of its content of trace metals. Secondary smelting and refining became a profitable activity but it attracted competition. The Britannia Lead Co. Ltd, originally set up in the early 1930s by the Australian Mt Isa lead producers as a primary smelter at Northfleet, added secondary smelting, while Enthovens also expanded from primary to secondary smelting. Also in the field are Chloride Metals Ltd (a subsidiary of Chloride Group Ltd which had purchased a further secondary smelter, Lead & Alloys Ltd of Wakefield), which clearly had an advantage over its competitors in obtaining supplies of used batteries which provided it with the raw material for new production. Four[6] large and very powerful companies had proved too large a capacity for the available supplies of secondary lead, which have declined especially as batteries have lasted longer, and ALM, finding itself with excess capacity, has had to close plant. In 1978 smelting capacity at Millwall was closed and the plant became a collection point for smelting materials from London and the south-east which could then be transported by road to Elswick or Chester for treatment. Even this move, however, left the company with excess capacity and in 1980 smelting capacity at Chester was closed and the company's secondary smelting was concentrated at Elswick, with collection depots being set up at Birmingham, Chester and in the south-east.

Declining levels of activity in some areas have not however been the whole story of development since 1949. Largely as a result of the research and development work undertaken at Perivale a number of new uses for lead has been developed. Perhaps the most significant of these has been the use of white lead as a stabiliser to give thermal stability to PVC. Initially an American development, the basic technical knowledge came to ALM through the exchange of information agreement with National Lead and the Perivale research centre then began experiments with white lead as a stabiliser. Production commenced at Hayhole in 1942 and sales of white lead stabiliser exceeded 1,000 tons p.a. by 1960. As earlier, with the development of the battery industry, there were some problems with lead absorption by employees in plastics firms manufacturing rigid PVC (which had to take its lead stabiliser in powder form since there was neither plasticiser nor oil with which to blend it). As the output of rigid PVC rose rapidly in the later 1960s there were two responses. The first was the development of the 'small packs' business (initiated by Novadel and followed by Mersey), where the lead manufacturer combined the stabiliser with other ingredients required by the plastics manufacturers in packs of 2-10 kg, which eliminated the need for handling the stabiliser. Secondly came the introduction of granulation of the stabiliser which reduced the likelihood of dust in works' atmospheres. As the demand from the plastics manufacturers increased there came the development of new lead stabilisers of which the most significant was tri-basic lead sulphate, first produced in America and developed in the UK from 1949 by ALM.[7] This was to become much the most significant of the lead stabilisers, overtaking basic lead carbonate in the mid-1960s, and in the late 1970s, with an output of more than 5,000 tons per annum, it accounted for about two-thirds of ALM's total output of stabilisers.

Another new product, an original development from the Perivale research centre based on the work of N.J. Read, was calcium orthoplumbate, a rust-inhibiting pigment which was to make inroads into the traditional market for red lead. In 1949 a plant was erected on Tyneside to manufacture this material by heating lead oxide with calcium hydroxide. Other developments in the pigment field have included the development and improvement of various flame-retardant paints, using antimony as a base. Not all the new developments have, however, been on the chemical side and it is clear that lead's use as a metal has not been exhausted. In particular its resistance to corrosion, despite the growing importance of plastics, has attracted attention. From the late 1940s there were attempts both by Cox Bros at Derby (who licensed the Schori process) and ALM at Chester, to develop a satisfactory process for spraying steel sheets with powdered lead and then fusing it with a flame gun to produce a weatherproof construction material, although they were not entirely satisfactory. By the early 1970s, however, cold-roll bonding techniques, developed by ALM from initial research by the British Non-ferrous Metals Association, were being used to produce a lead-clad steel, which has since been used as a protective and decorative material on a number of buildings, bringing up to date a traditional use of lead which goes back many centuries. With these and other developments the scope for a continued role for lead in modern society has been considerable despite the rise of new materials and the range of products in which lead plays a part is both wide and sometimes surprising.[8]

Closely associated with the need to take decisions over the closure of plant for products the demand for which was declining and the development of new products was the relationship between ALM and its competitors. As in the inter-war years there were two solutions, cartelisation and takeover, although neither bulk so large in the more buoyant markets post-1945. The international cartels became moribund with the onset of the War and were not renewed when it was over. International trade in lead products was of insignificant proportions compared with the levels of the 1920s and there was no incentive to cartelisation. With a resurgence of foreign competition from the later 1950s it became necessary to defend the home market, however, and price agreements were once again made. European agreements for both white lead and lead oxides were established which were subsequently converted into tonnage and quota agreements when price agreements became illegal. Further legislation led to the conversion of such agreements to an exchange of statistics on tonnage and these now exist on a European basis for lead oxides and lead stabilisers. The subject of international agreements between companies has been one which has caused considerable public interest and disquiet. In 1977, however, the EEC Commission instituted an investigation into the way in which the white lead agreement had worked out but, although it found ALM and Continental manufacturers technically guilty of infringement of EEC legislation, no fine was imposed. Current arrangements for the exchange of statistical information on market size may clearly be used either to encourage or discourage international competition dependent on the individual company's assessment of its market power.

The UK conventions had continued in existence despite the War but neither of those for red or white lead appears to have been of much significance as a result of the declining level of competition in the products. ALM heavily dominated red lead production leaving the lead oxide convention with little significance but two new firms were attracted by the high prices and began to contest ALM's monopoly. These firms were Deanshanger Oxide Ltd of Milton Keynes (a company now 50 per cent owned by Morris Ashby Ltd, which in its turn is owned by Billiton (UK) Ltd) and Bourne Chemicals Ltd of Welwyn Garden City. They soon built up a useful share of the UK market and each now has about one-sixth, with ALM having the remainder. In 1972 these companies were brought into a resuscitated lead oxide association although Deanshanger subsequently resigned. ALM was instrumental in the formation of this association, since by this means the company obtained not only a united voice for the interests of the oxide manufacturers but also access to international organisations and the exchange of statistics which were not open to it as an independent firm.[9] The association is (in 1982) largely inactive, holding only an annual meeting with exchange of statistics.

While the lead oxide market was becoming more competitive, that in white lead was becoming less so as outside competitors declined in number. At the end of the 1940s and in the early 1950s, with the fall in white lead consumption as more and more paint manufacturers turned to the use of titanium dioxide, several firms had closed their white lead departments. In 1950, for instance, Champion, Druce, largely owned by ALM, did so and in the same year Cox Bros closed its stacks and, having considered erecting a quick process plant, decided against doing so and arranged to take its supplies of white lead from ALM. In addition, competition was further restricted by the market sharing agreement between ALM and the most significant remaining firm, Mersey. As we have seen this had commenced in 1927 but in the 1960s Novadel, the only other remaining white lead producer, was brought into the agreement. Together with the fact that other outside firms, such as Robinson Bros in 1954, had given up the production of white lead and agreed to take their supplies from pool members, this meant that white lead was produced under virtual monopoly conditions and the White Lead Convention had little relevance. Following the passage of the 1956 Monopolies and Restrictive Practices Act the pooling agreement between ALM and Mersey, like several thousand others, was made a registerable restrictive practice. There was at first some discussion about the possibility of defending the agreement before the Restrictive Practices Court but, having seen the general tenor of that institution's decisions, it was decided to wind up the White Lead Marketing Co. and the intention of the Act was evaded by the signing of an exchange of information agreement between the companies by which price stability was retained. Control over white lead was further ensured when Novadel's white lead interests were purchased by ALM from 1 January 1971 and its manufacturing plant closed down. When, in 1972, ALM took over Mersey it became the single supplier of white lead in the UK, although a subsequent investigation by the EEC Commission did not find it guilty of using this position against the public interest.

Unlike the white and red lead conventions, that for blue lead had a more positive role to play after the War, purely because ALM was a much less dominant supplier and there existed a number of small competing firms. It was estimated in the late 1940s that there were 29 large and 10 small lead rolling mills and 164 pipe presses shared between about 70 firms seeking a part of a total UK output of from 70,000 to 80,000 tons p.a. For comparison the number was down to fewer than ten firms in 1981, contesting a total market of about 50,000 tons p.a. The reduction in numbers has been the result of the disappearance of the majority of the small firms which had survived from the inter-war period. Many of them, which had originally been builders' merchants and had taken up the manufacture of sheet and pipe to meet their own

requirements, gave up production as uneconomic and took their requirements from the big manufacturers. These included firms such as Stocks
Sons & Taylor Ltd of Birmingham, Rosewall (Southampton) Ltd and
Robert Fell & Sons Ltd of Skipton. Yet other firms gave up sheet and
pipe manufacture to concentrate on more profitable products, such as
Grey & Marten Ltd of Lambeth, which has concentrated on the manufacture of solders and, as a constituent of E.G.M. Solders Ltd, is yet
another part of the Billiton (and, therefore, Shell) Group.

Yet others of the small sheet and pipe producers have been taken
over by the larger companies, several of them, as we shall see, by ALM.
As a result the manufacture of sheet and pipe is now heavily oligopolistic
with the majority of firms belonging to the British Lead Manufacturers
Association (BLMA) having less than 50 employees and producing little
in the way of sheet and pipe. Approximately 90 per cent of output is
produced by three members of the BLMA, of which two, ALM and the
Firland Group are considerably larger than the third, Sheldon, Bush.
The latter company we saw in Chapter 3 to have remained independent
and, as the successor to William Watts, the firm celebrated its bicentenary
in December 1982. Sheldon, Bush has recently expanded by the purchase of a share in Highfield Lead Mills Ltd of Liverpool. This was a
company controlled by a consortium of builders' merchants, including
Rowe Bros & Co. Ltd, whose share was sold in 1972 to the Tilling
Group, Baxendale & Co. Ltd, whose share has been sold to Sheldon,
Bush and Giddings & Dacre Ltd, another old-established sheet and pipe
manufacturer of Manchester, now owned by Whitecroft Ltd.

British Lead Mills Ltd of Welwyn Garden City, originally set up in
1932, was attracted by the higher prices which resulted from the
increased control of the UK market by the UKLMA and for a long time
it was a serious price competitor. Gradually, however, it became part of
the establishment. In 1949 British Lead Mills Ltd was purchased by
Firth Cleveland Ltd and became the founder member of the Firland
Metals Group. That group steadily expanded to become the UK's
second largest producer of sheet and pipe and has taken over a number
of smaller producers, including R.E. Roberts & Son Ltd of Bolton,
Holman, Michell & Co. Ltd of St Helens and Brunton & Co. Ltd of
south London. Firth Cleveland Ltd, inclusive of the Firland Metals
Group, was acquired by GKN in 1972 and the Firland Group was sold
to Billiton (UK) Ltd in 1977. It thus means that the Lead Division of
the latter group, with its involvement in lead smelting, and the manufacture of lead oxides as well as sheet and pipe, has become the major
competitor to ALM in lead manufacture.

After the War the dominant blue lead organisation was the Lead Sheet and Pipe Manufacturers Federation[10] which had about 80 per cent of the total UK trade represented in its membership. The Federation came to a formal end in the Restrictive Practices Court in May 1962 when it was decided not to defend the agreement.[11] Price-fixing was then replaced by information agreements among the members of the BLMA, which still exists to provide its members with information and links to the CBI, government bodies and institutions in the building industry. As a result of the research and publicity work carried out by the various blue lead organisations and by the Lead Development Association[12] the consumption of lead sheet and, to a much lesser extent, pipe has remained at a much higher level in the UK than in any other western country. Indeed, in 1980 output exceeded 50,000 tons, the highest level since 1968, while profit margins were high, but this has had the disadvantage of attracting the first new firms to be set up in this area of lead production for many years. These firms, two of which were set up during 1981, are Avomet Ltd of Ebbw Vale (in which the A. Cohen Group and a South African company are involved), Dumetco Ltd of Dublin (a firm set up by Dutch interests) and George Royston Ltd of Barnsley, a subsidiary of Central and Sheerwood Ltd. None of the three has joined the BLMA and it is clear that each intends to carve out a slice of the UK market and the blue lead trade is obviously in for one of its periods of increased competition. Of the three newcomers Royston is probably the least significant and is unlikely to have any serious impact. A supplier of chemical lead, it has decided to attempt to make its own lead on a second-hand mill but it is always possible that it will be tempted to use spare capacity to enter the general sheet lead market. The other two companies represent potential competition of a totally different order. While Avomet has a traditional rolling mill, both companies have installed DM casting machines. This is a method of manufacturing sheet lead, chiefly for soundproofing, which has been around for some time, although attempts are now being made to use it for weather proofing for which it appears to be unsatisfactory. However, the capital cost of a DM machine is perhaps one-tenth of that of a new rolling mill and it will be interesting to see whether the problems can be overcome and the new technique used to obtain a significant share of the sheet lead market. In addition to these companies, the Dutch firm Hoboken has set up a British subsidiary, Sogemin Ltd, with a London warehouse for the supply of lead sheet to the British market and it seems likely that other continental producers will show interest in the currently buoyant UK market.

While agreements of various kinds between companies have had some impact on the lead market, the second method of controlling competition lay in takeover and, after two quiescent decades in the 1930s and 1940s, when the earlier mergers were being digested, ALM again followed this path after its reorganisation in 1949. The pace of takeover was much less hectic than it had been in the 1920s and it has been suggested by members of the company that its policy was not to pursue a policy of takeover but merely to respond to opportunities which were offered to it. While this would appear a rational policy to be adopted by much the largest organisation in an industry in which the demand for several of its traditional products was falling, it does have elements of a *post hoc ergo propter hoc* argument. It was inevitable that small producers of sheet and pipe or white lead would be squeezed out of the industry by the inexorable pressures of declining demand which meant that their own individual outputs would eventually be forced below an economic level, since the functioning of convention and marketing agreements meant that they could not increase their market share. Meanwhile ALM could respond to a declining level of output by reducing capacity and diversifying – opportunities which were limited for the small single-product firms – and wait for competition to be reduced by the closure of smaller competitors or their offer for sale at a price much lower than ALM would have had to offer if it were making the running. Undoubtedly this position did occur but it is also clear from the board minutes of other companies that ALM did from time to time make the initial overtures for takeover.

In 1954 ALM had bought up the outside shareholdings in Champion, Druce and thus brought the Islington company, which had left the Walkers partnership in 1825, back into the fold. Similarly in 1961 Cox Bros & Co. (Derby) Ltd was taken over by ALM, although the latter had first made an offer for the company in 1959. Again this was a case of the repurchase of an old possession, since the Walkers partnership had commenced the Mill Hill, Derby works in 1792 and sold it to Cox in 1839. In neither of these takeovers is it possible to see much economic logic. Champion, Druce had made good profits during and after the War and was financially viable on the basis of its paint output, having closed its white lead department in 1950 and agreed to draw supplies from other convention members. It may be that ALM at this time felt that control of one of the declining number of users of white lead for paint was beneficial and that its aim of expanding its own output of paint would be assisted if it bought out the shareholdings in Champion, Druce belonging to other convention members. The Islington works was closed and ALM's London paint output concentrated at Millwall. When Cox Bros was taken over in 1961 it had already ceased white lead production and was producing sheet and pipe, red lead and some paint. The firm appears to have been an example of the 'small is beautiful' thesis – providing a very good service to its customers by tailoring production promptly to their requirements. As such it would not have seemed a suitable purchase for ALM which could not have hoped to retain much of the customer loyalty. Otherwise Cox's only attraction was that it had the largest rolling mill in the country – of 12 ft width – largely wasted since it only rarely rolled sheets to that size. Although Cox Bros had made comfortable profits for most of the post-war period it was in difficulties in the late 1950s with steadily declining reserves. A purchase price of 35s per £1 ordinary share seems a high price to have paid for poor quality assets. The paint department at Derby was closed in 1962 and although sheet and pipe output was retained that was subsequently closed down as well.

It seems probable that both of these takeovers were initiated by ALM and there is clear evidence of interest in other companies. The Managing Directors Committee (Lead Side) of GWLI occasionally had discussion of competing firms on its agenda. For example in February 1950 Quirk, Barton & Co. Ltd was mentioned as follows:

There is no management there at all except for G.H. Quirk. The other two or three Directors are old men between 70 and 75 years of age and it is said they have not visited the Works for many years and know next to nothing about the Company's affairs. . .
Here again I have no idea whether they are open for purchase. . . .
The question of taking them over was mooted many years ago.

In 1957 the company was taken over by ALM, its Rotherhithe factory closed and output concentrated at Millwall to which works Quirk, Barton's tea lead rolling mills were transferred. In 1959 the sale of the Rotherhithe property of Quirk, Barton realised £75,000 but this would not have gone far towards repaying ALM for its takeover. In ALM discussions in the 1950s there is a frequently expressed fear that a weak competitor might be taken over by an outsider to the industry who would then compete vigorously for trade and it seems likely that several of ALM's takeovers may be attributed to the need to quell this particular fear.

There have also been a number of takeovers by ALM among the sheet and pipe manufacturers. Among these was Henry Halmshaw Ltd of Hull, taken over in 1966, a firm that had been a thorn in ALM's flesh in that area in competition with the Hull branch which had once belonged to Walkers, Parker. The Hull works of Halmshaw was closed and the firm amalgamated with Polans of Sheffield, which had been taken over by ALM in 1965. In 1968 the Mining Company of Ireland and Strachan Bros Ltd was taken over, although it was made part of the Fry's Metal Group rather than ALM, because the former already had a subsidiary in Southern Ireland. Subsequently, in 1971, the sheet and pipe manufacturing interests of James & Rosewall (Plymouth) Ltd were also taken over by ALM as were those of Glynwed Ltd[13] in 1979. In these later instances ALM had no interest in retaining the manufacturing capacity but was concerned to retain the goodwill. The manufacturing works were closed and the rolling mills cannibalised for spare parts to ensure that they did not fall into an outsider's hands. There was also at least one abortive takeover. In 1966 ALM showed interest in purchasing the Liverpool works of Highfield Lead Mills Ltd with its 'well equipped and well run mill'. It offered 'another important step towards rationalising the Industry' and caused ALM's managing directors to ponder the possibility of not replacing the Chester and Elswick mills which were 'pretty old' and concentrating output on the new acquisitions. As it happened, however, Highfield was not taken over since it proved impossible to agree to acceptable terms on which ALM would subsequently supply the consortium of builders' merchants which owned Highfield.

The remaining takeover of significance, worth exploring in some detail since it throws light on the way in which a small company could be competitive, was that of Mersey in 1972, which made ALM the sole UK producer of white lead. From the 1920s Mersey had had an almost incestuous relationship with ALM, with joint directors and marketing activity but yet maintaining a stubborn independence. Bound as it was by the pooling agreement, Mersey offered no threat to ALM and could be tolerated and even encouraged. In 1949, for instance, Mersey adopted the Librex quick process, details of which were obtained from ALM, whose engineers drew up plans for and supervised the installation of the plant. Although this was a satisfactory development while demand was high, Mersey began to feel the pinch of a declining total market for white lead in the 1950s and the pooling agreement effectively prevented the company from increasing its market share. Concerned about the future, C.F. Hill tried to sell Mersey to ALM but could not agree terms and after his death in 1958 the company was reorganised under the leadership of John Day, grandson of one of the founders.

About this time there was further discussion of the possibility of the sale of the company to ALM and the latter could undoubtedly have bought Mersey for a very low price but presumably felt that such a purchase was not worthwhile. Meanwhile Mersey had decided to develop in the stabiliser market as a means of maintaining its white lead output. It did so by the simple expedient of copying existing products (after much trial and error) — since stabiliser prices were high it was not difficult to persuade the various plastics companies to produce samples of their existing consumption by holding out the offer of producing a cheaper product. Since stabilisers were not included in the exchange of information agreement, Mersey was in a position to attack ALM's market and this it did very effectively. By the time that an exchange of information agreement on basic lead carbonate stabilisers was signed with ALM in 1963 Mersey had rather more than 60 per cent of the total UK sales of such products of the two companies.

From Mersey's viewpoint the signing of the agreement was an expression of the fact that the company could now hold its own with ALM but for the latter it merely recognised the fact that white lead stabilisers were of declining importance. ALM was already expanding sales of a more sophisticated plastics stabiliser, tri-basic lead sulphate, and in 1965 its total sales of this product, at over 2,000 tons, exceeded those of white lead stabilisers for the first time. Once again Mersey had to copy the new technology and, after many difficulties, did so and as a result of price competition had obtained about one-third of the total UK sales of the two companies by the early 1970s, although it had a much smaller share of export markets. Apart from price competition Mersey's success came from the flexibility and adaptability possessed by a small company in meeting consumer demand. For instance, Mersey recognised the opportunities offered by a rapidly expanding rigid PVC market in the later 1960s more quickly than did ALM and it was very successful in pioneering small packs of stabilisers blended with customers' requirements of other raw materials.

Mersey had made considerable steps from the rather somnolent company of the 1950s. In the years 1955 to 1958 its annual output of white lead was running at a little over 2,000 tons but in 1968 deliveries of white lead equivalent, at over 4,000 tons, were the highest for 30 years. Having obtained a respectable share of the market for tri-basic lead sulphate Mersey accepted the extension of the exchange of information agreement to that product, which led to a rise in price of about £25 per tonne. Competition between the two companies, it was recognised, had benefited only the consumer. From the point of view of both companies a closer arrangement appeared beneficial. ALM had discovered that Mersey could offer serious competition, albeit in a very small part of ALM's total product range. Mersey, on the other hand, had established a satisfactory base of profitability, which justified a reasonable sale price for the company. Since the latter's executive directors were approaching retirement age with no obvious young men to follow them (inevitably a problem for a small company anxious to keep down overheads) and there was considerable concern as to how entry to the EEC would affect the company's sales, it was with some gratitude that Mersey received an offer in July 1972 from ALM's parent company.[14] In the early years after the takeover there was little change at Mersey beyond the replacement of the outside directors with ALM representatives, since it had been agreed that Mersey would be kept in production under its own name. Gradually, however, the Warrington works was run down and closed in 1980 and production transferred elsewhere. Inevitably this meant some ill-feeling among Mersey men, although the case for the eventual closure of the Warrington works, given ALM's wide diversity of works and the considerable potential cost of replacing old and unsuitable buildings at Warrington, was a strong one.

The various takeovers outlined above gave ALM a marked increase in control over the output of a number of products, including a monopoly of the outputs of white lead and lead stabilisers after 1972, although, as we have seen, considerable competition has remained in blue lead products. Although many of the firms taken over in the 1950s and 1960s were rapidly closed down in order to reduce surplus capacity, the takeovers added to the problems of the overall organisational structure in ALM. This was increased by the fact that expansion came not only by takeover in traditional lead manufacturing fields but also by overseas developments and by diversification. The South African plant at Jacobs, which began production in 1950 has already been mentioned. It has subsequently developed a considerable output of lead products including stabilisers. In addition in 1948 a joint company with Enthovens, Enthoven Fry Ltd, was set up at Germiston (Transvaal) to manufacture lead and tin alloys. A second factory was commenced in 1953 at Port Elizabeth and in 1956 the Enthoven interest was bought out and the company became a wholly owned subsidiary of GWLI with the new title Fry's Metals (Pty.) Ltd. There were further developments in India based on the local company, Waldies, with which ALM had joined in the inter-war years. In Australia LIG has a substantial shareholding in Dulux (Pty.) Ltd,[15] the owner of Commonwealth Litharge and Red Lead, which manufactures lead oxides, pigments and stabilisers.

In 1977 Penalmex was established in works near Paris to manufacture lead stabilisers, jointly owned by ALM, which was to provide technical expertise, and Penarroya, which was to undertake administration and marketing. This arrangement proved unsatisfactory, which, together with the fact that the level of continental competition in stabilisers was greater than had been expected, led to the closure of the company in 1980. While these overseas activities were aimed at securing foreign markets for lead products and paints which had been built up on the basis of exports, it was clear that the overall scope for lead products was limited and that diversification was needed.

This need for diversification had clearly been recognised by the board of GWLI at least by the early 1940s and the first response was the purchase of Fry's Metal Foundries in 1943. Expansion was to take place in areas closely akin to the existing interests of GWLI rather than follow the path which would lead to a conglomerate.[16] Diversification, however, inevitably meant a decline in the significance of lead and, therefore, ALM within GWLI. As we have seen, up to the late 1940s the lead side made up virtually the whole of the GWLI board but, as a result of decisions following from its recognition of the need for diversification, the subsequent period has seen major growth of the non-lead side of GWLI (and later of LIG), although this growth has been largely directed by men who originally started in the lead side. Although this history has been concerned with the lead companies it is necessary to pay some attention to the diversification which took place, largely under the GWLI banner, in order to show the way in which that company has developed up to the present.

In 1954 Fry's Diecastings Ltd was purchased and made part of the newly established Fry Group of companies, which also has works in the USA, Ireland, India and South Africa.[17] This group also includes Durastic Ltd, a firm producing flooring and roofing materials, which has since 1950 occupied the Burdett Road premises which had previously belonged to LL & J. Also in 1950 the manufacture of zirconium products was commenced by ALM at Howdon on the site of the old Cookson blue lead works which had been unused since the end of the 1920s. Zircon (zirconium silicate) is a highly refractory mineral extracted from beach sands chiefly found in Australia and Florida. The most immediate link with ALM's existing business lay in the fact that zirconium oxide was challenging antimony oxide as an opacifier in vitreous enamels and was, therefore, a potential threat to part of the output of the Howdon antimony works. Moreover zircon was also taking over from tin oxide as an opacifier for glazes in the pottery industry, while zirconium oxide was an ingredient of ceramic stains. Since ALM was already heavily involved with the supply of lead frits as a medium for pottery glazes, the development of zircon presented an interesting possibility for diversification into a product which was closely connected with the company's existing activities. As with stabilisers the technology came once again from the USA and in 1947 R.A. Cookson went to that country and obtained a licence from TAM Ceramics Inc., a company subsequently purchased by LIG in 1979, to manufacture zirconium products in the UK. With the introduction of zircon at Howdon, ALM was able to extend its position as a supplier to the UK ceramics industry and, with considerable expansion of overseas demand for zirconium products, was able to set up factories abroad in, for example, Italy in 1963 and Spain in 1973. As a result ALM was probably the world's largest producer of both antimony and zirconium products, and a new company, Anzon Ltd, was set up from 1 January 1977 to take over the manufacture of these products. This followed an earlier reorganisation which had separated antimony and zircon from the lead selling organisation. Anzon thus became, like ALM, a subsidiary of LIG, rather than merely a works of ALM. Closely linked with this expansion in the field of ceramic glazes were the takeovers of Harrison & Sons (Hanley) Ltd and E.W.T. Mayer Ltd in May 1963 and August 1965 respectively and their merger as Harrison Mayer Ltd from 1 January 1967. This company, with a new manufacturing works in the Potteries completed in 1971, produces glazes and colours for the ceramics industry using products from ALM and Anzon.

While Fry's and the ceramics developments have been the most successful long term diversifications, there have also been other developments. During the War ALM had commenced at Elswick the production of aluminium by smelting secondary materials but after a period of some profitability this department was closed in 1975 as a result of increasing competition and narrowing profit margins. Also at Elswick there was a brief flirtation with the manufacture of zinc and, at research and development level, with the production of copper from the copper drosses produced as a by-product of lead smelting and previously sold to copper manufacturers abroad. In 1956 ALM became the only UK producer of lithium, the lightest solid and third lightest element by atomic weight, with plant established at Bootle. As a material in many modern processes[18] lithium and its products offered a diversification of considerable significance which proved profitable to ALM. The main raw material source, however, the mineral petalite (which contains about 4 per cent lithium oxide), came in suitable quality and quantity only from Rhodesia and supplies dried up after the declaration of independence by Ian Smith. Briefly, thereafter, production of lithium and lithium salts was continued at Bootle from materials obtained from both the USA and USSR. This proved commercially unsatisfactory but although manufacture at Bootle soon ceased, trading continued successfully until the mid-1970s on the basis of lithium products purchased from a manufacturer in the USA.

The overall result of diversification into new products, takeover of additional lead companies and diversification by takeover of non-lead companies was considerable pressure on administrative structures. In 1949, when the administrative problems inherited from the 1920s' mergers at last came to be dealt with, ALM had been constituted with a regional structure, the Southern Area based on London, the Northern Area based on Newcastle upon Tyne and the North-western Area based on Chester. This was obviously a rationalisation based on existing works and aimed at dividing up a large organisation into more manageable units. It also made considerable sense for ease of administration and delivery of heavy products at a time when the UK road network was unsuitable for long distance, especially cross country, journeys. But although regionalisation may have been a suitable way of dealing with overlapping sales structures and offered some opportunity for product rationalisation within the regions, it did little to encourage elimination of excess capacity within the group as a whole or recognition of the most efficient plants for particular products. Most of the major lead products continued to be produced in all three regions, although white lead was no longer produced in the Southern Area, only the North-western Area produced grey oxide and a Special Chemicals Division, producing among other things stabilisers, was set up at Hayhole in the Northern Area.

The regional structure could, therefore, only be a temporary solution to administrative problems and, indeed, diversification was going to increase the necessity for further administrative reorganisation. From 1949, when ALM was set up with a regional structure, it becomes increasingly difficult to continue to make a clear product distinction between ALM and its holding company, GWLI (and then LIG). Between 1930 and 1949 one may reasonably make the distinction that ALM was the lead arm of GWLI, while Goodlass, Wall was the paint arm, although most of the lead companies had expanding interests in lead paint. But the takeover of Fry's in 1943 meant that GWLI had a third major area of interest while a further problem was created in 1949. Although ALM took over most of the lead companies, GWLI retained as direct investments those lead companies which were not wholly owned, including Harburger, Alexander, Fergusson,[19] T.B. Campbell, Champion, Druce and the shareholdings in BALM and also British Titan Products. Subsequent takeovers, although always undertaken by GWLI, or later LIG, led to additions to ALM and other subsidiaries of the holding company. In one of these instances, the takeover of the Mining Co. of Ireland, a lead manufacturing company was added to the Fry Group. While the most significant diversification of product range came in those areas controlled by GWLI, ALM was not merely a lead company. It controlled the manufacture of antimony because of the historical accident that this product had been developed by a lead company, Cooksons, and, because zircon was seen as complementary to the lead companies' supplies to the pottery industry, the manufacture of zirconium products after 1950 was undertaken by ALM.

The addition of new companies with a consequent increase in inter-company trading at non-market prices compounded the problem which already existed in the 1949 ALM regional structure. While the regions created profit centres and therefore broke up the monolith, it was impos-sible to allocate accurate costings even in such a specialised area as stabiliser production, where the profit margins in the 1960s were regarded as very much 'back of the envelope' calculations. By the early 1970s it was coming to be felt within LIG, whose board was by now much less dominated by men who had grown up in the lead industry and had an affection for it, that the lead parts of ALM were no longer pulling their weight in profit terms. Moreover as a result of the developments of the previous two decades ALM found itself a much smaller proportion of LIG (by the early 1970s ALM accounted for little more than one-third of the turnover of LIG, which was around 150th in *The Times*' lists of the largest UK companies for most of the late 1960s and 1970s). ALM had traditionally dominated the holding company but its significance by the early 1970s depended heavily on its non-lead interests. Moreover the share of LIG's profits contributed by ALM was declining[20] and in 1974 profits accruing to LIG from associated com-panies actually exceeded group profit from subsidiaries. It was clear that the relative profitability of particular parts of the group had to be established before further investment took place.

For ALM, however, the problem of isolating the profitable sectors could only be overcome by an administrative reorganisation from a geographical to a divisional structure. Although few large companies had followed the ICI example before the War[21] and in 1950 only 13 of the 100 largest UK companies had done so, by 1960 the number had increased to 30 and by 1970 72 had adopted a divisional organisation.[22] The result, as one recent survey of the British economy has put it, was that

> Day-to-day operations and the formulation of many investment pro-
> posals were the responsibility of several specialist autonomous
> divisions, leaving to the main board of directors the responsibility for
> choosing from amongst the investment proposals put up to them by
> the divisions, and for choosing the senior managers of the divisions.[23]

ALM was late to follow this trend but it was commenced with the setting up of Anzon Ltd in 1977, stripping the antimony and zircon interests out of ALM and making the new company a direct subsidiary of LIG. This was an obvious and easy step to take since there was no inter-company trading between antimony and zircon products and the lead interests, but to go further with the formation of individual companies within lead would have encountered this problem. The removal of antimony and zircon, however, made it clear how little contribution to overall profit was being made by the lead interests and in 1978 ALM finally adopted a divisional structure.

Within the old area structure there had been moves in this direction from 1965 with marketing and some accountancy undertaken in three divisions, Building Materials, Pigments and Chemicals, and Smelting and Alloys. While there was a commercial director for each marketing division, who was a member of the ALM board, the allocation of over-heads to each division was done, within each region, on a very sketchy basis.[24] Stemming from the earlier developments the company was sub-divided as from 1 January 1979 into three formal divisions, Chemicals, Metals and Paints, representing the three clear product divisions and regional structure was abolished. Paints division had its head office at Millwall and the paints factory there was divided completely from the lead works and was responsible for all ALM's output of paint. Chemicals and Metals divisions had several works each, including one joint works, Chester. Chemicals division had its head office at Elswick (where it was completely separate from the lead works) and was responsible for Hayhole, Bootle and the Chester output of chemicals. Metals division had its head office in London at Clements House but its accounting organisation at Chester and it controlled works at Millwall, Elswick, Sheffield (Halmshaw/Polan), Glasgow (Alexander, Fergusson) and the metal output of Chester. As a result lines of responsibility were made much clearer, with a single director in ultimate charge of purchasing of materials and production and marketing of the final products of each division, and it became possible to allocate overheads more effectively and assess the profitability of particular products.

The move to a divisional structure came too late to give ALM the benefits, which might have accrued had it come earlier, of more effective rationalisation of output and within two years of the introduction of divisionalisation further changes proved necessary because of intolerable pressures from excess capacity and the high fixed costs of running a considerable number of geographically diversified works. The changes experienced by ALM within the last few years have no doubt been bewildering to many of its employees but it is merely a microcosm of what has been happening to British industry as a whole. Dependent on when this last chapter was written one might within that time have chosen any one of several different points at which to end. In 1977 Walkers, Parker & Co. Ltd was struck off the Companies Register after a period of 199 years in which the name of Walker had been part of the lead industry. In 1978 ALM celebrated the bicentenary of the founding of the first of its constituent companies, Walkers, Fishwick & Co. From the beginning of 1979 came the implementation of divisionalisation with its promise of a more efficient structure and more effective growth in the future. Also in 1979 LIG announced that it was expanding its existing operation in the USA by taking over part of the lead interests of NL Industries Inc. (formerly National Lead) for the sum of $40m. Such an expansion, backed up by the discovery that productivity in the American plants was certainly no higher than that in the UK, might have been a suitable point at which to end, emphasising the technical competence of ALM and the fact that it still had a future in lead even in the most advanced country in the world. In the following year, 1980, ALM's board of directors was moved from London to Newcastle, an unfashionable step wisely based on the desire to avoid high London office rents and no doubt influenced by the fact that the newly-appointed chairman was a Geordie. Again this might have been a pleasant note on which to end, emphasising a return of the power centre to Elswick, where it had all begun more than two centuries earlier. Unfortunately, however, the researching and writing of history takes time and during that time events do not stand still but in themselves become the stuff of history. Rather than ending on a pleasant note of expansion it is necessary to end on contraction, although at the time of going to press in 1983 the change of name of the holding company from LIG to Cookson Group marks the special role that that lead company played in the creation of the whole structure.

In May 1981, however, proposals were announced for the closure of the Millwall and Chester[25] works of ALM as a response to the deterioration in the company's profitability. These proposals are being carried through and will be completed during 1983, with a considerable reduction in employment and the severing of links with the long-established works at Millwall. As the chairman put it in a notice informing the employees of the company of the proposed closures, 'In the view of the Board such dramatic changes represent the only means of providing a sound basis for the long-term survival of the Company' since 'without a major reduction in the present high level of fixed overheads, there can be no realistic prospect of adequate profitability in the foreseeable future'. In such circumstances there is no scope for platitudes about the past. Mistakes have been made and the failure to rationalise production sooner is going to have disastrous effects on the lives of many employees. However having had the problem forcibly brought to its attention by declining profitability it is better that the company should take severe action now to prune its unwieldy limbs while it still has the resources to invest in expansion in those works which are retained. In that way the company should have a much more rational and secure base from which to face what appears to be an unstable industrial future in Britain as a whole.

Notes

1. For technical reasons at the end of 1948 Walkers, Parker formally took over the other lead companies and then changed its name to ALM while the old inactive ALM changed its name to Walkers, Parker. One of the reasons for this complex manoeuvring was that Walkers, Parker had an advantageous arrangement with the Inland Revenue with regard to the valuation of its base stock of lead, which would have been lost had the company been taken over. The companies which were taken into ALM were Cooksons; LL & J; Walkers, Parker; Librex; and London Lead Oxide. ALM of course remained a subsidiary of GWLI. The value of the assets officially taken over by Walkers, Parker was £1,617,534, made up of Cooksons £733,508, LL & J £679,905, London Lead Oxide £50,842 and Librex £153,279 (a far cry from its purchase in 1924 for the sum of £350,000).

2. Growing concentration in the lead industry between 1935 and 1951 is discussed in R. Evely and I.M.D. Little, *Concentration in British Industry* (Cambridge, 1960), pp. 53, 119, 152, 219-24 and 299. The figures from the Censuses of Production, on which that work is based, are not completely reliable however because of the difficulties involved from the fact that some lead products (e.g. white lead) were raw materials for non-lead products (e.g. paint) made by the lead manufacturers. Such products were not therefore traded. Secondly, GWLI was effectively larger than its returns would suggest because of its control of such firms as Alexander, Fergusson and T.B. Campbell, which were not wholly-owned subsidiaries and were therefore returned separately.

3. Metallgesellschaft, *Metal Statistics*. In addition to the figures given, 10,000-15,000 tons of lead annually was used for solders, between 12,000 and 20,000 tons annually for alloys and around 20,000 tons for miscellaneous products.

4. For discussion of consumption patterns and the outlook for the future see Zinc Development Association and Lead Development Association, *Into the 80's – the Critical Factors* (n.d.).

5. ALM's home trade in white lead (divided roughly in thirds between dry white lead, ground white lead and white lead sold in the form of ready mixed paint) was 19,464 tons in 1948 but only 15,918 in 1949. ALM's total capacity was an annual output of 23,500 tons of dry white lead (18,000 tons at Bootle and 5,500 tons from the Hayhole chamber plant), to which the Millwall plant would have added a further 7,000 tons.

6. A fifth secondary lead smelting company also exists. This is Capper Pass Ltd of North Ferriby, a company originally set up in Bristol. Much smaller than the other four in lead smelting (its chief activity is tin smelting) the company buys in special lead drosses, especially copper-rich ones and does not compete for battery lead. See B. Little, *150 The First Hundred and Fifty Years. A History of Capper Pass and Son Ltd* (1963).

7. A considerable number of lead salts (organic such as lead stearates, as well as the standard inorganic carbonate and sulphates) was manufactured as stabilisers. Apart from basic lead carbonate and tri-basic lead sulphate output of these stabilisers was small, running at a few hundred tons or less of each particular type per annum. In nearly every instance the initial development of new types of stabiliser was American and ALM's research laboratory merely responded to developments from abroad chiefly as a result of requests from the plastics companies for the new products.

8. Aside from the obvious, the range includes electric light bulbs, many types of floor tiles, gramophone records, ceramics, brake linings, rubber manufacture, thermo-electric and semi-conductor devices, while lead-covered tumblers are used on safes to prevent the combinations from being read by X-ray equipment.

9. For similar reasons ALM had been instrumental in 1967, at the instance of its then managing director, A.S. Davies, in obtaining the formation of the Lead Smelters and Refiners Association and the White Lead Manufacturers Association. A chartered accountant, Davies joined ALM in 1936. After the War he became assistant to J.H. Stewart and then commercial director of ALM. He was made a director of GWLI in 1955, managing director in 1965 and chairman of LIG in 1973, a post he held for only a short period until his early death in 1976. He played a leading part in bringing many of the smaller companies into the ALM merger in the post-war years.

10. The UKLMA remained in existence as a holding company for the shares in the continental works, which had been purchased in the inter-war years, but active control of UK quotas, etc. passed to the LSPMF.

11. The significance of the restrictive practice for price control may be seen from the fact that in 1958 Federation members had annual pipe capacity on single shift working of 91,250 tons against output of 15,000 and rolling mill capacity of 75,850 tons against output of 30,000. This was after the closure of eight mills and 29 presses between 1949 and 1958, the relevant firms having agreed to take their supplies from other Federation members.

12. See for instance Lead Development Association, *Lead Cable Workshop* (June 1978). In 1981 the BLMA contributed £50,000 to the LDA, to which the latter added a similar sum for promotional purposes for lead sheet and pipe.

13. Glynwed Ltd was formed in 1939 as an amalgamation between Glynn Bros Ltd, lead sheet and pipe manufacturers of Park Royal, Middlesex and

Wednesbury Tube Co. Soon after the War F.A. Clark & Son Ltd of Hammersmith, another lead sheet and pipe manufacturer, was added to the group but Glynwed's diversification took it away from lead and by 1979 the lead interests were small when they were sold to ALM.

14. The offer was of two LIG shares for each Mersey share, which put a value of £2.70p. on each Mersey share of nominal value of 50p.

15. In 1927 ICI had taken a 40 per cent share in BALM as a means of establishing itself in the Australian paints market. At about the same time BALM, in conjunction with Broken Hill Associated Smelters Pty Ltd, had set up Commonwealth Litharge and Red Lead Pty Ltd. By 1945 GWLI owned 90 per cent of the British holding in BALM and in 1947 it sold sufficient shares to ICI to give that company control of BALM, since it was largely on the basis of acting as the Australian arm of ICI in paints that BALM had flourished. By 1955 ICI owned 70 per cent of the shares in BALM (with GWLI owning 29 per cent) and, reflecting the greater role in paints, the name was changed to BALM Paints (Pty) Ltd, to which company in 1957 were sold both GWLI's and Alexander, Fergusson's Australian paint companies. Subsequently the name of the company has been changed to Dulux (Pty) Ltd in order to reflect its major activity and ownership. See Barncastle, British Australian Lead Manufacturers Pty Ltd.

16. In the annual report for 1968, for instance, R.A. Cookson commented on the 'vigorous policy, to which I have referred before, of extending our manufacturing activities into other expanding fields which are associated with our established business by common technological ground, for example, ceramic and zircon products'.

17. Brief historical details of the Fry Group are given in its house journal, *The Fry Record*, 181 (Sept. 1981), pp. 17-28.

18. Among other uses lithium products contribute to the production of synthetic vitamins, artificial rubber and lubricants since they combat the tendencies for normal greases to melt, freeze and become waterlogged, while lithium carbonate is used as a flux in enamels, an obvious link with another of ALM's activities.

19. In 1974 the Clyde Lead Works of Alexander, Fergusson at Kinning Park, was brought into ALM, although until the end of 1978 it continued trading as Alexander, Fergusson. From 1 January 1979, however, the name was dropped for the lead manufacturing operation, although at Ruchill Alexander, Fergusson continued to manufacture and trade in paints as a subsidiary of LIG.

20. It was 25 per cent in 1970 and only 15 per cent in 1975 despite capital expenditure of £4.6m at ALM in the years 1968-72 (out of a total LIG capital expenditure of £12.6m).

21. Reader, *ICI*, vol. II and Hannah, *Rise of the Corporate Economy*, pp. 91-7.

22. Hannah, *Rise of the Corporate Economy*, p. 173, quoting D.F. Channon, *The Strategy and Structure of British Enterprise* (1973), p. 67.

23. J.F. Wright, *Britain in the Age of Economic Management* (Oxford, 1979), p. 76.

24. The overall structure was chaotic with separate sales, production and accounting functions for each region, most of the products being duplicated in each area, some of the products (lithium for instance) being produced in one area but sold from another and antimony being produced in one but sold by all three areas.

25. As a result of subsequent decisions a small part of the Chester site has been retained and output of special products, such as lead-clad steel, will be continued, as will shot production.

STATISTICAL APPENDICES

APPENDIX I: Pig Lead Prices, 1783-1980 (£ per long ton, from 1970 per metric ton)

(1785 = 100)

Year	Price	Index	Year	Price	Index	Year	Price	Index	Year	Price	Index
1783	18.71	104	1810	36.91	206	1837	19.50*	123*	1864	21.60	121
1784	17.50	98	1811	30.71	172	1838	19.00*	120*	1865	20.10	112
1785	17.91	100	1812	29.50	165	1839	19.00*	120*	1866	20.50	115
1786	17.79	99	1813	29.83	166	1840	18.50*	117*	1867	19.55	109
1787	20.23	113	1814	31.88	176	1841	19.50*	123*	1868	19.33	108
1788	23.10	129	1815	26.45	148	1842	17.75*	112*	1869	19.08	107
1789	21.44	120	1816	20.91	117	1843	13.50*	85*	1870	18.65	104
1790	18.80	105	1817	19.41	109	1844	17.50	98	1871	18.20	102
1791	19.67	110	1818	25.50	142	1845	19.50	109	1872	20.00	112
1792	20.75	116	1819	24.79	138	1846	18.50	103	1873	23.30	130
1793	20.29	113	1820	23.50	131	1847	18.75	105	1874	22.10	123
1794	19.00	106	1821	21.12	118	1848	16.75	94	1875	22.50	126
1795	18.15	102	1822	22.67	127	1849	15.95	89	1876	21.65	121
1796	21.31	119	1823	23.50	131	1850	17.51	98	1877	20.56	115
1797	19.50	109	1824	23.25	130	1851	17.17	96	1878	16.70	93
1798	19.15	107	1825	27.75	155	1852	17.78	99	1879	14.25	80
1799	21.15	119	1826	22.12	124	1853	23.40	131	1880	16.31	91
1800	21.91	123	1827	20.59	115	1854	23.65	132	1881	14.96	84
1801	25.91	145	1828	19.33	108	1855	23.15	129	1882	14.37	80
1802	30.50	171	1829	16.00*	101*	1856	24.00	134	1883	12.90	72
1803	33.25	186	1830	13.50*	85*	1857	23.84	133	1884	11.13	62
1804	33.00	185	1831	13.00*	82*	1858	21.58	121	1885	11.50	64
1805	37.84	211	1832	12.75*	80*	1859	22.30	125	1886	13.22	74
1806	39.91	223	1833	13.00*	82*	1860	22.31	125	1887	12.85	72
1807	35.04	196	1834	17.00*	108*	1861	21.45	120	1888	13.91	78
1808	31.90	178	1835	19.50*	123*	1862	20.81	116	1889	13.05	73
1809	39.75	222	1836	26.00*	164*	1863	20.80	116	1890	13.39	75

APPENDIX I: Contd

Year	Price	Index	Year	Price	Index	Year	Price	Index	Year	Price	Index
1891	12.43	69	1914	18.69	104	1937	23.30	130	1960	72.15	403
1892	10.74	60	1915	22.88	128	1938	15.33	86	1961	64.21	359
1893	9.93	55	1916	30.98	173	1939	15.29	85	1962	56.29	314
1894	9.58	53	1917	30.00	168	1940	25.00	140	1963	63.44	354
1895	10.62	59	1918	30.13	168	1941	25.00	140	1964	101.25	565
1896	11.30	63	1919	28.20	157	1942	25.00	140	1965	115.00	642
1897	12.42	69	1920	38.23	213	1943	25.00	140	1966	93.15	520
1898	13.09	73	1921	22.73	127	1944	25.00	140	1967	83.76	468
1899	15.07	84	1922	23.74	133	1945	27.78	155	1968	101.80	568
1900	17.18	96	1923	26.82	150	1946	48.04	268	1969	122.70	685
1901	12.70	71	1924	33.70	188	1947	85.05	475	1970	126.43	706
1902	11.26	63	1925	35.86	200	1948	95.50	533	1971	103.79	580
1903	11.74	66	1926	31.11	174	1949	103.19	576	1972	102.58	573
1904	12.14	68	1927	24.40	136	1950	106.48	595	1973	174.34	973
1905	13.88	77	1928	21.16	118	1951	162.20	906	1974	252.48	1410
1906	18.20	102	1929	23.25	130	1952	136.75	764	1975	185.38	1035
1907	19.62	110	1930	18.07	101	1953	91.50	511	1976	250.44	1398
1908	13.74	77	1931	13.03	73	1954	96.45	539	1977	353.79	1975
1909	13.08	73	1932	12.04	67	1955	105.86	591	1978	342.50	1912
1910	12.95	72	1933	11.80	66	1956	116.31	649	1979	567.09	3166
1911	13.96	80	1934	11.05	62	1957	96.64	540	1980	391.27	2129
1912	17.79	99	1935	14.28	80	1958	72.78	406			
1913	18.31	102	1936	17.63	98	1959	70.78	395			

Note: The figures are all London prices with the exception of those marked with an asterisk (for which dates the London prices are unavailable), which are Hull prices. There was no close correlation between the two sets of prices (before 1829 Hull prices were above London ones while after 1843 they were below London ones). The figures given here for the years 1829-43 cannot therefore be taken as reflecting the course of the missing London prices.

Sources: 1783-1874, L. Willies, 'A Note on the Price of Lead, 1730-1900', Bulletin of the Peak District Mines Historical Society, vol. 4, part 2 (1969), pp. 179-91; 1875-1980, Metal Bulletin Handbook.

APPENDIX II: Walkers, Fishwick & Co., Elswick — Financial Statistics

Period to	Wages £	%[b]	Factory expenses[a] £	%[b]	Lead £	%[b]	Turnover £	Profit £	%[b]
8 May 1780	469	12.2	288	7.5	1,802	47.0	3,834	712	18.6
18 Nov. 1782	1,398	9.8	766	5.4	6,755	47.2	14,307	2,839	19.8
Nov. 1783	614	6.5	1,070	11.4	4,754	50.6	9,396	2,180	23.2
Aug. 1784	502	6.7	705	9.4	3,200	42.7	7,489	2,434	32.5
Aug. 1785	741	7.3	987	9.7	4,676	46.2	10,127	3,089	30.5
Aug. 1786	732	7.0	969	9.3	4,256	40.8	10,424	2,990	28.7
July 1788	1,833	3.3	4,227	7.7	35,703[c]	64.8[c]	55,070	9,499	17.2
Aug. 1789	960	3.8	2,915	11.4	17,453	68.5	25,475	2,025	7.9
Aug. 1790	959	3.4	2,565	9.1	16,418	58.2	28,202	6,183	21.9
Aug. 1791	1,127	2.4	3,723	7.8	30,645	64.2	47,720	10,314	21.6
Aug. 1792	1,304	2.3	4,372	7.7	38,330	67.8	56,560	10,067	17.8
Aug. 1793	1,278	2.9	5,041	11.3	27,726	62.2	44,602	7,740	17.4
Mar. 1794	500	2.7	2,665	14.5	11,510	62.5	18,417	2,850	15.5
Aug. 1795	1,113	2.4	4,203	9.0	27,440	59.0	46,541	11,397	24.5
Aug. 1796	2,095	2.4	7,620	8.7	47,056	54.0	87,190	27,319	31.3
Apr. 1797	962	2.5	4,878	12.7	23,451	60.8	38,549	7,354	19.1
Apr. 1798	1,357	2.8	6,673	13.6	30,606	62.5	48,946	8,975	18.3
Apr. 1799	1,832	3.6	7,237	14.1	26,264	51.3	51,209	11,992	23.4
May 1800	2,638	3.6	8,847	12.1	40,890	55.8	73,218	15,680	21.4

Notes: a. Since figures (not quoted here) are given for variable costs other than wages and labour (e.g. acid, colours, litter and bark, utensils and bad debts) it seems clear that factory expenses referred to repairs and extensions to buildings and plant.   b. Each item is given as a percentage of turnover.   c. It is possible that this increase was caused not only by expansion at Elswick but also by the purchase of pig lead on account of the Islington works.

APPENDIX III

From 1817 to 1881 the minutes (John, *Walker Family*, pp. 36-49) have figures for the distribution of the partnership capital of Walkers, Parker & Co. among the various works. These figures are presented below in tabular form, and include both fixed and variable capital. They are not always comparable since throughout the period the concept of capital was the partners' holdings in the firm and these included cash deposits which are sometimes included in the valuation of each branch and sometimes shown separately for the partnership as a whole.

Year	Total capital	Elswick	Islington	Lambeth, Red Bull Wharf	Orange St Southwark	Derby	Chester, Liverpool, Newcastle-under-Lyme	Dee Bank	Glasgow	Deposits
1817	456,729	128,797	53,325	100,930		35,772	137,905			
1824	376,225e	111,215		71,158		30,371	138,559			
1830	354,399	110,415		53,024		21,444	112,428			
1841	270,312	111,478		64,932	24,922		63,254	20,648		25,000a
1849	464,997	80,919		64,433	34,088		101,011	49,142		10,000b
1856	493,926	104,025		36,174			75,446	81,401		169,492
1860	519,849	129,729		42,097			101,226	105,092		196,880
1863	503,959	148,370		47,755			107,565	112,064		141,705
1866	520,616	149,674		40,681			121,390	120,147		88,205
1869	572,396	159,637		42,486			162,305	118,835	2,385	88,724
1876	466,704	114,452		73,937			169,429	100,113	8,773	86,748
1881	473,065c	146,994		60,938			174,300	85,232	5,612	74,271d

Notes: a. This sum is specifically attributed to William Parker.   b. This sum was a loan to 'J.W.' (probably Joshua Walker II).   c. The addition in the original is incorrect and this sum should be £473,075 or else one of the constituent figures should be £10 smaller.   d. This figure is included in the individual works figures.   e. Of the total capital in 1824 of £376,225, a sum of £62,000 was attributable to fixed capital, which conforms fairly well to the usual suggestion in this period that fixed capital accounted for about 20 per cent of total capital employed. The figure of £62,000 does, however, relate to the old partnership including Islington while the total capital of £376,225 relates

APPENDIX III: Contd

to the new partnership, excluding Islington, but including the less valuable Orange Street, Southwark premises. The fixed capital was divided as follows, Red Bull Wharf £3,900, Chester £10,600, Full Street, Derby £2,500, Windmill Hill, Derby £4,000, Elswick £19,000, Lambeth £14,900, Islington £5,200, Newcastle-under-Lyme £1,900.

While there was little change in capital employed in the period from 1817 (although it may be argued that this figure is inflated by high Napoleonic War prices and subsequently falls, the argument remains true if one begins with 1849) until 1881 it is worth noticing that a fairly constant level of capital input was financing an increasing level of output which rose from 24,000 tons in 1842 to 43,000 tons in 1880. This was chiefly the result of more efficient use of money capital — through more productive machines, better factory layout and an increase in the ratio of annual sales to working capital by more efficient means of financing credit.

APPENDIX IV: Deliveries of Lead Products, Walkers, Parker & Co., 1854-81 (tons)

(a) by product[a]

Year	Sheet	Pipe	Shot	White lead	Paint	Red lead	Litharge	Pig lead	Total	Monthly average stock of pig lead
1854[c]	11,006	4,874	2,419	2,414	3,094	1,424	700	1,598	27,529	6,691
1855	9,871	5,365	2,729	2,969	3,674	1,836	871	1,453	28,768	5,804
1856	10,579	4,316	2,562	3,492	3,909	1,936	1,185	1,302	29,281	6,091
1857	11,687	3,754	2,702	3,208	4,132	2,221	1,076	368	29,148	5,386
1858	10,099	3,557	2,613	3,837	4,022	1,957	838	1,102	28,025	5,865
1859	11,228	4,371	3,388	4,175	4,257	2,195	1,004	601	31,219	5,985
1860	11,569	3,800	2,658	3,342	4,170	2,015	1,141	2,073	30,768	7,174
1861	11,180	3,555	2,683	3,625	4,204	2,093	976	2,265	30,581	5,843
1862	10,801	3,431	2,821	3,357	3,971	2,044	786	3,415	30,625	6,789
1863	11,593	4,139	2,708	2,652	3,886	2,152	919	3,280	31,329	6,426
1864	12,279	4,698	2,971	3,260	3,831	1,967	808	2,994	32,808	7,217
1865	13,335	4,979	3,452	3,912	4,013	2,068	754	3,076	35,589	6,358
1866	14,856	4,893	3,498	3,699	4,070	1,835	788	1,657	35,296	7,475
1867	15,430	6,112	3,529	4,407	4,743	2,321	1,068	3,107	40,717	8,334
1868	15,377	6,701	3,473	4,390	4,846	2,177	1,226	2,591	40,781	6,282
1869	12,701	5,948	4,111	4,767	4,898	2,104	1,361	2,842	38,732	7,293
1870	12,967	6,517	3,748	4,220	4,589	2,239	1,246	2,688	38,214	6,636
1871	15,803	7,168	3,942	4,183	5,102	2,546	1,106	4,403	44,253	6,559
1872	15,890	7,487	3,743	4,232	4,304	2,120	1,131	1,973	40,880	8,726
1873	16,179	7,207	3,216	4,013	4,368	1,711	910	3,122	40,726	8,184
1874	16,802	7,789	3,446	4,046	4,704	1,967	1,013	1,749	41,516	6,125
1875	16,229	6,860	3,955	3,595	4,156	1,936	886	1,903	39,520	7,226
1876	17,228	7,158	3,173	3,370	4,133	1,632	850	1,452	38,996	6,441
1877	16,216	7,059	3,043	3,609	4,684	1,828	733	1,993	39,165	6,348

APPENDIX IV: Contd

(a) by product[a]

Year	Sheet	Pipe	Shot	White lead	Paint	Red lead	Litharge	Pig lead	Total	Monthly average stock of pig lead
1878	16,002	7,283	3,222	3,181	5,080	1,956	747	907	38,378	7,111
1879	16,916	8,255	3,151	3,417	4,676	2,084	856	2,779	42,134	8,137
1880	17,965	9,298	3,061	3,442	4,369	2,133	600	2,251	43,119	8,823
Jan.-Oct. 1881	16,375	8,812	3,258	2,869	4,362	2,098	555	1,714	40,043	7,815

(b) by selling branch[a]

Year	London	Elswick	Chester	Liverpool	Glasgow	Total
1854[c]	10,063	6,162	5,916	5,335		27,476
1855	11,056	6,520	5,787	5,436		28,799
1856	9,463	7,540	6,546	5,911		29,460
1857	9,730	7,195	6,425	5,862		29,212
1858	8,996	7,045	7,321	4,789		28,151
1859	10,354	8,353	6,616	5,817		31,140
1860	9,322	9,133	7,189	5,315		30,959
1861	8,425	9,809	7,365	5,044		30,643
1862	8,782	10,256	6,717	4,952		30,707
1863	8,682	10,107	6,996	5,656		31,441
1864	9,462	10,772	6,975	5,527		32,736
1865	10,563	11,022	8,279	5,821		35,685
1866	11,750	10,267	7,144	5,914	404	35,479
1867	11,266	10,994	8,064	8,448[b]	1,852	40,624
1868	10,739	10,279	8,907	8,623	2,253	40,801
1869	9,874	10,672	8,146	7,722	2,223	38,637

APPENDIX IV: Contd

Year	London	Elswick	Chester	(b) by selling branch[a] Liverpool	Glasgow	Total
1870	9,760	10,695	7,876	7,726	2,239	38,296
1871	11,753	13,553	8,226	8,339	2,507	44,378
1872	11,829	10,601	7,841	8,222	2,623	41,116
1873	11,967	11,194	6,982	7,389	3,139	40,671
1874	12,719	10,787	7,507	7,870	2,398	41,281
1875	10,627	11,245	6,845	7,765	2,891	39,373
1876	10,363	9,909	7,394	8,212	3,297	39,175
1877	9,011	11,189	8,197	7,922	3,049	39,368
1878	9,124	10,578	7,722	8,374	2,676	38,474
1879	10,475	13,245	7,875	8,249	2,352	42,196
1880	12,383	12,217	7,088	8,528	2,992	43,208
Jan.-Oct. 1881	12,070	10,635	6,695	8,511	1,971	39,882

Notes: a. These figures are based on the monthly delivery books of the partnership. The failure of the totals columns to agree reflects (i) small sales of lead ore which are not included in the first set of figures but are included in the second (ii) small arithmetical errors, but since the range of difference between the totals is only +236/−235 it has not been considered worth the effort of recalculating them.   b. The sudden increase in Liverpool deliveries between 1866 and 1867 is probably a result of sales through that branch of pipe and sheet produced at the Bagillt works which were taken over in the former year.   c. The figures for 1854 are for eleven months (Feb.-Dec.) only.

APPENDIX V: Cookson & Co., Deliveries of Lead Products (tons)

Year	Pig lead	Sheet	Pipe	Howdon total	Red lead	Litharge	Orange lead[a]	Dry white lead	Paint	Hayhole total
1872	4,547			4,547	900	23		1,001	161	
1873	5,004			5,004	799	38	168	877	92	1,974
1874	3,930			3,930	1,064	11	104	699	283	2,161
1875	2,971			2,971	1,470	13	17	854	233	2,587
1876	?	10			1,313	66	43	775	314	2,511
1877	?	1,002			1,366	36	48	987	220	2,657
1878	6,313	2,957		9,271	1,582	31	26	1,041	160	2,840
1879	4,699	3,987		8,687	1,339	36	23	1,161	124	2,683
1880	4,429	3,440		7,869	1,356	75	30	1,172	14	2,647
1881	3,309	4,154		7,463	1,422	24	69	1,296	81	2,892
1882	2,940	3,979	296	7,215	1,508	136	45	1,266	61	3,016
1883	?	4,176	510		?	?	?	?	?	
1884	?	4,182	573		?	?	?	?	?	
1885	?	3,490	555		?	?	?	?	?	
1886	2,051	3,703	430	6,184	1,600	224	64	1,300	138	3,327
1887	2,366	4,166	636	7,168	1,646	173	54	1,244	98	3,216
1888	3,410	5,066	559	9,035	1,822	75	54	1,054	205	3,209
1889	3,453	6,206	703	10,367	1,851	206	80	1,264	140	3,541
1890	9,636	6,676	710	17,022	1,863	219	18	1,538	116	3,755
1891	15,940	6,753	773	23,467	2,009	257	51	2,443	181	4,940
1892	15,612	6,229	703	22,544	1,645	261	76	2,086	349	4,417
1893	11,819	7,304	736	19,909	1,631	202	101	2,059	1,535	5,528
1894	9,914	7,959	709	18,582	1,909	261	49	2,233	1,038	5,490
1895	11,936	6,863	927	19,726	1,867	127	40	1,687	2,170	5,891
1896	11,402	7,010	1,207	19,620	1,604	56	45	2,525	1,942	6,172
1897	15,051	5,938	951	21,940	1,500	37	29	2,567	1,754	5,887
								2,876[s]	1,603[s]	
1898	13,549	6,178	1,279	21,006	1,562	56	48	331[c]	40[c]	6,516

APPENDIX V: Contd

Year	Pig lead	Sheet	Pipe	Howdon total	Red lead	Litharge	Orange lead[a]	Dry white lead	Paint	Hayhole total
1899	13,707	6,299	1,288	21,294	1,522	94	92	3,043	1,167	9,576
1900	12,901	5,970	1,321	20,192	1,763	66	94	2,571 / 2,406	1,087 / 1,248	10,312
1901	11,775	6,921	1,592	20,288	1,734	—	34	2,892 / 2,140	1,843 / 1,727	9,175
1902	11,419	5,223	1,431	18,073	1,867	37	48	2,546 / 1,980	994 / 1,106	8,812
1903	12,531	5,203	1,341	19,075	1,875	66	71	2,383 / 2,797	1,391 / 1,202	11,710
1904	13,501	5,496	1,550	20,547	2,038	70	65	3,407 / 2,914	2,292 / 1,001	11,734
1905	16,886	4,154	1,354	22,394	2,016	42	94	4,419 / 3,428	1,227 / 966	12,126
1906	20,963	3,619	1,181	25,763	1,661	50	65	4,619 / 3,719	961 / 905	13,215
1907	19,589	2,898	1,149	23,636	1,639	65	79	5,544 / 4,442	1,271 / 615	14,190
1908	?	7,076		?	?	?	?	4,437 / ?	2,913 / ?	?
1909	?	6,334		?	?	?	?	4,316 / 4,408	572 / 4,284	?
1910	25,265	6,415		31,680	2,163	123	—	4,936 / 3,700	924 / 4,265	16,111

Notes: a. There is no evidence from these figures to support the claim frequently made in Cooksons' own and other literature that the firm commenced the manufacture of orange lead about 1900 in order to challenge what had, up to that time, been a German monopoly.
s., c. From 1898 the top figure in the dry white lead and paint columns is for stack manufacture and the bottom figure for chamber manufacture.

APPENDIX V: Contd

Analysis of Sales by Markets (tons)

Year to	Total sales	London	Quay[a]	Continent	Country[a]	Scotland	Ireland	Canada	Australia	USA
30.6.1887	8,781	3,642	2,322		1,449	829	355	636	188	
1888	9,455	3,054	2,538		1,897	1,149	195	295	327	
1889	10,730	4,152	2,654		1,863	1,360	209	79	413	
31.12.1889	12,569	3,974	3,872		2,284	1,627	246	186	380	
1890	17,667	5,627	2,131	797	4,953	2,673	284	1,074	128	
1891	26,423	8,259	3,421	1,117	7,616	4,151	338	1,516	5	
1892	22,429	5,957	4,923	748	5,759	3,398	370	1,248	26	
1893	20,803	6,117	4,350	452	5,478	3,585	438	357	26	
1894	20,648	7,239	6,415	678	2,999	2,670	421	226	–	
1895	19,820	5,922	5,803	806	2,895	3,848	211	262	36	37
1896	20,256	4,768	6,175	1,801	2,756	4,271	334	144	7	–
1897	23,111	5,082	4,847	2,649	2,360	4,743	268	131	5	–
1898	22,545	5,940	4,455	2,098	4,035	5,282	86	644	5	–
1899	25,412	7,784	2,841	1,448	6,179	5,860	72	1,213	15	–
1900	17,907	5,040	2,349	2,252	3,559	4,339	42	316	10	–
1901	23,505	6,279	4,379	1,445	5,127	4,858	207	1,204	5	2
1902	19,799	6,549	2,486	1,808	3,960	3,887	211	893	5	–
1903	22,454	8,897	2,534	1,356	4,413	3,778	245	1,191	–	40
1904	22,521	5,137	4,119	2,215	5,428	3,391	225	1,896	–	110
1905	24,178	5,254	4,926	2,532	4,521	3,910	121	2,854	–	60
1906	28,335	6,285	3,902	1,765	4,885	2,995	123	4,664	–	3,716
1907	25,707	8,915	3,078	1,526	5,009	3,459	209	3,511	–	–

Note: a. Quay and country were, presumably, respectively deliveries to English (excluding London) customers by sea and by rail and road.

APPENDIX VI: UK Exports of Lead (tons)

Year[a]	Pig and rolled lead and shot	Red lead	White lead	Total (including litharge) and lead ore	Re-exports[c]
1816	17,445	395	527	19,298	
1817	17,742	267	694	19,548	
1818	13,028	403	579	14,586	
1819	14,068	215	462	15,249	
1820	18,299	229	602	19,776	
1821	15,645	242	652	17,399	
1822	13,784	441	572	15,828	
1823	11,042	280	549	12,913	
1824	10,833	338	885	12,991	
1825	8,616	338	616	10,540	
1826	10,217	408	629	12,404	
1827	13,261	534	1,012	16,203	
1828	10,001	382	1,133	13,256	
1829	6,832	382	750	8,647	
1830	7,442	520	662	9,309	861
1831	6,777	281	434	7,932	1,234
1832	12,181	396	652	13,898	957
1833	9,015	565	917	11,145	857
1834	8,672	412	781	10,412	865
1835	11,082	562	978	13,372	1,275
1836	9,769	427	819	11,418	913
1837	7,864	401	710	9,561	1,520
1838	7,351	542	774	9,404	3,440
1839	10,469	730	1,230	12,991	3,736
1840	13,223	876	1,264	16,071	2,530
1841	12,690	648	1,106	14,980	947

APPENDIX VI: Contd

Year[a]	Pig and rolled lead and shot	Pig	Sheet[d]	Pipe[d]	Shot[b]	Red lead	White lead	Total (including litharge and lead ore)	Re-exports[c]
1842	20,205					577	1,066	22,653	1,836
1843	14,611					707	1,225	17,098	2,440
1844	14,696				968	957	1,172	18,372	3,200
1845	10,485				1,092	1,033	1,353	14,533	3,241
1846	6,422				1,073	812	1,437	10,547	4,700
1847	8,259				1,177	840	1,389	12,079	3,462
1848	4,977				1,152	842	1,168	8,557	3,747
1849	15,277				1,799	1,621	1,676	20,985[e]	5,161
1850	20,166				1,750	2,112	2,043		3,217
1851	18,029				1,459	1,500	1,840		4,288
1852	18,641				1,355	1,182	1,731		2,967
1853	14,935				1,307	1,026	1,528		1,439
1854	17,857				1,748	1,103	1,815		199
1855	20,466				1,681	1,782	3,009		98
1856	20,868				2,266	1,846	2,819		566
1857	19,272				2,816	2,540	2,875		240
1858	17,645				1,910	2,293	2,684		203
1859	18,414				2,157	2,641	3,624		N
1860	21,986				1,811	2,455	2,813		E
1861	17,500				1,795	2,474	2,694		G
1862		28,591	3,966	1,447	2,136	2,396	3,175		L
1863		26,758	4,725	2,144	2,425	2,704	2,910		I
1864		27,868	4,103	1,610	2,186	2,874	3,023		G
1865		18,441	4,690	1,695	2,452	3,423	3,338		B
1866		20,563	4,728	2,097	2,653	2,918	4,798		L
									E

APPENDIX VI: Contd

Year[a]	Pig	Sheet[d]	Pipe[d]	Shot[b]	Red lead	White lead	Total (including litharge and lead ore)
1867	19,726	4,958	2,047	2,388	3,963	5,026	
1868	33,697	5,577	2,281	2,330	3,800	5,193	
1869	40,242	6,254	1,869	3,323	3,309	5,790	
1870	39,299	6,000	2,503	3,347[f]	3,147[f]	6,201[f]	
1871	34,425	10,044					
1872	33,403	10,927					
1873	21,906	10,104					
1874	25,872	10,481					
1875	24,271	11,127					
1876	24,459	11,462					
1877	32,008	10,457					
1878	20,864	13,521					
1879	22,780	13,996					
1880	21,797	11,754					
1881	28,002	14,992					
1882	22,299	15,076					
1883	23,583	15,732					
1884	17,464	16,087					
1885	23,545	14,981					

Notes: a. Up to and including 1854 the statistics are for the year to 5 January of the year following that for which they are given.
b. Before 1843 included in the total for 'pig and rolled lead and shot'.   c. Foreign lead brought into this country and then exported without further processing.   d. Previously included in the total of 'pig and rolled lead'.   e. After 1849 no total figure is given for lead exports.   f. After 1870 the returns cease so to include figures for shot and red and white lead and after 1884 they cease altogether. From 1893, however, a comprehensive collection of statistics is available in the Metallgesellschaft A.G., *Metal Statistics* (Frankfurt, various years), from which the following figures (on p. 402) for exports of pig lead and sheet and pipe are taken.

Source: *BPP*, annual returns of exports of lead, from 1828. The figures for 1816-19 are contained in *BPP*, 1828 (90) XIX, 405 and for 1820-6 in *BPP*, 1826-7 (520) XVIII, 157.

APPENDIX VI: Contd

Year	Pig lead	Sheet and pipe	Year	Pig lead	Sheet and pipe
1893	49,145	19,645	1904	37,703	19,043
1894	42,303	19,566	1905	43,031	18,360
1895	40,900	19,452	1906	41,530	17,835
1896	31,483	22,897	1907	40,954	16,612
1897	27,785	20,161	1908	42,532	18,043
1898	35,429	19,720	1909	36,505	17,895
1899	38,107	19,839	1910	43,675	17,659
1900	31,151	18,806	1911	44,862	18,111
1901	31,546	19,739	1912	41,601	18,000
1902	24,408	18,625	1913	46,500	n.a.
1903	29,121	18,183			

APPENDIX VII: Company Financial Statistics

1. Associated Lead Manufacturers Ltd (£,000)

Year to	Capital	Reserves	Loans	Creditors	Property & plant	Goodwill	Investments	Stocks	Debtors	Cash	Trading profit	Net profit	Return on capital employed (net profit as % capital plus reserves)
31 Dec. 1925	2,005	226	259	237	803	366	294	857	406	1	132	83	3.7
1926	2,412	123	394	226	989	369	523	833	428	12	65	9	0.4
1927	2,412	34	326	160	967	357	572	670	353	15			3.0
1928	2,412	66	245	181	973	355	561	638	378	10	117	74	
1929	2,551	84	254	417	1,029	347	578	874	469	9			1.6
30 Nov. 1930	2,860	(65)	842	86	3,596 (combined Property & plant, Goodwill, Investments, Stocks)				8	114	68	46	

Year to	Capital	Reserves	Loans	Creditors	Property	Plant & Goodwill	Investments	Stocks	Debtors	Cash	Sales	Trading Profit	Net Profit	Dividend (free of tax)	Net profit as % of capital employed	Trading profit as % turnover
31 Dec. 1949	1,878	2,215	426	1,694	731	688	1,908	738	1,313	835		1,383	522	282	12.8	
1950	1,878	2,804	428	1,833	792	806	1,735	1,450	1,988	171		1,641	704	376	15.0	
1951	1,878	3,629	397	2,339	865	1,001	1,149	3,073	1,923	231		2,141	728	400	13.2	
1952	1,878	2,866	687	1,548	926	1,121	179	1,789	1,068	1,896		1,026	583	350	12.3	
1953	1,878	2,889	2,458	541	914	1,066	992	1,243	1,128	2,423		1,670	713	450	15.0	
1954	1,878	3,158	2,351	1,195	917	1,063	1,846	1,735	1,618	1,403		1,520	643	450	12.8	
1955	1,878	3,136	2,150	1,403	948	1,151	2,004	2,439	1,705	319		1,355	546	450	10.9	
1956	1,878	3,095	1,987	1,449	1,044	1,243	1,735	2,542	1,664	181		933	426	450	8.6	
1957	1,878	2,680	2,026	1,050	1,167	1,358	1,844	1,823	1,251	191		840	478	450	10.5	
1958	1,878	2,940	1,289	614	1,220	1,490	947	1,656	1,183	225		927	411	400	8.5	
1959	1,878	3,013	1,754	1,083	1,300	1,457	1,434	1,714	1,478	344		743	488	350	10.0	
1960	1,878	2,895	2,313	1,064	1,443	1,504	1,249	2,145	1,504	304		794	354	250	7.4	
1961	1,878	3,107	2,580	1,020	1,494	1,588	1,491	2,079	1,517	416		674	356	250	7.1	
1962	1,878	3,072	2,368	964	1,490	1,558	1,580	2,139	1,278	236		522	338	150	6.8	
1963	1,878	3,288	1,474	946	1,442	1,491	503	2,011	1,741	398		1,069	241	215	4.7	
1964	3,000	3,215	1,984	1,647	1,435	1,297	606	3,592	2,899	16		1,471	413	400	6.6	
1965	3,000	3,224	1,473	1,876	1,405	1,257	687	3,215	2,362	649		1,242	882	400	14.2	
1966	3,000	3,368	844	2,006	1,427	1,338	775	3,195	2,097	386	16,936	1,065	742	400	11.7	6.3
1967	3,000	3,887	1,314	1,284	1,390	1,342	730	3,299	2,245	478	19,455	1,266	609	400	8.8	6.5
1968	3,000	4,499	1,764	1,943	1,422	1,702	1,066	3,968	2,849	199	24,131	1,420	696	500	9.3	5.9
1969	3,000	4,077	2,125	3,535	1,406	1,919	1,017	4,901	3,495	–	27,504	1,506	761	500	10.8	5.5
1970	3,000	4,896	2,243	3,859	1,540	2,474	1,386	5,324	3,235	38	19,922	1,186	999	400	12.7	6.0
1971	3,000	5,696	1,979	2,247	1,658	2,985	996	4,408	2,815	60	22,534	863	829	400	9.5	3.8
1972[a]	3,000	5,996	4,558	2,036	1,593	3,040	1,220	5,856	3,557	323			568	400	6.3	

1. Associated Lead Manufacturers Ltd (£,000): Contd

Year to 31 Dec.	Capital	Reserves	Loans	Owing to Holding Co.	Tax & Grants Equalisation a/c	Tax Payable	Creditors	Fixed Assets	Holdings in subsidiaries	Stocks	Debtors	Cash	Sales	Trading Profit	Net Profit	Dividend	Net Profit as % capital employed	Trading Profit as % turnover
1972[a]	3,000	5,094	297	4,261	902	432	1,524	4,633	1,220	5,856	3,477	323	22,534	863	568	400	7.0	3.8
1973	3,000	5,157	297	5,217	1,301	2,217	4,076	5,068	1,965	8,873	5,215	144	32,485	1,081	577	400	7.1	3.3
1974	3,000	5,702	379	4,348	3,511	3,419	5,095	5,830	3,343	10,191	6,024	66	47,095	2,495	1,270	880	14.6	5.3
1975	3,000	5,164	173	6,330	5,382	2,069	4,261	7,689	886	10,400	5,975	1,322	37,650	1,837	962	1,500	11.8	4.9
1976	3,000	6,054	—	13,020	10,710	(246)	7,075	9,682	1,740	17,032	10,027	1,132	50,908	1,732	799	560	8.8	3.4
1977	3,000	10,303	812	5,357	6,956	(48)	3,367	6,904	1,195	14,135	7,478	35	44,339	544	72	64	0.5	1.2
1978	3,000	15,547	975	4,591	3,382	657	3,276	7,854	1,115	14,143	8,167	149	43,773	96	113	50	0.6	0.2
1979	3,000	19,743	41	8,921	4,151	89	5,429	9,493	702	18,648	12,307	224	65,922	(707)	(311)	—	-1.4	-1.0
1980	3,000	16,557	287	5,060	662	(1,035)	5,893	11,294	471	11,216	7,371	72	48,179	(844)	1,752	1,400	9.0	-1.8

Note: a. After 1972 the method of presentation of the accounts was changed. To facilitate comparison the 1972 figures are presented on both bases.

2. Cookson Lead and Antimony Co. Ltd (CLACO) (£,000)

Year to	Capital	Reserves	Loans	Creditors	Property	Plant	Investments	Stocks	Debtors	Cash	Sales	Trading Profit (£)	Net Profit (£)	Dividend (£)	Net profit as % capital employed	Trading Profit as % turnover
31.12.1923	400	396	110	47	358			391	154	24		87,869	71,850		9.0	
1924	400	490	142	43	359		25	512	161	18		61,670	45,142		5.1	
1925	400	622	181	38	481		25	398	148	1		60,365	37,327		3.7	
1926	400	541	131	66	469		213	365	128	5		(1,380)	(24,078)	27,047	2.6	
1927	400	500	139	26	454		171	270	109	1		16,082	(25,561)	22,587	−2.8	
1928	400	505	70	30	436		231	214	102	1		72,858	40,601	40,601	4.5	
1929	400	510	163	25	428		252	252	119	1		121,819	49,709	49,709	5.5	
30.11.1930	400	522	66	55	422		298	165	97	1		67,586	10,745	8,000	1.2	
31.12.1931	400	366	105	44	306		356	181	68	1		116,898	51,949	57,397	6.2	
1932	400	347	89	30	302		342	154	68	1		104,120	47,622	68,566	7.0	
1933	400	355	141	35	296		408	152	74	1		186,972	76,483	75,085	10.1	
1934	400	371	128	36	289		406	155	84	1		151,887	82,066	74,565	10.6	
1935	400	438	204	43	284		489	216	94	1		181,949	101,793	53,005	12.1	
1936	400	392	149	50	279		423	189	100	1		169,937	90,981	74,305	11.5	
1937	400	405	153	56	277		389	236	109	3		135,567	70,082	68,336	8.7	
1938	400	437	172	51	196	79	450	226	108	—		156,558	74,343	70,048	8.9	
1939	400	534	180	90	193	78	544	212	151	24		247,107	108,319	54,088	11.6	
1940	400	545	223	62	191	80	528	300	130	1		173,075	67,339	51,812	7.1	
1941	400	611	141	158	189	81	600	184	177	79		225,121	90,853	68,523	9.0	
1942	400	690	252	48	187	76	619	161	347	1		205,135	65,166	52,269	6.0	
1943	400	629	227	121	184	60	873	106	153	1		165,347	35,454	31,364	3.4	
1944	400	606	235	99	182	57	790	113	169	28		186,490	85,632	88,719	8.5	
1945	400	577	138	273	179	57	726	225	200	1		218,693	109,265	110,106	11.2	
1946	400	888	390	93	177	57	907	226	297	107		449,817	114,529	73,275	8.9	
1947	400	1,261	392	154	178	75	1,077	320	555	1	3,310	722,293	403,816	233,820	24.3	21.8
1948	400	1,215	306	394	201	123	861	398	587	145	3,716	463,370	240,778	205,319	14.9	12.5

3. Goodlass Wall & Lead Industries Ltd (from 1966 Lead Industries Group Ltd), Consolidated Group Figures (£,000)

Year to	Capital	Reserves	Loans	Creditors	Goodwill	Land, buildings, plant	Company Investments	Other investments	Stocks	Debtors	Cash	Net profit	Dividend (%)	Net profit as % capital employed
31.12.1931	3,003	86	831	48	434	207	2,804	143	99	233	49	219	2	7.1
1932	3,003	103	528	402	499	1,299	538	336	804	499	60	199	3	6.4
1933	3,003	158	550	463	499	1,290	605	255	868	574	82	237	5	7.5
1934	3,003	193	528	520	499	1,280	630	261	906	572	96	239	6	7.5
1935	3,003	224	525	560	499	1,251	617	248	973	638	86	249	6	7.7
1936	3,003	336	579	697	497	1,216	668	214	1,219	730	71	275	7	8.2
1937	3,003	441	449	613	497	1,178	793	153	869	735	280	267	7	7.8
1938	3,003	474	428	733	497	1,126	713	143	962	654	542	275	7	7.9
1939	3,003	496	425	967	497	1,065	700	368	883	825	552	268	7	7.7
1940	3,003	479	425	1,002	497	1,070	781	424	1,324	680	132	266	6	7.6
1941	3,003	487	427	1,229	497	1,064	735	402	1,072	907	469	305	6	8.7
1942	3,003	508	427	1,339	497	1,054	695	421	1,128	946	535	288	7	8.2
1943	3,003	519	423	1,526	449	1,012	681	539	1,055	784	950	364	7	10.3
1944	3,003	683	554	1,798	660	1,118	765	226	1,134	1,214	950	572	9	15.5
1945	3,003	745	551	2,060	610	1,155	793	196	1,420	1,235	950	551	10	14.7
1946	3,003	1,031	536	3,034	610	1,298	850	61	1,369	1,649	1,766	1,114	15	28.0
1947	3,003	1,764	539	4,698	610	1,454	1,153	51	1,754	2,627	2,354	1,463	15	33.8
1948	3,003	5,179	913	4,190	641	2,333	853	67	3,491	3,619	2,281	501	15	9.7
1949	3,003	5,885	953	3,309	644	2,694	961	48	2,879	3,011	2,912	473	15	8.6

3. Goodlass Wall & Lead Industries Ltd (from 1966 Lead Industries Group Ltd), Consolidated Group Figures (£,000): Contd

Year	Capital	Reserves	Minority holders and loans	Tax reserve	Goodwill	Creditors	Land & buildings	Plant	Company investments	Other investments	Stocks	Debtors	Cash	Sales	Trading profit	Gross profit	Net profit	Dividend (%) (b)	Net profit as % capital employed	Trading profit as % turnover
1950	3,003	5,696	1,162	1,186	642	4,011	1,742	1,279	867	1,182	4,412	4,333	600		2,597	2,806	764	20	8.8	
1951	4,207	5,619	1,494	1,870	644	4,935	1,902	1,524	879	835	7,409	4,403	528		3,542	3,686	768	12	7.8	
1952	4,207	6,493	1,106	407	644	3,407	2,026	1,710	1,074	132	4,375	3,153	2,508		1,679	1,793	515	12	4.8	
1953	4,207	7,368	930	1,100	644	2,436	2,026	1,650	1,074	910	3,252	3,206	3,281		2,584	2,841	1,061	16	9.2	
1954	5,412	7,077	1,081	1,232	644	3,462	2,174	1,795	1,099	1,801	4,292	4,226	2,232		2,830	3,083	1,288	14(10.7)	10.3	
1955	5,412	8,203	1,246	1,194	642	4,001	2,352	1,987	1,258	2,317	5,627	4,859	1,014		2,636	2,984	1,451	16	10.7	
1956	5,412	9,115	1,376	1,205	644	4,128	2,753	2,192	1,631	2,103	6,121	5,134	661		2,424	2,826	1,281	18	8.8	
1957	5,412	10,067	1,026	786	644	4,059	2,984	2,326	1,996	2,917	4,871	4,743	868		2,566	3,037	1,410	13(12)	9.1	
1958	7,219	9,641	1,030	1,148	642	3,433	3,245	2,600	2,478	3,275	4,807	4,643	780		2,869	3,452	1,650	16	9.8	
1959	7,219	10,850	1,205	1,179		4,513	3,414	2,752	2,549	3,821	5,373	5,228	829		2,885	3,636	1,863	19	10.3	
1960	9,026	10,571	1,260	926		4,390	3,752	2,954	4,189	2,758	5,968	5,830	722		2,283	3,207	1,653	15(14.25)	8.4	6.7
1961	9,026	11,384	1,320	987		4,520	3,923	3,302	4,323	2,931	5,906	5,964	889	33,332	2,218	3,252	1,626	15	8.0	6.6
1962	9,026	12,937	1,377	831		4,545	4,060	3,392	5,006	3,556	6,190	5,820	691	33,213	2,182	3,234	1,626	16	7.4	7.1
1963	10,833	12,298	1,276	1,047		5,082	4,304	3,549	5,350	2,545	6,473	7,126	1,188	36,595	2,586	3,601	1,610	14(12.8)	7.0	7.4
1964	10,833	13,401	1,514	1,999		6,310	4,356	3,409	5,643	1,417	8,831	9,719	681	46,728	3,461	4,471	1,972	16	8.1	8.1
1965	14,447	12,504	2,105	1,620		6,854	4,566	3,515	7,255	3,192	8,872	9,242	889	50,603	4,097	5,141	2,699	12.5(11.4)	10.0	7.1
1966	14,447	12,858	1,932	1,402		6,706	4,708	3,567	7,291	3,568	8,080	8,338	889	45,738	3,252	4,385	2,347	12.5	8.6	6.7
1967	14,447	14,258	2,511	1,729		6,964	5,082	3,872	7,489	2,996	9,157	10,000	1,315	49,512	3,335	4,219	2,482	12.94	8.6	6.7
1968[a]	14,447	19,349	2,680	2,093		8,178	5,332	4,487	11,035	1,902	10,548	12,433	1,010	55,820	3,761	6,116	3,265	13.39	9.7	6.9
1969	19,507	15,904	3,822	2,448		9,606	6,059	5,341	11,083	1,191	12,276	14,419	918	64,520	4,478	6,937	3,703	10.5(9.56)	10.5	6.2
1970	19,507	18,521	4,780	2,582		10,792	7,109	7,258	11,432	411	13,481	15,643	848	73,585	4,554	7,331	3,970	11	10.4	6.8
1971	19,507	20,773	5,422	1,754		10,815	7,537	8,470	12,080	907	13,213	14,785	1,289	63,774	4,326	6,585	3,829	11.5	9.5	6.8
1972	19,846	25,295	10,371	2,864		11,515	8,280	9,157	17,221	230[d]	16,138	17,461	1,404	74,688	4,512	7,040	3,829	12.08	8.5	6.0
1973	19,846	34,680	12,919	3,006		19,145	9,245		20,540	1,266[c]	22,651	24,272	1,622	95,173	7,369[d]	12,204	5,933	12.68	10.9	7.7
1974	19,846	42,952	12,854	7,678		22,117	20,708		24,079	—	28,302	27,785	4,573	146,027	11,889	21,322	9,816	7.13p net	15.6	8.1
1975	19,911	51,770	13,098	5,596		20,174	24,008		27,107	—	28,505	27,252	4,677	131,842	9,614	13,650	6,513	7.83p gross	9.1	7.3
1976	19,926	57,880	19,964	18,042		26,194	28,689		31,713	—	42,645	36,831	2,128	168,862	11,682	20,546	9,302	8.60p gross (5.59p net)	12.0	6.9
1977	22,418	78,384	21,680	20,575		29,874	29,683		35,177	—	57,265	41,045	9,761	198,531	15,789	20,911	9,553	7.37p net	9.5	8.0
1978	22,497	106,496	22,427	10,308		31,372	34,122		48,945	—	58,155	43,246	8,632	195,268	12,702[e]	14,815[e]	7,662[e]	8.23p net	5.9	6.5
1979	22,543	123,794	58,645	6,267		42,994	51,489		49,844	—	86,947	59,874	6,089	269,737	14,837	20,341	12,035	9.66p net	8.2	5.5
1980	22,600	160,580	61,177	(46)		33,643	59,963		90,789	—	68,156	49,212	9,834	277,047	12,146	10,314	7,610	9.66p net	4.2	4.4

Notes: a. Figures for 1968 are not directly comparable with those for 1967 because of a change in presentation. b. After 1951 there was no fall in the effective rate of dividend paid – where an increase in capital, by capitalisation from reserves, caused a reduction in the actual dividend rate, a figure (in brackets) is given for the rate of dividend on the new capital had the previous year's actual dividend been maintained. c. On a comparable basis the trading profit for 1972 was £5,192,000. d. After 1973 'other investments', chiefly short-term deposits, are included in 'cash'. e. From 1978 profit figures have been struck after a change in depreciation policy. The comparable 1977 figures are: trading profit £14,594,000, net profit £11,355,000 and gross profit £19,716,000.

4. Walkers, Parker & Co Ltd (£,000)

Year to	Capital	Reserves	Loans	Creditors	Property	Plant	Investments	Stocks	Debtors	Cash	Sales	Trading profit (£)	Net profit (£)	Dividend (%)	Net profit as % capital employed	Trading profit as % turnover
31.12.1889												58,476	38,033	5		
1890												23,628	2,117	—		
1891												2,283	(15,774)	—		
1892												11,920	(5,228)	—		
1893												32,773	3,656	—		
1894												16,602	(9,438)	—		
1895	275	2	380	42	343	60	15	182	90	9		33,176	8,145	—	2.9	
1896												32,813	5,460	—		
1897												22,480	(1,196)	—		
1898												30,815	5,783	—		
1899												43,884	18,697	—		
1900												27,539	2,548	—		
1901												(24,721)	(44,315)	—		
1902												34,013	13,031	—		
1903												37,538	16,202	—		
1904												31,092	3,099	—		
1905												29,033	6,644	—		
1908	275	23	323	32	313	71	10	144	108	7		18,986	(1,547)		-0.5	
1909	275	25	319	34	312	70	10	146	104	10		30,976	7,900	3	2.6	
1910	275	28	319	41	312	68	10	145	110	17		31,163	8,380	3	2.8	
1911	275	28	323	53	311	69	10	152	122	15		29,362	6,871	3	2.3	
1912	275	46	319	44	310	72	10	130	133	28		46,900	23,615	5	7.4	
1913	275	50	319	34	309	70	10	127	137	24		40,518	16,999	5	5.2	
1914	275	76	319	50	308	66	—	142	167	26		59,314	37,409	6	10.7	
1915	275	195	302	81	307	66	—	188	210	83		161,782	60,771		12.9	
1916	275	261	302	94	308	66	151	216	270	72			115,678	10	21.6	
1917	213	445	302	500	237	183	197	181	244	464			41,076	10 (£20,000)	6.2	
1918	223	300	302	204	237	171		193	177	55			40,165	10	7.5	
1919														12%		
1920	223	386	377	106	237	181	207	185	196	85			104,179	15 (£28,000)	17.1	
1921	260	267	302	109	237	186	52	169	147	147			58,015	10	11.0	
1922	260	197	302	103	237	128	141	161	138	57			26,168	10	5.7	
1923	260	205	300	108	237	128	120	153	162	71			33,534	11%	7.2	
1924	260	217	299	102	252	83	102	164	214	64			38,186	11%	8.0	
1925	260	227	299	85	252	83	98	151	230	56		65,671	39,184	11%	8.0	
1926	260	231	299	84	252	80	102	147	207	84	1,644	55,342	33,376	11%	6.8	3.4
1927	260	194	298	53	229	78	125	146	158	66	1,400	39,676	20,992	10	4.6	2.8
1928	260	202	298	54	229	78	118	146	161	81	1,292	46,336	29,455	11%	6.4	3.6
1929	260	207	298	71	229	69	119	142	189	79	1,354	51,586	34,007	11%	7.3	3.8
30.11.1930	260	235	298	74	229	63	114	132	170	152	1,028	79,365	51,925	12%(£32,500)	10.5	7.7
31.12.1931	260	252	298	79	226		212	139	149	100	937	72,009	50,248	£50,248	9.8	7.7

Statistical Appendices 409

4. Walkers, Parker & Co Ltd (£,000): Contd

Year to	Capital	Reserves	Loans	Creditors	Property	Plant	Investments	Stocks	Debtors	Cash	Sales	Trading profit (£)	Net profit (£)	Dividend (%)	Net profit as % capital employed	Trading profit as % turnover
1932	260	252	298	86	224	67	235	156	125	89	826	61,637	48,926	£51,988	9.6	7.5
1933	260	254	298	86	222	64	218	138	155	101	940	88,003	55,186	£55,186	10.7	9.4
1934	260	256	298	99	228	65	209	170	156	85		83,578	56,791	£56,791	11.0	
1935	260	241	297	110	225	61	143	240	195	44		69,518	47,057	£47,057	9.4	
1936	260	251	297	134	221	61	96	296	242	27		88,352	56,486	£56,486	11.1	
1937	260	232	297	80	218	61	88	187	197	119		62,065	33,734	£33,734	6.9	
1938	260	256	297	183	215	69	63	184	194	271		87,913	52,610	£52,610	10.2	6.8
1939	260	234	297	189	217	73	63	149	175	304	1,302	59,491	16,856	£16,856	3.4	
1940	260	208	297	187	212	74	63	309	283	12		42,022	13,555	£13,555	2.9	
1941	260	227	297	177	209	77	98	216	148	248		79,530	32,920	?		
1942	260	253	297	220	206	79	92	182	145	319		155,585	58,663	£58,663	11.4	
1943	260	262	297	277	203	79	198	178	172	376		146,812	66,801	£66,801	12.8	
1944	260	264	297	315	201	71	221	185	234	246		165,066	68,781	£68,781	13.1	
1945	260	244	297	519	202	71	140	270	223	332		145,990	49,617	?		
1946	260	207	297	613	209	107	45	210	205	507		222,475	104,203	£104,203	22.8	
1947	260	225	297	1,041	212	111	—	254	429	772		427,673	210,178	£91,864	39.6	
1948	260	365	297	1,052	213	135		349	574	703		485,915	195,191	?		

BIBLIOGRAPHY

Books and articles

No attempt is made to list all the sources which have been used. This bibliography contains references only to works which have significant material on lead, the lead industry and business history.

Armstrong, H.E., and C.A. Klein, 'Paints, Painting and Painters, with Special Reference to Technical Problems, Public Interests and Health', *Journal of the Royal Society of Arts*, no. 3588 (26 Aug. 1921), pp. 655-85

Associated Lead Manufacturers Ltd, *Southern Area News Bulletin*, 1-126 (Jan. 1964-June 1974), and *News Bulletin*, 1-74 (July 1974-May/June 1981)

Baillie, Rev. J., *An Impartial History of the Town and County of Newcastle upon Tyne* (Newcastle, 1801)

Bainbridge, H.C., *Twice Seven* (1933)

Beckwith, F., 'Fishwick and Ward', *The Baptist Quarterly* (April 1954), pp. 249-68

Benwell Community Project, *The Making of a Ruling Class* (Benwell, 1978)

Booth, C., *Life and Labour of the People in London* (1895)

Brook, F., and M. Allbutt, *The Shropshire Lead Mines* (Cheddleton, Staffs., 1973)

Buchanan, R.A., and N. Cossons, *The Industrial Archaeology of the Bristol Region* (Newton Abbot, 1969)

Burt, R., 'Lead Production in England and Wales 1700-1770', *Economic History Review*, XXII, 2 (1969), pp. 249-68

— The Lead Industry of England and Wales 1700-1880 (University of London, PhD, 1971)

Chandler, A.D., 'The Growth of the Transnational Industrial Firm in the United States and the United Kingdom. A Comparative Analysis', *Economic History Review*, XXXIII, 3 (1980), pp. 396-410

Chemical Trade Journal, 'The Lead Industries of the Tyne', 19 Jan. 1895

Church, R.A., (ed.), *The Dynamics of Victorian Business. Problems and Perspectives to the 1870s* (1980)

Clow, A. and N.L., *The Chemical Revolution* (1952)

Cocks, E.J., and B. Walters, *A History of the Zinc Smelting Industry in Britain* (1968)

Cohen, J.M., *Life of Ludwig Mond* (1956)

Cookson, C., 'Antimony' in G.B. Richardson and W.W. Tomlinson (eds), *Official Handbook to Newcastle and District* (Newcastle, 1916), p. 127

Cookson, N., 'Lead' and 'Antimony' in J. Wigham Richardson (ed.), *Visit of the British Association to Newcastle-upon-Tyne, 1889, Official Local Guide. Industrial Section* (Newcastle, 1889), pp. 115-20 and 129

—— 'On Rozan's Process for Desilverising Lead', *Trans. Newcastle Chemical Society*, iv (1877-80), pp. 171-6

Department of Health and Social Security, *The Report of a DHSS Working Party on Lead in the Environment* (1980)

Elliott, W.Y., *et al.*, *International Control in the Non-ferrous Metals* (New York, 1937)

Fâche, E.C., *Lancaster & Locke. A Pedigree* (n.d.)

Forster, Westgarth, *A Treatise on a Section of the Strata from Newcastle upon Tyne, to the Mountain of Cross Fell . . .* (3rd edn, 1883)

Glendenning, I., 'Shot Making and the Shot Tower at Elswick', *Proceedings of the Society of Antiquaries of Newcastle upon Tyne*, 5th series, 1 (1951-6), pp. 351-60

Goadby, Sir Kenneth, 'Immunity and Industrial Disease', *Journal of the Royal Society of Arts*, lxix, no. 3580 (1 July 1921), pp. 523-39

Greninger, D., *et al.*, *Lead Chemicals* (New York, n.d.)

Hall, F., *An Appeal to the Poor Miner* (1818)

Hannah, L., 'Mergers in British Manufacturing Industry 1880-1918', *Oxford Economic Papers*, 26 (1974), pp. 1-20

—— *The Rise of the Corporate Economy* (1976)

—— and J.A. Kay, *Concentration in Modern Industry* (1977)

Harn, O.C., *Lead the Precious Metal* (New York, 1924)

Hedley, W.P., and C.R. Hudleston, *Cookson of Penrith, Cumberland and Newcastle upon Tyne* (Kendal, n.d.)

Higgins, R.M., 'Lead Shot Tower at Elswick, Newcastle upon Tyne', *Bulletin of the North East Industrial Archaeology Society*, 10 (1970), pp. 9-16

Hofman, H.O., *The Metallurgy of Lead* (New York, 1899)

Holley, C.D., *The Lead and Zinc Pigments* (New York, 1909)

Hughes, M., Lead, Land and Coal as Sources of Landlord Income in

Northumberland between 1700 and 1850 (University of Durham, PhD, 1963)

Hunt, R., (ed.), *British Mining* (1884)

— *Mineral Statistics* (various years)

— *Ure's Dictionary of Arts, Manufactures and Mines* (7th edn, 1878)

Hutchins, B.L., and A. Harrison, *A History of Factory Legislation* (2nd edn, 1911)

International Labour Office, *White Lead* (Geneva, 1927)

International Lead and Zinc Study Group, *Lead and Zinc. Factors Affecting Consumption* (New York, 1966)

Jefferys, J.B., Trends in Business Organisation in Great Britain since 1856 . . . (University of London, PhD, 1938)

John A.H., (ed.), *The Walker Family, Iron Founders and Lead Manufacturers 1741-1893* (1951)

Johnson, W., *et al.*, 'Small Spherical Lead Shot Forming from the Liquid, Using a Shot Tower', *Metallurgia and Metal Forming* (March 1976), pp. 68-72

Krysko, W.W., *Lead in History and Art* (Stuttgart, 1979)

Lead and Zinc Development Association, *Into the 80's – the Critical Factors* (n.d.)

Leathart, T.H., 'Lead and Copper', in G.B. Richardson and W.W. Tomlinson (eds), *Official Handbook to Newcastle and District* (Newcastle, 1916), pp. 123-6

Lee, G.A., 'The Concept of Profit in British Accounting 1760-1900', *Business History Review*, xlix, 1 (1975), pp. 6-36

Legge, T.M., *Report on White Lead Works* (1898)

— and K.W. Goadby, *Lead Poisoning and Lead Absorption* (1912)

Mackinlay, W.M., 'History of Glasgow Paint Manufacturers', *Oil and Colour Trades Journal* (6 Feb. 1948), pp. 314-18

Mess, H.A., *Factory Legislation and its Administration* (1926)

Metal Bulletin Handbook previously *Quin's Metal Handbook and Statistics* (various dates)

Metallgesellschaft, *Metal Statistics* (Frankfurt, various dates)

Mitchell, B.R., and P. Deane, *Abstract of British Historical Statistics* (Cambridge, 1962)

Miall, S., *A History of the British Chemical Industry* (1931)

Mosse, J., 'Redcliff Shot Tower', *Bristol Industrial Archaeology Society Journal*, 2 (1969), pp. 4-5

Nicholls, D.A., 'Chester's Shot Tower', *Cheshire Life*, 24 (May 1958), pp. 63-5

Oils, Colours and Drysalteries (15 June 1895), pp. 412-17 (on

Alexander, Fergusson & Co.). This journal and its contemporaries, *Oil and Colourman's Gazette* and *Oil and Colourman's Journal*, contain numerous details of relevance

Oliver, T., *Lead Poisoning in its Acute and Chronic Forms* (1891)

Pattinson, H.L., 'On a New Process for the Extraction of Silver from Lead', *British Association Eighth Report* (1838), pp. 50-5

Payne, P.L., 'The Emergence of the Large-scale Company in Great Britain, 1870-1914', *Economic History Review*, XX, 3 (1967), pp. 519-42

—— 'Industrial Entrepreneurship and Management in Great Britain', in P. Mathias and M.M. Postan (eds), *Cambridge Economic History of Europe*, vol. VII, part I (Cambridge, 1978), pp. 180-230

Percy, J., *The Metallurgy of Lead* (1870)

Plummer, B., *Newcastle upon Tyne: its Trade and Manufactures* (Newcastle, 1874)

Pulsifer, W.H., (comp.), *Notes for a History of Lead* (New York, 1888)

Raistrick, A., *Quakers in Science and Industry* (1968 edn, Newton Abbot)

—— *The Lead Industry of Wensleydale and Swaledale*, 2 vols. (Buxton, 1975)

—— 'The London Lead Company 1692-1905', *Trans. of the Newcomen Society*, XIV (1933-4), pp. 67-81

—— and B. Jennings, *A History of Lead Mining in the Pennines* (1965)

Reader, W.J., *Imperial Chemical Industries. A History*, 2 vols. (Oxford, 1970 and 1975)

Rees, A., *The Cyclopaedia* (1819)

Report of the Departmental Committee on Industrial Paints (1923)

Rhodes, J.N., The London Lead Company in North Wales 1693-1792 (University of Leicester, PhD, 1970)

Ritchie, E.J., *Lead Oxides, Chemistry-Technology-Battery Manufacturing Uses-History* (Largo, Florida, 1974)

Rowe, D.J., A History of the Lead Manufacturing Industry in Great Britain 1778-1980, with Special Reference to the Constituent Companies of Associated Lead Manufacturers Ltd (University of Newcastle upon Tyne, PhD, 1982)

Savary des Bruslons, J., *The Universal Dictionary* 2 vols. (2nd edn, 1757)

Schumpeter, E.B., (ed.), *English Overseas Trade Statistics 1697-1808* (Oxford, 1960)

Sherard, R.H., 'The White Slaves of England. V. – The White-lead Workers of Newcastle', *Pearson's Magazine*, II (1896), pp. 523-30

Singer, C., *et al.*, *A History of Technology,* vol. III, *From the Renaissance to the Industrial Revolution* (Oxford, 1957) and vol. IV, *The Industrial Revolution c.1750-c.1850* (Oxford, 1958)

Smythe, J.A., *Lead* (1923)

Sopwith, T., and T. Richardson, 'On the Local Manufacture of Lead . . .', in Sir W.G. Armstrong, *et al.* (eds), *The Industrial Resources of . . . the Tyne, Wear and Tees* (2nd edn, 1864), pp. 126-43

Stone, G., *Laws and Regulations Relating to Lead Poisoning* (1922)

Tylecote, R.F., *A History of Metallurgy* (1976)

—— 'Lead Smelting and Refining during the Industrial Revolution, 1720-1850', *Industrial Archaeology*, 12, 2 (1977), pp. 102-10

Watson, R., *Chemical Essays*, vols. III-V (various editions, 1787-9)

White, W., *Northumberland and the Border* (1859)

Williams, C.R., A Dissertation on the Industrial Changes and Developments in the County of Flint from 1815 to 1914 (University College of Wales, Aberystwyth, MA thesis, 1951)

Willies, L., 'A Note on the Price of Lead', *Bulletin of the Peak District Mines Historical Society*, 4, 2 (1969), pp. 179-91

—— 'Gabriel Jars (1732-1769) and the Derbyshire Lead Industry', ibid., 5, 1 (1972), pp. 31-9

—— 'Lead Poisoning in the Derbyshire Lead Industry', ibid., 5, 5 (1974), pp. 302-11

—— Technical and Organisational Development of the Derbyshire Lead Mining Industry in the 18th and 19th Centuries (University of Leicester, PhD, 1980)

Parliamentary Papers

Annual returns of imports and exports of lead and lead products 1816-85, commencing with *BPP* 1826-7 [520] XVIII, 157 and ending with 1886 LX, 213. Annual returns of mining and mineral statistics (from 1882), commencing with 1884 [C.3869] LXXXV, 535.

Reports of the Chief Inspector of Factories and Workshops (from 1878), commencing with 1878-9 [C.2274] XVI, 439.

1842   [382] XVII, Royal Commission on the Employment of Children, Appendix to First Report of Commissioners (Mines)

1843   [432] XV, Royal Commission on the Employment of Children, Appendix to Second Report of Commissioners

(Trade and Manufactures)

1862   [486] XIV, Report of the Select Committee of the House of Lords on Injury from Noxious Vapours

1876   [C.1443] XXIX-XXX, Report of the Royal Commission on the Working of the Factory and Workshops Act

1878   [C.2159] XLIX, Report of the Royal Commission on Noxious Vapours

1882   [C.3263] XVIII, 957, Report by Alexander Redgrave ... [on] White-lead Works

1883   [C.3516] XVIII, 929, Communications to the Secretary of State with Report by Alexander Redgrave, CB

1893-4 [C.6894-XXIII] XXXVII, part 1, pp. 151-5, Reports from the Lady Assistant Commissioners on the Employment of Women to the Royal Commission on Labour; Report by Miss May E. Abraham on the Conditions of Work in the White Lead Industry

1893-4 [C.7239 and C.7239-I] XVII, 717, Report and Minutes of Evidence of the Departmental Committee on the Various Lead Industries

1896   [C.8149] XXXIII, 1897 [C.8522] XVII, 1899 [C.9073] [C.9420] and [C.9509] XII, Interim, Second Interim, Third Interim, Fourth Interim and Final Reports of the Departmental Committee of the Home Office on Certain Miscellaneous Dangerous Trades

1899   [C.9207] XII, 277, Report of Prof. T.E. Thorpe and Prof. T. Oliver on the Employment of Compounds of Lead in the Manufacture of Pottery

1901   [Cd 527] X, 721 and [Cd 679] X, 683, Reports by T.E. Thorpe on the Use of Lead in the Manufacture of Pottery

1905   [Cd 2466] X, 727, Report on the Manufacture of Paints and Colours Containing Lead

1908   [Cd 3793] XII, 765, A.M. Anderson and T.M. Legge, Report on Dangerous or Injurious Processes in the Coating of Metal with Lead

1910   [Cd 5152] XIX, 51, E.L. Collis, Report on Dangerous or Injurious Processes in the Smelting of Metals Containing Lead and in the Manufacture of Red, etc. Lead

1910   [Cd 5219] XXIX, 85, Report of Departmental Committee on the ... Use of Lead ... in the Manufacture of Earthenware [Cd 5278] XIX, 245, Appendices to the Report

1914-16 [Cd 7882] XXIV, 901, 1920 [Cmd 630] XX, 1, [Cmd 631]

XX, 49 and [Cmd 632] XX, 125, Reports of the
Departmental Committees on the Use of Lead Compounds
in Painting. Vol. 1, Report on Painting of Buildings; vol. 2,
Report on Painting of Coaches and Carriages; vol. 3,
Appendices to both Reports; vol. 4, Minutes of Evidence
taken by both Committees

1918   [Cd 9236] XIII, 789, Report of the Committee of
Reconstruction on Trusts

Manuscripts

The basic source has been the uncatalogued papers of Associated Lead
Manufacturers Ltd, deposited in Tyne and Wear Archive Department,
accession number TWAD 1512, minute books, account books, sales
ledgers, day books, technical specification books, of the constituent
companies of ALM from many different places (including London,
Tyneside, Chester, Derby, Hull, Warrington and north Wales). Records
of the UK and British Lead Manufacturers Associations, held by Mr M.R.
Perrott, 68 High Street, Weybridge, Surrey, KT13 8BL. H.J. Barncastle,
British Australian Lead Manufacturers Pty. Ltd, History of the Company,
3 Vols (Ts., 1955), ICI Library, Millbank, London, Cookson Mss, Dept
of Palaeography, University of Durham. Blackett (Wylam) Papers,
Northumberland Record Office. Johnson papers, held by Mr R.J.
Johnson, Threadgolds Farm, Great Braxted, Witham, Essex. Company
House, Cardiff and London, company files. Public Record Office,
Home Office Papers and Board of Trade, files of deceased companies.
Material on various companies has also been obtained from Bristol,
Chester, Clwyd, Derbyshire, Northumberland, Norwich and Tyne and
Wear Record Offices and from a number of libraries in many parts of
the country.

INDEX